BIODIVERSITY, NATURAL PRODUCTS
AND CANCER TREATMENT

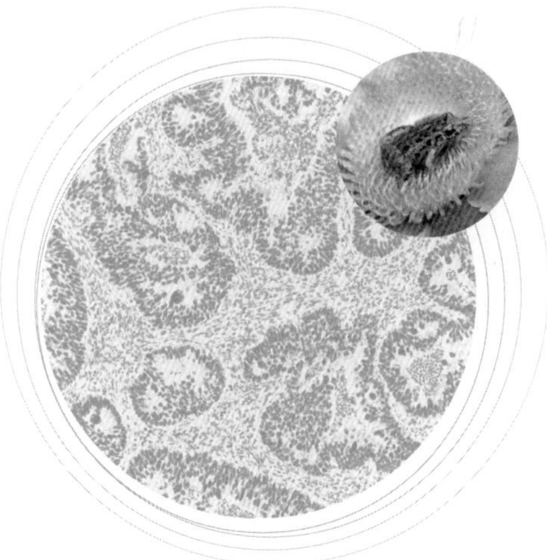

BIODIVERSITY, NATURAL PRODUCTS AND CANCER TREATMENT

Edited by

Victor Kuete (University of Dschang, Cameroon)

Thomas Efferth (University Mainz, Germany)

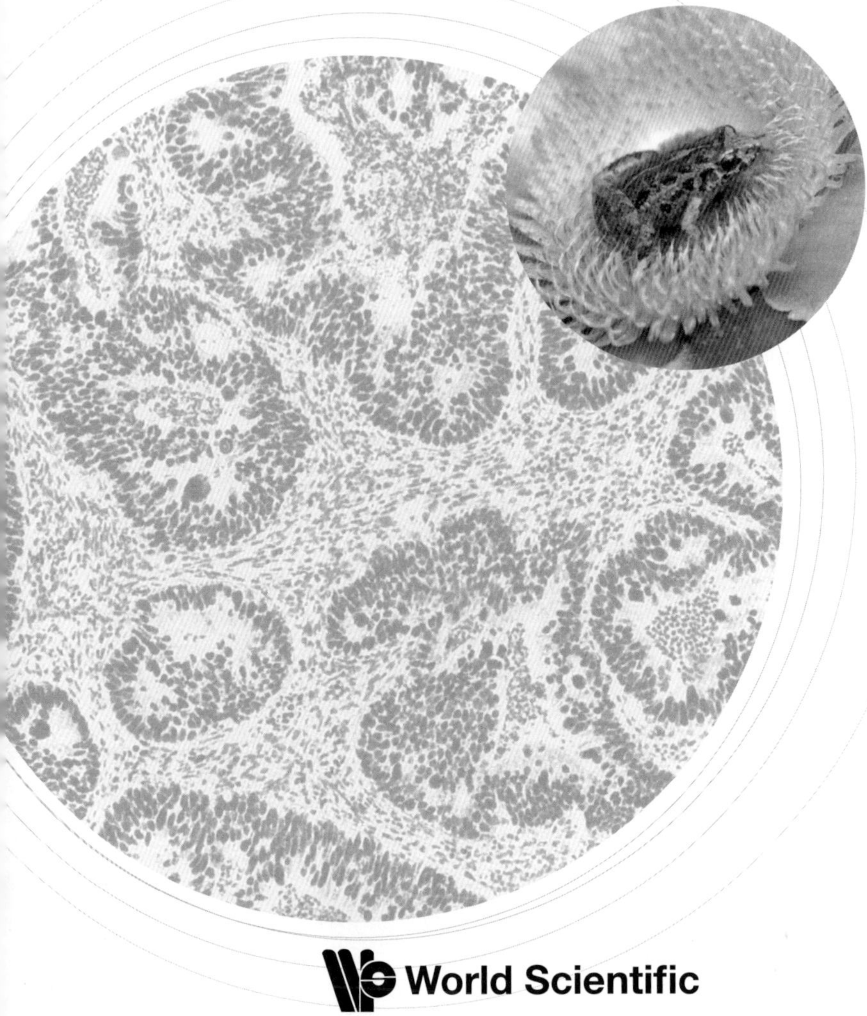

World Scientific

NEW JERSEY • LONDON • SINGAPORE • BEIJING • SHANGHAI • HONG KONG • TAIPEI • CHENNAI

Published by

World Scientific Publishing Co. Pte. Ltd.
5 Toh Tuck Link, Singapore 596224
USA office: 27 Warren Street, Suite 401-402, Hackensack, NJ 07601
UK office: 57 Shelton Street, Covent Garden, London WC2H 9HE

Library of Congress Cataloging-in-Publication Data
Biodiversity, natural products and cancer treatment / [edited by] Victor Kuete, Thomas Efferth.
p. ; cm.
Includes bibliographical references and index.
ISBN 978-9814583503 (hardcover : alk. paper) -- ISBN 9814583502 (hardcover : alk. paper)
I. Kuete, Victor, editor. II. Efferth, Thomas, 1960– editor.
[DNLM: 1. Neoplasms--drug therapy. 2. Biodiversity. 3. Biological Products--therapeutic use. 4. Ethnopharmacology--methods. 5. Materia Medica--therapeutic use. 6. Phytotherapy--methods. QZ 267]
RC262
616.99'4061--dc23

2014018400

British Library Cataloguing-in-Publication Data
A catalogue record for this book is available from the British Library.

The cover image shows a frog of the genus *Laptodactylus* sitting in a *Clusia* blossom. *Clusia* species produce antibacterial compounds, some of which are also cytotoxic towards cancer cells. The tissue in the background represents a colon carcinoma immunohistochemically stained for the proliferation marker Ki-67.

Printed in Singapore by FuIsland Offset Printing (S) Pte Ltd

Preface

Major advances in modern medicine and pharmacology were made possible by the presence of chemical structures in nature, i.e. molecules that have developed during millennia of evolution of life in terrestrial and marine microorganisms, plants and animals. The majority of clinically established anticancer drugs are either natural products, derivatives of them, or molecules that are based on mechanistic principles taken from nature. This seemingly unexpected high number of molecules from nature in cancer therapy is not peculiar to the field of oncology. In medicine in general, a considerable proportion of all drugs across all the different therapeutic fields have their origins in natural compounds.

Bearing in mind that the very first chemical used for therapeutic purposes was morphine isolated from *Papaver somniferum* in 1804 by the German pharmacist Friedrich Wilhelm Adam Sertürmer, it should be realized that modern pharmacotherapy is not much older than two centuries. On the other hand, the various traditional medicines all around the world are probably as old as mankind. While the elaborate forms of traditional medicines such as traditional Chinese medicine, Ayurveda and others are well documented with century- or millennia-old written documents still accessible, other traditional forms of medicine are handed down from generation to generation by oral tradition, thus losing their very first roots with the passage of time. Therefore, traditional medicines are not only a valuable cultural heritage of mankind, but also a rich resource for modern drug development. Through the ages, nature has been a source and continues to be the provider of drugs for a variety of ailments.

To understand the link between chemicals from nature and human health, it is important to know the ecology of the area of the drug source.

Plants as well as many marine organisms such as sponges, corals and tunicates are sessile and live in the same environment with pathogenic bacteria, viruses and herbivores. To defend themselves, they produce phytoalexins or antibiotics, providing a possible link between chemical defense against such harmful organisms and drugs for use in humans. Marine products have long been used in traditional medicine in Taiwan, Japan, China and India and it has been known for centuries that sponges are a source of medicinally important compounds. Plants such as *Glycyrrhiza glabra*, *Commiphora* species, and *Papaver somniferum*, used today for medicinal purposes, were already used in Mesopotamia 2600 years before the Christ era. The World Health Organization (WHO) estimated in 1985 that approximately 65% of the population of the world predominately relied on plant-derived traditional medicines for their primary healthcare; meanwhile, phytochemicals or their synthesized derivatives still play an important role in drug discovery.

Examples of pharmaceuticals include the bronchodilator, chromolyn (which originated from the phytochemical, khellin, a furanochromone from *Ammi visnaga* (L) Lamk.); the antihypertensive verapamil (which originated from papaverine, an opium alkaloid antispasmodic drug from *Papaver somniferum*); the well-known antimalarial drug, quinine from Cinchona trees); and the sesquiterpene lactone artemisinin from *Artemisia annua.* Plants have also been used in the treatment of cancer and have afforded anti-proliferative drugs in clinical use, such as the *Vinca* alkaloids, vinblastine and vincristine from the Madagascar periwinkle, *Catharanthus roseus*; and etoposide and teniposide, which are semi-synthetic derivatives of the natural product, epipodophyllotoxin, from *Podophyllum peltatum.* There is also a large chemical and biological diversity of the different marine evolutionary group, making them a remarkable resource for the discovery of new drugs, including anti-neoplastic agents. Products from marine flora and fauna have in the past been used in the treatment of various human ailments. Several compounds from marine organisms such as sponges, coelenterates and microorganisms as well as echinoderms, tunicates, mollusks, and bryozoans are a source of anticancer drugs. Examples of established drug from marine organisms include the well known Ara-A and Ara-C (respectively, antiviral and anticancer from marine sponge) and the cephaolosporins antibiotics from marine fungi. Other anticancer drugs in the clinical phase include LAF389 from sponge (phase I), Yondelis^{TM} from sea squirt (phase II/III), Cemadotin from sea slug (phase II), and ILX651and Dolastatin-10 from sea slug (phase II).

From our point of view, it makes much sense to bring together scientists from different disciplines, e.g. medicine, pharmacology, organic chemistry, molecular biology, gene and biotechnology, botany, ethnobotany, phytochemistry, microbiology, oceanography, and other related fields to work on an interdisciplinary field such as biodiversity. This point of view is supported by the fact that biodiversity, natural products and cancer therapy have become thriving fields of research attracting numerous scientists. A comprehensive survey of published papers from 1990 to 2012 shows that such fields of research have been growing over time (Fig. 1).

The advent of molecular biology has revolutionized almost all fields in the life sciences. This is a very illustrative example that inter- and cross-disciplinary approaches do not only cross-fertilize the various disciplines, but also can lead to scientific revolutions and major breakthroughs with gain of new knowledge. At our universities, we are frequently faced with tightly packed curricula, resulting in a lack of room for topics outside the

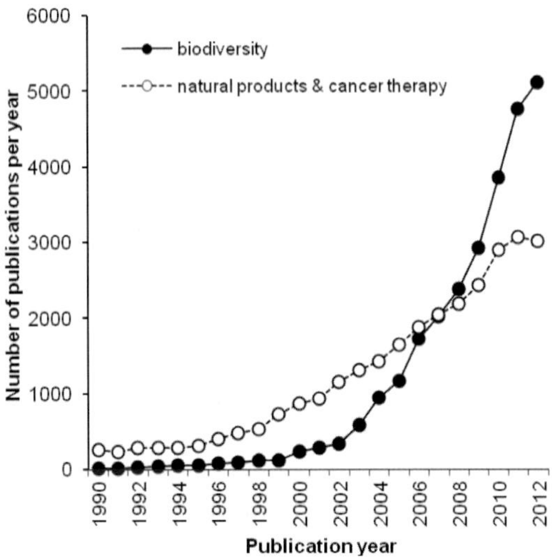

Figure 1 Survey of the literature documented in the PubMed database from 1990 to 2012 with the keywords "biodiversity" and "natural products and cancer therapy."

classical teaching content. Inter- and cross-disciplinary topics are mostly not sufficiently represented in academic teaching. Therefore, we felt that it is necessary to further strengthen inter- and cross-disciplinary research and teaching and try to span a bow from biodiversity to cancer therapy with several of its major topics.

The highlight of this book is an exhaustive compilation of scientific data on Biodiversity of medicinal plants (Chapter 1), Biodiversity and metagenomics (Chapter 2), Chemical ecology of medicinal plants (Chapter 3), Chemical ecology of marine organisms (Chapter 4), Natural products from terrestrial microbial organisms with activity towards cancer cells (Chapter 5), Marine organisms (Chapter 6), Ethnopharmacology and phytotherapy (Chapter 7), Contribution of African flora in the worldwide fight against cancer (Chapter 8), Natural products derived from terrestrial plants with activity towards cancer cells (Chapter 9), and Established anticancer drugs from natural origin (Chapter 10).

These topics have been chosen for seminars in the Pharmacy and Molecular Biotechnology programs at the Universities of Mainz and Heidelberg, Germany during the past few years. Students have chosen topics of their interests, prepared papers based on up-to-date knowledge taken from literature and presented them to the plenum. Hence, the existing curricula have been used to establish possibilities for inter- and cross-disciplinary topics, which are of high relevance not only for research and education in pharmacy and biotechnology, but also in related fields such as molecular medicine, pharmacology, drug development, medicinal chemistry, molecular biology etc.

In doing so, students acquired not only knowledge but also a sense how fertile inter- and cross-disciplinary research can be. This might be helpful in their later professional life.

The book discusses the state-of-the-art of each documented topic to serve as a reference resource tool for scientists and scholars in pharmaceutical sciences, pharmacology, organic chemistry and biochemistry, pharmacognosy, phytochemistry, ethnomedicine and ethnopharmacology, complementary and alternative medicine, medical and public health sciences and others.

In 2007, one of the editors (TE) met Dr. Ben ter Welle (Georgetown, Guyana) during an expedition in the rainforests of Guyana. Dr ter Welle's excellent knowledge of the plants and ecology of the rainforest was impressive and gave the motivation for a book which brings together biodiversity, chemical ecology, phytochemistry and cancer therapy. Without his inspiring love of nature, the idea to write a book like this would not have been born. Thank you, Ben!

We are also grateful to the Scientific Editor, Sook-Cheng Lim and her team from World Scientific Publishing. The realization of this book would not have been possible without their support and patience. Furthermore, we thank Karen Duffy (Cornell University, Ithaca, NY, USA) for reading and correcting the manuscripts as a native speaker. A special thanks also to Ilona Zirbs for her secretarial support in preparing the manuscripts.

Victor Kuete and Thomas Efferth
Mainz, October 2013

Contents

List of Contributors

André Antunes
Institute of Pharmacy and Molecular Biotechnology,
University of Heidelberg, 1m Neuenheimer feld 364,
69120 Heidelberg, Germany

Clara Becker
Institute of Pharmacy and Biochemistry,
Johannes Gutenberg University,
Staudinger Weg 5,
55128 Mainz, Germany

Katharina Cibis
Institute of Pharmacy and Biochemistry,
Johannes Gutenberg University,
Staudinger Weg 5,
55128 Mainz, Germany

Theresa Dreis
Institute of Pharmacy and Biochemistry,
Johannes Gutenberg University,
Staudinger Weg 5,
55128 Mainz, Germany

Karen Duffy
Cornell University,
Ithaca, NY, USA

Thomas Efferth
Institute of Pharmacy and Biochemistry,
Johannes Gutenberg University,
Staudinger Weg 5,
55128 Mainz, Germany

Sonia Falenska
Institute of Pharmacy and Biochemistry,
Johannes Gutenberg University,
Staudinger Weg 5,
55128 Mainz, Germany

Franziska Faulstich
Institute of Pharmacy and Biochemistry,
Johannes Gutenberg University,
Staudinger Weg 5,
55128 Mainz, Germany

Caroline Gartner
Institute of Pharmacy and Biochemistry,
Johannes Gutenberg University,
Staudinger Weg 5,
55128 Mainz, Germany

Kerstin Hoffmann
Institute of Pharmacy and Biochemistry,
Johannes Gutenberg University,
Staudinger Weg 5,
55128 Mainz, Germany

Laura Hoffmann
Institute of Pharmacy and Biochemistry,
Johannes Gutenberg University,
Staudinger Weg 5,
55128 Mainz, Germany

Jennifer Honek
Institute of Pharmacy and Molecular Biotechnology,
University of Heidelberg, 1m Neuenheimer feld 364,
69120 Heidelberg, Germany

Irini Karagianni
Institute of Pharmacy and Biochemistry,
Johannes Gutenberg University,
Staudinger Weg 5,
55128 Mainz, Germany

Christian Kersten
Institute of Pharmacy and Biochemistry,
Johannes Gutenberg University,
Staudinger Weg 5,
55128 Mainz, Germany

Tanya King
Institute of Pharmacy and Biochemistry,
Johannes Gutenberg University,
Staudinger Weg 5,
55128 Mainz, Germany

Ina Kirmes
Institute of Pharmacy and Biochemistry,
Johannes Gutenberg University,
Staudinger Weg 5,
55128 Mainz, Germany

Stephanie Kletting
Institute of Pharmacy and Biochemistry,
Johannes Gutenberg University,
Staudinger Weg 5,
55128 Mainz, Germany

Julia Krebs
Institute of Pharmacy and Biochemistry,
Johannes Gutenberg University,
Staudinger Weg 5,
55128 Mainz, Germany

Victor Kuete
Department of Biochemistry,
Faculty of Science,
University of Dschang,
P.O. Box 67, Dschang, Cameroon

Stephanie Lenz
Institute of Pharmacy and Biochemistry,
Johannes Gutenberg University,
Staudinger Weg 5,
55128 Mainz, Germany

Ariane Löhnert
Institute of Pharmacy and Biochemistry,
Johannes Gutenberg University,
Staudinger Weg 5,
55128 Mainz, Germany

Susanne Löhnert
Institute of Pharmacy and Biochemistry,
Johannes Gutenberg University,
Staudinger Weg 5,
55128 Mainz, Germany

Viktoriya Mogilevskaya
Institute of Pharmacy and Biochemistry,
Johannes Gutenberg University,
Staudinger Weg 5,
55128 Mainz, Germany

Corinna Risch
Institute of Pharmacy and Biochemistry,
Johannes Gutenberg University,
Staudinger Weg 5,
55128 Mainz, Germany

Sandra Schick
Institute of Pharmacy and Biochemistry,
Johannes Gutenberg University,
Staudinger Weg 5,
55128 Mainz, Germany

Mathias Schneider
Institute of Pharmacy and Biochemistry,
Johannes Gutenberg University,
Staudinger Weg 5,
55128 Mainz, Germany

Eva-Maria Surmann
Institute of Pharmacy and Molecular Biotechnology,
University of Heidelberg,
Im Neuenheimer feld 364,
69120 Heidelberg, Germany

Katrin Viertel
Institute of Pharmacy and Biochemistry,
Johannes Gutenberg University,
Staudinger Weg 5,
55128 Mainz, Germany

Janina Wich
Institute of Pharmacy and Biochemistry,
Johannes Gutenberg University,
Staudinger Weg 5,
55128 Mainz, Germany

Kirsten Yacoub
Institute of Pharmacy and Biochemistry,
Johannes Gutenberg University,
Staudinger Weg 5,
55128 Mainz, Germany

CHAPTER 1

Biodiversity of Medicinal Plants

Kirsten Yacoub, Katharina Cibis and Corinna Risch

ABSTRACT

Biodiversity describes the universality of all life forms, i.e. plants, animals and microorganisms, and the variety of ecosystems they live in. There are three classes of biodiversity, namely genetic diversity, species diversity and ecosystem diversity. Biodiversity is associated with chemodiversity, which plays a significant role in the discovery and development of drugs and medical pharmacotherapies. In addition to medical applications, biodiversity of microorganisms plays a crucial role in biotechnology and industrial biocatalytic processes. It is expected that in the future, biotechnology will produce high-quality bioproducts, which could replace fossil fuels. In this context, it is important to note that biodiversity is being threatened by the human species. Biodiversity is being destroyed by global environmental changes, such as changes in the atmospheric composition land degradation, depletion of fisheries and shortage of freshwater. Reduced biodiversity may lead to fewer natural chemical substances and hence reduced potential for medical and biotechnological applications.

1. INTRODUCTION

The term "biodiversity" describes the universality of all life forms, i.e. plants, animals and microorganisms, and the variety of ecosystems they live in. It also encompasses the differences in the genetic make-up and the numerous ecological processes therein.[1] Biodiversity is divided into three classes (Fig. 1). The first class is referred to as genetic diversity, and describes the variability of the genetic make-up between and within species. The

Institute of Pharmacy and Biochemistry, Johannes Gutenberg University, Staudinger Weg 5, 55128 Mainz, Germany.

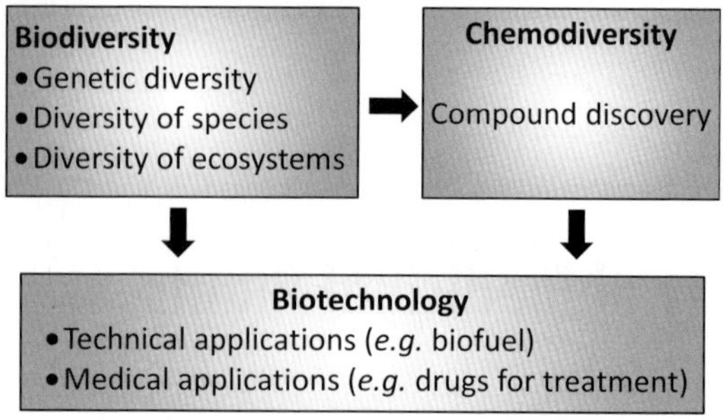

Figure 1 From Bio- and chemodiversity to biotechnology.

second class is the diversity of species and the third class is the diversity of ecosystems.[1] Biodiversity contributes greatly to the chemodiversity of pharmaceutical substances, many of which originate in nature and are significant for human health.[1] Modifications and changes in biodiversity can have an extensive impact on humans and their environment.[2]

2. CLASSIFICATION OF BIODIVERSITY

There is a widely held belief that biodiversity refers to the enormous number of plant and animal species. However, biodiversity in fact is much more than that. It corresponds to the complex natural living system, its structure and diverse levels of organization.[1] The three classes of biodiversity mentioned earlier i.e. genetic diversity, species diversity and ecosystem diversity are described in detail in the following sections.

2.1. Genetic Diversity

The genetic code is universal; it applies to all species. Genetic diversity does not only refer to individual variation in genomes within a species, but also to genetic variation between different species.[1] To date, scientists have identified about 1.7 million species on our planet.[3] The genomes of various species, including both eukaryotic and prokaryotic organisms, have been sequenced. There are currently fully-sequenced genomes of 2927 different species. One hundred and forty-six of them are archeal, 2615 are bacterial and 166 are eukaryotic genomes.[4] Among these sequenced genomes are those of the archea *Methanococcus maripaludis X1*,[5] the bacteria *Escherichia coli*[6] and the eukaryote *Homo sapiens*.[7]

Based on the presence of homologous genes in different organisms, it is possible to estimate how closely related two species are from a phylogenetic point of view. In this way scientists have been able to prove that humans, yeasts and sunflowers are closely related. The use of genetic data has revolutionized the evaluation of phylogenetic relationships, as using phenotypic details sometimes leads to inaccurate assumptions of evolutionary relationships.[8] Archaeal and eubacterial genomes can be sequenced much faster than eukaryotes, because of their smaller size.[8] Information about complete genomes, chromosomes or large segments of DNA is collected in databases (*e.g.* Ensembl, NCBI (National Center for Biotechnology Information), GOLD (Genomes OnLine Database)).[4]

The benefits of DNA sequencing are numerous. Taxonomists do not have to identify species by morphological features any more, which are not always indicative of relationships between species. Using cutoffs in genetic variation provides a more subjective way to define a species.

2.2. Species Diversity

The second kind of biodiversity is species diversity.[1] At least half a billion species have existed on earth. Currently, there are probably about 20–80 million species, with many of them listed as endangered.[9] About 1.4 million species have been morphologically described,[8] 265,000 of which are flowering plants.[1] The extinction rate at present is estimated to be about three species per hour.[9] Natural selection has resulted in a huge genetic variation favoring the generation of new species.[9] The actual number of species in a given geographic region depends on speciation, extinction, and immigration and emigration rates of the species in the area.[10]

2.3. Ecosystem Diversity

The third kind of biodiversity is ecosystem diversity, which refers to the variety of habitats that exist. Ecosystem diversity refers not only to abiotic conditions such as climate, soil, etc., but also to the communities of organisms and ecological processes, which play a major role in defining an ecosystem.[1]

The fact that many species are still unknown and have not been morphologically or genetically described necessitates further systematics and taxonomy. Biodiversity research is important not only for the discovery and classification of new species, but also for the preservation of species.[9]

3. BENEFITS OF BIODIVERSITY

Humanity benefits greatly from biodiversity in many ways. Three of the most important advantages derived from the earth's vast biodiversity are the use of chemical substances from nature, innovations in biotechnology and the impact of biodiversity on nutrition and health. Each of these three advantages is explained in the following sections.

3.1. Chemical Substances from Nature

The natural compounds produced by many plants play a significant role in drug discovery and development. Roughly half (49 percent) of the new small-molecule chemical entities introduced between 1981 and 2002 were chemical substances of natural origin, semi-synthetic or synthetic substances based on natural products.[11]

3.1.1. *Definition*

Natural products include extracts, mixtures of many compounds, and their isolated chemical constituents that are naturally biosynthesized.[12] Valuable natural products can come from terrestrial or marine life forms, e.g. plants, fungi, bacteria, protozoans, insects, other animals or humans.[12] Natural products can be subdivided into primary and secondary metabolites. Primary metabolites are indispensable for nutrition of the organism that produces them, and are universal in nature. On the contrary, secondary metabolites have beneficial features such as protection against parasites or predators or attracting pollinators, but are not essential to growth and reproduction of the producer. In general, secondary metabolites display larger chemical diversity.[12]

3.1.2. *Advantages and disadvantages of natural products over synthetic substances*

Chemical substances of natural origin differ from synthetic ones in that they display much greater structural diversity. Natural products typically have an increased steric complexity and a higher number of chiral centers than synthesized chemicals. They frequently bear a higher number of oxygen atoms and solvated hydrogen-bond donors and acceptors. Natural products have greater molecular rigidity, a broader distribution of molecular mass, a higher octanol-water partition coefficient, and a greater diversity of ring systems.[11]

The great chemical diversity of natural products is a result of evolutionary pressure, which acts on every organism. The biosynthesis machinery of an organism is subject to continuous mutation, which allows it to adapt to changing environments. In the course of evolution, an enormous variety of chemical substances has developed.[12] The architectural complexity of natural products compared to synthetic substances makes subsequent modifications of natural molecules difficult and expensive.[11] Furthermore, this complexity results in difficulties in the procedures for total chemical synthesis. As a result, many commercially available natural products are biosynthesized in the living organism or, if possible, in cell cultures. For example, isolation from *Papaver* species is still the most economical way of obtaining alkaloids such as morphine or codeine.[12] Another benefit of natural products is their increased compatibility and stability in biological milieus.[11]

In summary, chemical substances of natural origin are preferable molecules for the development of new drugs.[11] The plant kingdom still holds huge potential because only 10 percent of higher plants worldwide have been analyzed for chemical ingredients and biological activity. Furthermore, only half of the about 100,000 secondary metabolites that were discovered to have biological activity have already been investigated.[11]

3.1.3. *Examples of drugs from nature*

Alkaloids such as morphine, narcotine, emetine, strychnine, chinine, cinchonine, caffeine, atropine, etc., represent a highly potent class of drugs derived from nature.[12] The antimalarial drugs quinolin and related aryl alcohols are based on quinine, an ingredient of *Cinchona* bark.[11] Artemisin, an ingredient of the herb *Artemisia annua*, is effective against fever. Several semi-synthetic derivates of artemisin are used as antimalarials.[11] Macrocyclic spermine alkaloids isolated from the stem bark and leaves of *Albizia adinocephola* interact with the antimalarial target plasmepsin II.[11]

Vinblastin and vincristin from *Catharanthus roseus* are important anticancer drugs. Taxol is an anticancer drug that is extracted from the dried bark of the pacific yew tree (*Taxus brevifolia*). As this tree is slow-growing, and moreover is home to the endangered spotted owl, scientists prefer the semi-synthetic way to produce taxol from 10-deacetyl baccatin extracted from the needles of more common yew species.[13]

3.1.4. *Methods for discovering bioactive natural substances*

The search for new drugs is characterized by a search for new lead compounds for further derivatization. In addition to target specificity, considerations such as bioavailability, biological half-life, and tolerable side effects are also important features of drugs derived from natural products.[12]

Several approaches can be adopted for the development of drugs based on natural products:

(1) Accidental discovery, which has no rational basis.
(2) Structure activity relationships: Many chemical substances show structure activity relationships (SAR), which means that their chemical structure contributes to their biological function. This means that structurally similar molecules have similar effects.

    When a chemical structure with a certain biological effect is well-established, one can screen for similar chemical entities using this SAR. However, this method does not lead to the identification of new lead compounds.[12]
(3) The ethnopharmacological approach is based on the analysis of plants that have been used as medical plants by ancient cultures.[12]
(4) Screening of extracts: Typically, whole or roughly fractionated extracts, containing a large number of chemical substances are screened to discover new lead compounds. The target is carefully defined. If the

extract manifests an interaction with the target, then it is subjected to bio-guided fractionation. Ideally, the chemical substance responsible for the activity are isolated and characterized. One difficulty in bioguided fractionation is that the activity may not increase during separation; instead it may be spread among many fractions and compounds. This phenomenon may result from synergistic effects, where different chemical substances contribute to the biological effect. It is also possible that isolated chemical substances may lose their biological activity due to the loss of stabilizing co-metabolites, pH-changes, solvation, etc.[12]

(5) Computational approach: Using large compound libraries, a limited number of chemical molecules are chosen by evaluating appropriate pharmacophores and virtually fitting 3D structures. The selected structures are further analyzed in terms of their interaction with specific targets. This method is fast and inexpensive, but its major weakness lies in the inaccuracy of the functions that are used to estimate affinity of the ligand-target interaction.[12]

3.2. Potential Contribution of Biodiversity to Biotechnological Innovations

The success and potential of biotechnology relies heavily on the genetic resource of microorganisms' molecular biodiversity.[14] As a group, microorganisms display the greatest amount of biodiversity, and thus are the main focus of this article.[15]

The starting point for biotechnology development is the search for and the discovery of useful biological phenomena. Therefore, the establishment of efficient drug screening programs is fundamental. This process begins with sampling biological material; it is followed by screening for a desired property and selecting the best candidates among the screened samples; and it ends with the development of a final product or process.[16] Innovative selection and screening methods increase the chances of detecting new organisms and biochemical reactions, which can then lead to the discovery of specific pharmacological activities through rational target-directed screening.[16]

3.2.1. *Collecting genetic material to identify new organisms*

It was recently discovered that 99 percent of microorganisms cannot be cultivated on conventional nutrient solutions, and therefore could not be detected in the past.[14] Thus, the entirety of microorganisms is considerably greater and there are more rich species than was previously imagined. Because the diversity among the microorganism offers great potential, there have been numerous innovations in biotechnology exploring new organisms, processes, and products.[14]

One challenge in researching the chemical and genetic diversity of microorganisms is the difficulty of cultivating organisms under controlled conditions. In some cases, it is difficult to achieve sufficient growth for thorough investigation. In other cases, growth can be achieved in a lab setting, but the expression of desired genes does not take place.[13] Many technologies have been established to facilitate the cultivation of microorganisms.

Isolation and cloning of all genomes present in an environmental sample such as a drop of pond water allows access to the genetic information of microorganisms living in a certain habitat. The sum of this genetic information is called "metagenome."[15]

Metagenomics promises to yield new biocatalysts, diverse enzymes and other bioactive molecules from cultivable as well as uncultivable microorganisms. Compared to conventional cultivation methods, metagenomics provides an advantage in that it is not necessary to isolate and cultivate individual microorganisms.[15]

An ordinary soil sample of 30 to 100 cm^3 can contain 10^4 different bacterial species and one sample of marine prokaryotic plankton from the Sargasso Sea provides more than one million new open reading frames.[15,17] One can only imagine what an inexhaustible resource the entirety of microorganisms is, particularly for the discovery of new biomolecules with potential applicability.

The chemical and pharmaceutical industry deals with huge numbers of structurally different molecules, and the production of each requires individual enzymatic solutions. Hence, the value of biocatalysts is overwhelming and the popularity of microbial resources continues to increase in the chemical industry.[15]

3.2.2. *Analysis of genetic material*

The exploitation of microbial diversity was considerably fostered by the introduction of recombinant DNA techniques such as DNA-DNA hybridization, nucleic acid fingerprinting, methods of assessing the outcome of DNA probing, and rRNA sequencing.[16]

RNA-based methods for genetic analysis include the isolation, separation and sequencing of 16S rRNA molecules to decipher the phylogenetic relationships among organisms. With appropriate probes for certain rRNA sequences, members of communities at domain-, kingdom-, phylum-, genus- or species-specific levels of resolution can be analyzed.[16] Sequence analysis has been made easy by development of polymerase chain reaction (PCR).[16]

Another DNA-based method is DNA fingerprinting by restriction fragment length polymorphism (RFLP), which can be used to determine relationships between organisms, and is especially useful in distinguishing closely related organisms.[16]

DNA sequencing techniques achieved importance in bacterial systematics and taxonomy, to distinguish phenotypically similar, but genetically unrelated, bacterial strains. It is generally accepted that microorganisms with GC content differing by more than 5 percent in their DNA do not belong to the same species, and those with more than 10 percent differences do not belong to the same genus.[16]

3.2.3. *DNA databases*

Isolation and screening of biological material for new organisms and metabolites unavoidably leads to expensive rediscovery of already identified components. Hence, the generation of high quality databases is indispensable.[16] Huge data sets obtained via molecular analysis can be collected and stored in databases.[14]

DNA collections are becoming an important resource in the global effort to conserve biodiversity and to preserve genetic resources. Plant breeders and biotechnologists depend highly on DNA sequence collections for genetic resource characterization and exploitation.[18]

The main steps for the establishment of a DNA bank are collection of plant material, storage of DNA-rich material in a tissue bank, DNA extraction, DNA purification and quantification as well as storage of the DNA in TE buffer (10 mM Tris-HCl, 1.0 M EDTA) at −80°C in ultracold freezers. After extraction, the quantity and purity of the DNA can be detected by UV spectrophotometry or fluorescence-based assays. DNA absorbs ultraviolet light at 260 nm and 280 nm wavelengths, which can be measured with an UV spectrophotometer. Fluorescence-based methods are preferable, if the samples are impure or at low concentrations. Once DNA has been purified, it can be stored for decades.[18] DNA databases provide a fast and easy way to search for data and information associated with a particular sequence.[18]

3.2.4. *Biocatalytic processes*

Some microorganisms can exist under extreme physical and chemical conditions such as extreme temperatures and/or pH values, high salinity, high pressure or the presence of radiation. Such extremophiles contain a panel of enzymes that catalyze specific biochemical reactions enabling the organism to thrive even under these harsh conditions.

Such conditions are also common in industrial biocatalytic processes. Therefore, robust biocatalysts are of utmost value. A famous example of such an enzyme used in industrial applications is Taq polymerase, which is derived from a thermophilic bacterium and used to catalyze polymerization at high temperatures in PCR. The number of enzyme-driven reactions in chemical, pharmaceutical, food and other industrial applications is constantly increasing. The advantage of biocatalysts compared to chemical catalysts is their specificity, which leads to a more efficient production of single stereoisomers, less side reactions and a decreased environmental burden. Extremophilic organisms are a valuable source of novel and robust enzymes.[14]

In the future, biotechnology may replace fossil fuels with biofuels, and conventional chemical processes with bioprocesses; and produce high-quality bioproducts such as nutritionals, valuable chemicals and bioactive substances. Besides the food, detergent, and biofuel industries, numerous chemical and pharmaceutical industries are active players in industrial biotechnology. In these times of soaring energy costs, dwindling fossil fuel reserves and increasing environmental pollution, biotechnology is expected to play a major role in the near future.[15]

3.2.5. *Biofuels*

At present, scientists recognize three distinct generations of biofuels. First-generation biofuels were derived from crops such as sugarcane, corn and soybeans. However, the cultivation of these plants for fuel extraction led to water scarcity and deforestation. Second-generation biofuels originated from lignocellulosic biomass and forest residues. These biofuels required large tracts of land, but the increased demand led to shortages of agricultural land and increased food prices.[19] The third generation of biofuels is derived from microalgae.[20]

Microalgae are an energy source that has the potential to make biofuels widely accessible without the drawbacks of first and second generation biofuels. Compounds from microalgae can be easily converted into fuels without excessive environmental pollution.[20] The conversion from biomass to fuel takes place via thermo-chemical or biological methods. Thermo-chemical methods consist of direct burning, which can provide heat, electricity and mechanical power. Biological conversion methods include fermentation of biomass producing energy carriers such as hydrogen, biogas, ethanol and biodiesel. Fatty acids obtained from microalgae can be extracted and converted into biodiesel, forming a renewable, non-toxic, biodegradable and environmentally friendly fuel. Ethanol is obtained by fermentation of algal biomass, and biogas is the product of anaerobic digestion of biomass.[20]

Algal species used for biofuel include *Dunaliella* sp. and *Chlorococum* sp. for ethanol production, *Neochlorosis oleabundans* and *Chlorococum* sp. for the production of biodiesel, *Chlamydomonas reinhardtii* and *Spirulina platensis* for the recovery of hydrogen, and *S. platensis* UTEX 1926 and *Spirulina* Leb 18 for methane production.[20]

Compared to terrestrial plants, microalgae are quite efficient in their process to convert solar energy. About 10 percent of the solar energy they receive is converted into biomass. Microalgae present further advantages compared to terrestrial plants including continuous biomass production, rapid growth,[19] high energy content,[19] simple cell cycle, high tolerance to changing environmental conditions, the opportunity to utilize waste and brackish water for cultivation, and the use of land, which cannot be used for agriculture. Furthermore, when subjected to physical or chemical stresses, algae can produce specific compounds in high concentrations.[20] The development of transgenic algal strains with increased efficiency thanks to recombinant protein expression, an engineered photosynthesis pathway, and enhanced metabolism are the hope for the biofuel industry.[19]

3.3. Impact of Biodiversity Related to Food Products and Nutrition

There are approximately 30,000 edible plant species, but only about 7,000 of them are cultivated as food crops. Four plant species are of particular importance: rice, wheat, corn, and potato. Together these account for more than 75 percent of human calorie intake.[21]

3.3.1. *Evolution, breeding and genetic erosion*

The biological diversity in the plant kingdom is a result of natural evolution, which has enabled plants to adapt to diverse climatic conditions and to develop certain resistances against parasites and disease.[21] Human agricultural practices have intervened in this natural process by selecting certain plant types that are particularly suitable for cultivation. The aims of plant breeding are increased production, improved product quality, and resistance to physical threats and biological pests.[21]

Earlier, many of the local landraces that were used and eaten by humans were easy to maintain and produced relatively stable harvests, even under extreme conditions. However, the variability between and within the plant populations meant a poor annual production compared to that of modern homogenous crop varieties.[22] However, the combination of several factors that led to a diversity of cultivated plants meant that at least farmers would not lose the whole harvest.[22] Nowadays, there is a marked loss of genetic diversity in crop plants, which has led industrial agriculture to be dominated by a small number of largely homogenous landraces.[22] Farmers no longer save their seed for the next harvest, but prefer to buy it from plant breeders. In this system, the process of evolution and development of unique plant varieties disappears.[22] Cultivation of a limited number of genotypes causes genetic erosion, a term referring to a reduction of a species' gene pool.[21]

3.3.2. *The importance of agro-biodiversity*

There is an important correlation between agro-biodiversity, dietary diversity, and health, especially in developing countries.

A healthy diet is dependent on the availability and accessibility of multi-nutritional food sources, which were historically part of many traditional food systems. Many traditional well-balanced food systems, including several plant and animal varieties, are still a source of important nutrients and micronutrients.[23] However, contemporary trends tend to disregard biodiversity.[23]

Various epidemiological studies underline the advantages of versatile nutrition, particularly when including fruits and vegetables for longevity, and the correlation between increase in food quality and a reduction of chronic degenerative diseases.[23] Well-balanced nutrition often contains probiotic, immune-stimulant and antibiotic compounds, all of which reduce one's susceptibility to infectious diseases such as diarrhea, a common but debilitating disease prevalent in developing countries. To some extent, there are also anti-cancer benefits attributable to antioxidants, flavonoids, carotenoids, lutein and phytoestrogens.[23]

In developed countries, availability and accessibility of a variety of plant and animal food sources may not be a concern for the population, but in developing countries it certainly is. In many countries, the poor subsist on limited, unbalanced and unhealthy diets, leading to malnourishment and undernourishment. Undernourishment is often not a caloric problem, but rather an inability to consume certain important nutrients. Although traditional food systems in developing countries are characterized by their nutrient diversity, globalization of culture and commercialization has lead to a westernization of food systems and diets with well-established consequences such as obesity, type 2 diabetes mellitus, cancer, cardiovascular disease, and others.[23]

Nowadays, modern plant breeders and molecular biologists make an effort to ensure food security, for example by breeding high-yield crop plants with resistance against several infections to feed the rapidly growing world population.[22]

Another modern strategy to improve the nutrient potential of agricultural crops is to increase the nutrient content or other functional components to address micronutrient deficiencies. However, this may lead to further simplified food systems and a decrease in dietary diversity.[23] The conservation of diversity must not be disregarded in the context of plant breeding, because it is precisely this diversity that contributes to food quality and the plurality of food products.[22]

4. LOSS OF BIODIVERSITY

Biodiversity is being threatened by the human species. Many different ecosystems exist on earth and they interact in many complex ways. If just one aspect of an ecosystem is changed, it is likely to have far-reaching impacts on many other ecosystems.[2] It is important to keep in mind that humans depend heavily on the world's diversity of ecosystems, not to mention all the other species on earth that have evolved to survive in their unique environments.

One obvious example of how ecosystems are affected by human behavior is global climate change and its impact, especially on tropical rain forests. Tropical forests cover only 10 percent of the earth's surface, mainly around the equator. These forests are home to up to two-thirds of all species on earth. It is unknown exactly how many species live there. The tropical rain forests are also important for human beings. They offer homes for various tribes such as the Yanomami of Amazonia. Moreover, they offer food, building materials and medicines on a global scale. These forests have the ability to store and process enormous quantities of carbon dioxide via photosynthesis and respiration. Their capacity to bind and process carbon is about six times the carbon amount produced by fossil fuel use by humans.[24] This ecosystem is being threatened by deforestation, and many species have already gone extinct. It is impossible to count the number of species that disappear because there is no complete inventory of all species in tropical forests. Therefore, the extent of species loss is difficult to estimate. Deforestation also means a loss of carbon-dioxide processing potential. This is problematic, because the carbon dioxide levels in the atmosphere are higher today than they have ever been in the last sixty million years.[24] Furthermore, slash-and-burn agriculture increases carbon dioxide levels in the atmosphere.[2] In the past century, fossil fuel and changes in land use have led to an increase in carbon dioxide concentration by 25 percent and an average increase in temperature by 0.6°C.[24] Furthermore, increasing temperatures are responsible for rising sea levels.[2] This is an impressive example of how earth's different systems (the atmosphere, oceans, and ecosystems) and humans interact.

Another way in which humans affect biodiversity is in the transfer of species from their native habitats into new ones.[25] In the past, species have certainly moved between ecosystems, but the changes have been slow enough so that creatures were able to adapt to the new environment. Today, many changes are taking place far too fast to allow successful adaption.[2] As a consequence many species are threatened by extinction. Human activities are largely responsible for the rising rate of extinction and the increasing transformation and fragmentation of natural landscapes. The result will is likely to be gloomy: more and more species are expected to disappear and biodiversity will become reduced.[2]

5. CONSEQUENCES

Biodiversity is being destroyed by global environmental changes, such as changes in the atmospheric composition, land degradation, depletion of fisheries and freshwater shortages.[26] The carbon dioxide concentration is 30 percent higher than it was three centuries ago, and half of this increase has taken place in the last 40 years. The methane concentration has doubled. Altogether, this contributes to accelerated global climate warming. These changes are expected to lead to the most rapid climate change since the last 18,000 years. Human beings have converted roughly half of the earth's surface and use about 54 percent of the fresh water. By 2050, it has been estimated that humans will use 70 percent of the earth's fresh water.[25] Collectively, these recent changes have resulted in reduced biodiversity. The extinction rate for species is now 100 to 1000 times higher than it was before humans dominated earth.[25]

An ecosystem is not only defined by the species that live in it, but also by the interactions between those species. These interactions are very important; they can even be essential to the survival of a species. For example, nitrogen-fixing microorganisms and plants rely on one another. A change in one single aspect of such an interaction can trigger large-scale changes in the whole ecosystem. For example, boreal forests are highly dependent on the temperature of the soil. The moss in these forests keeps the soil temperature low and stabilizes the soil composition.[25]

Humans change ecosystems either for economic reasons, or for cultural, intellectual, aesthetic and spiritual reasons. These human ecosystem changes affect biodiversity and vice versa; changes in biodiversity can also affect ecosystems e.g. by invasion of a new species.[25] The pathways of these interactions are so complex that it is very hard to predict the effects of human intervention.

Another example of anthropogenic changes of ecosystems is nutrition by cultivation of plants in plantations. Plantations represent drastic interventions in earth's ecosystems. The reduction in biodiversity and extinction of nutrition-delivering species that accompany plantations may have global effects on long-term food supply. Humans must find a balance for sustainably taking advantage of biodiversity, especially concerning the fact that the world's population is still growing.[24]

Keeping in mind the endangered state of our current biodiversity, we now shift our focus to a tropical disease that is prevalent: malaria. Every year over 400 million people are diagnosed with malaria, of which 2 million, mostly children, die every year. To put malaria in the context of global climate change, it is important to recognize that both the *Anopheles mosquito* and the *Plasmodium* are affected by climate changes such as locally increased temperatures. Some evidence shows that higher incidence rates of malaria correlate with higher local temperatures. In addition, malaria should be considered in the context of recent changes in land use. Malaria is carried by the *Anopheles* mosquito as a vector. Hence, if the vector's habitat is changed, the pattern of disease occurrence will also change. Agricultural areas frequently have dams to direct water flow. Some *Anopheles* species benefit from this man-made alteration, whereas others cannot survive in the resulting environment.[26]

Plant and animal species can become displaced or extinct due to global and local changes, and human populations are by no means immune to such changes either. For example, some coastal regions can no longer be inhabited because of flooding resulting from rising sea levels. Various health problems may arise from malnutrition, geographically mobile vector-borne infections, or from consequences of extreme weather events.[26]

Human health can be seen as an indicator for global health and biodiversity. Strategies to protect biodiversity are sustainable development[1] and habitat preservation.[27] Economic independent unselfish international cooperation is required to reach these goals.[26]

First of all, the scientific community should try to ascertain more precisely the response of ecosystems to certain changes in biodiversity. Secondly, the general public, policy makers and land managers must be informed of changes of biodiversity, their extent and their irreversibility. Land managers should consider ecological and social consequences of actions that will interfere with biodiversity. Through this sort of planning, land managers have the ability to preserve a large amount of regional biodiversity. A third point is that scientists should collaborate with governmental organizations, both local and national ones. Furthermore, new international networks should be established, such as the Intergovernmental Panel on Climate Change (IPCC), focusing on changes in biodiversity and the associated global consequences. International politicians should work to reach agreements, such as the Convention on Biological Diversity. This convention establishes ways to reduce biodiversity losses, fossil-fuel emissions and land-use changes.[25]

In the long run, solutions that balance the need to save biodiversity and conduct ourselves sustainably with economically attractive uses of natural resources are most promising.

6. BIODIVERSITY OF MEDICINAL PLANTS

6.1. Procedure for the Production of Herbal Drugs and their Components

Pharmaceutical drugs are developed both by chemical synthesis in the laboratory and through the screening of natural products. "Natural products" are defined as products from natural sources, and may be produced from plants, animals, microorganisms or mineral sources. For this reason, it is promising to start the discovery for new pharmaceutically active compounds by investigating natural habitats.[28]

Whereas microorganisms can be cultivated in small quantities and cultured in laboratories, the discovery of natural plant drugs is more complex. The isolation, characterization and synthesis of complex natural chemicals are time-consuming and expensive. New chemical and analytical techniques for extraction, isolation, screening and identification are needed that are quick, inexpensive and require smaller sample sizes.[29] Techniques such as high performance liquid chromatography and high field nuclear magnetic resonance spectrometry facilitate sample fractionation and characterization quickly and with small sample volumes.[29]

To search for new biologically active components of plants for drug development processes, plant material must first be collected, extracted and then screened biologically. Either small quantities of plant material (20–500 g dry weight)[28] or larger amounts of 1 to 10 kg material[29] are required, depending on the extraction and screening methods. If the extract or isolated compound shows promising activity or if synthesis of the active compound is difficult, larger sample sizes are required (5–100 kg dry weight).[28,29] Large quantities may also be necessary for bioactivity-guided fractionation, structure elucidation and for pharmacological and biological *in vitro* and *in vivo* studies.[28,29] Even larger quantities of plant material are required for *in vivo* experiments in animals and clinical trials in patients.[28]

With the technological progress in the past two decades, it has become possible to transfer genes from unrelated species into agricultural crops using genetic methods.[29] Tissue and cell culture techniques might be promising for natural product research, because for these methods only a little quantity of plants has to be collected and screened. Plant material such as seeds or leaves can be established in cell culture. Then, this culture can be exposed with different stresses (chemicals or environmental influences) to stimulate the biosynthesis of certain phytochemicals.[29]

6.2. Measuring Biodiversity

Correct identification of medicinal plants is very important and is essential for maintaining herbal drug quality, effective therapy, and patient safety.[30] In the 1990s, a mix-up between two plants in China led to a dramatic incident. The plant *Stephania tetrandra* ("han fung ji") was an ingredient in an herbal slimming mixture in Europe. However, *Aristolochia fangchi* ("guang fang ji") has a similar Chinese name and was accidently mixed with *Stephania tetrandra.* This led to severe nephrotoxic and carcinogenic side effects caused by the aristolochic acids in *Stephania tetrandra.* The confusion of the two plants caused severe renal failure in more than 100 cases necessitating renal transplants. Many affected patients also developed urothelial carcinoma.[31] This example demonstrates why correct identification and nomenclature of plants is crucial.

Three main methods are used to identify plants: morphology or histology, chemotaxonomy and DNA-based authentication.

6.2.1. *Identification of medicinal plants by morphology*

Traditionally, morphology has been used to distinguish plants from one another. Shape, color, texture, size, smell, flowers and floral formula are some of the characteristics used by experienced botanists for correct identification.[32,33] Histological techniques include microscopic studies of tissue structure and the arrangement of cortex, cork cells, sieve tubes, xylem vessels or cell compounds.[32] The identification and authentication of medicinal plants requires raw plant material and well-trained experts. The identification of plant mixtures and dried or processed plant parts is rather difficult.[30,34]

6.2.2. *Identification of medicinal plants by chemotaxonomy*

Chemical identification of medicinal plants is based on analysis of chemical constituents. Such chemical analysis commonly makes use of techniques such as thin layer chromatography (TLC) and high performance liquid chromatography (HPLC).[32]

The constituents of a solution can be separated by TLC. The stationary phase forms an absorbance layer. Adsorbent agents like silica-gel, cellulose or aluminium oxide are put on a plate made of glass, aluminium or a synthetic material. The samples are applied to one end of the plate, which is transferred to a jar with a solvent (known as the mobile phase).[35] Depending on the mobile phase, constituents of the original solution migrate at different speeds through the stationary phase.[36] At the end of the process, the solvent front is marked. Then, R_f values (retention factors) are calculated and compared with those of reference substances to identify the constituents.[35]

HPLC is a form of column chromatography. The sample and the mobile phase are driven through a thin column containing a stationary phase. Depending on the interaction between a given compound and the stationary phase, the compound will elute from the column at a different time. If the compound interacts vigorously with the stationary phase, it will remain for a longer time in the column as compared to a compound with less affinity for the stationary phase. A detector can verify the different compounds eluted. The primary function of HPLC is the separation, identification and quantification of samples. It requires only small sample volumes.[36] Examples of other chemical approaches include reversed-phase HPLC, ultra spectroscopy, infrared spectroscopy and gas chromatography[32] as well as liquid chromatography coupled mass spectrometry (LC-MS).

The composition and yield of plant chemicals depends on growing conditions, harvesting, post-harvest processing and storage. Variation in chemical composition can hinder identification. Furthermore, distinguishing between closely related species with similar chemical components is difficult.[32]

6.2.3. *DNA-based authentication of medicinal plants*

Morphological features and chemotaxonomy do not always lead to definite or correct authentication of a species. An alternative used to identify medicinal plants is DNA technology. DNA-based molecular marker techniques are useful for the identification of plants, which are morphologically and/or chemically indistinguishable. The advantage of DNA is its stability. DNA can be isolated from fresh, dried or processed material. In contrast to chemo-profiles, DNA markers are species-specific; yet they do not differ among tissues of the same species. Another advantage of DNA-based methods is that analysis can be performed on very small amounts of plant material.[30]

There are two general approaches for genome-based identification of medicinal plants. In the first approach, the nucleotide sequence of the plant of interest at one or more gene loci is determined to assign the test sample to its species.[33] Technical methods used in this approach include allele-specific diagnostic PCR, amplification refractory mutation system (ARMS), multiplex amplification refractory mutation system (MARMS), DNA microarray and DNA sequencing.

In allele specific PCR, primers with allele specific 3′ ends labeled with different fluorochromes at their 5′ ends are applied along with a general primer in a polymerase chain reaction (PCR). Analysis of the resulting amplified DNA can be performed by gel electrophoresis or capillary electrophoresis using an automated DNA sequencer.[33]

The principle underlying ARMS is that primers require complementarity at their 3′ ends to bind to target DNA. Oligonucleotides with mutated 3′ ends cannot bind to the normal target sequence and consequently DNA amplification will not take place. Thus, the presence or absence of certain PCR products translates to the presence or absence of the target allele. In multiplex ARMS (MARMS), several primer combinations are used simultaneously.

To accelerate the often slow and labor-intensive molecular analyses, miniaturized chip-based analytical tools have been developed.[33] Simultaneous analyses of multiple genes across many taxa are possible with a single DNA microarray. For authentication of medicinal plants, DNA sequences are identified that are unique to each species.[30] In DNA microarrays, DNA of known species is synthesized and immobilized on glass slides or on silicon or nylon membranes. Then, the DNA extracted from a test sample is labeled with a specific fluorescent dye and applied to the microarray. The sample DNA hybridizes with the complementary DNA on the microarray. After a washing step, the fluorescence on the microchip is analyzed by a microarray reader.[30,33]

DNA is sequenced using cycling process of polymerization. A heat stable DNA polymerase, nucleotides, and dideoxynucleotides labeled with a fluorescent dye are combined in a thermocycler. In capillary electrophoresis, the resulting PCR amplicons are divided. Detection is carried out using laser-induced fluorescence and analysis by computer software.[33] The DNA sequence that is obtained can be compared to known sequences in genome databases; for example using BLAST (Basic Local Alignment Search Tool) in the GenBank database of NCBI (National Center for Biotechnology Information).[30]

Another possibility is the use of sequence-specific markers for loci such as the internal transcribed spacer 2 (ITS2) region of nuclear ribosomal DNA. ITS2 has the advantage of having a highly conserved region that can be used for universal primers, yet amplification of this region also provides adequate variability to distinguish between species that are closely related.[37] Yao *et al.*[37] propose use of the ITS2 locus as a barcode for authentication of plant species. A DNA barcode is a short DNA segment that can be used to distinguish one species from another.[38] The Consortium for the Barcode of Life (CBOL) uses standard DNA markers from nuclear, plastidal and mitochondrial regions for correct taxonomic identification of species.[30]

In the second approach to DNA-based identification, species-specific variations (polymorphisms) of nucleotide sequences of entire genome are used for characteristic fingerprinting of genomic DNA.[33] Techniques for the generation of genomic fingerprints include restriction fragment length polymorphism (RFLP), randomly amplified polymorphic DNA (RAPD), arbitrarily primed PCR (AP-PCR), inter-simple sequence repeat-PCR (ISSR), amplified fragment length polymorphism (AFLP) and multiplex-PCR.

Determination of restriction fragment length polymorphisms is a non-PCR based method. Genomic DNA is digested with selected restriction endonucleases and the resulting DNA fragments are separated by gel electrophoresis. After transferring the digested sample to a matrix (such as a nitrocellulose or nylon membrane), the band pattern is hybridized with chemically labeled DNA. RFLP requires large amounts of high quality DNA and is time-consuming.[30,33]

Randomly amplified polymorphic DNA PCR uses short primers with arbitrary sequences of 10 nucleotides and results in DNA fragments of different lengths, which are analyzed by gel electrophoresis. In comparison to RAPD, arbitrary polymerase chain reaction uses two larger arbitrary primers (10–50 nucleotides) and the amplicons are analyzed by polyacrylamide gel electrophoresis.[30]

In inter-simple sequence repeat PCR, primers carry simple sequence repeats (SSRs). The primers are 1 to 7 nucleotides long and are ubiquitous and highly polymorphic.

For AFLP, genomic DNA is digested with restriction enzymes. The resulting restriction fragments are ligated with specific oligonucleotide adapters, which are amplified by PCR.[30,33]

Another method is multiplex-PCR. There are various sets of forward and reverse primers resulting in parallel amplification.[33]

The use of DNA-based authentication has some advantages, e.g. its requirement of only small quantities of plant material. Furthermore, DNA authentication is not impaired by environmental factors.[30] However, DNA methods are not without limitations; the success of DNA-based methods is contingent on DNA quality and quantity. Furthermore, high concentrations of secondary metabolites can interfere with DNA extraction and PCR. Additionally, molecular authentication methods cannot be used for plant extracts.[30]

6.3. Geographical Biodiversity of Medicinal Plants

6.3.1. *Medicinal plants of America*

There exists an enormous variety of medicinal plants in the Amazon rainforest. The medicinal plants of the United States are listed in the "United States Pharmacopeia"[39] and the medicinal plants of Brazil are listed in "Brazilian Pharmacopoeia."[40] *Euterpe oleracea* Mart. (Arecaceae, common name: Açai) is a palm tree that grows in the Amazon Basin. Its fruits exert antioxidant activity due to their anthocyanins.[41] Another plant of the Western Amazon Basin that grows in regions such as Peru is Camu-Camu (*Myrciaria dubia* (H.B.K.) McVaugh, Myrtaceae). The high levels of vitamin C in the fruit provide antioxidant effects.[41] The red viscous latex of the bark of *Croton lechleri* L. (Euphorbiaceae) is also known as Dragon's Blood. This latex is used in the Amazon Valley for wound-healing and as an anti-inflammatory, antiviral and antitumor drug. Bioactive compounds include the alkaloid taspine, the lignan 3′,4-O-dimethylcedrusin, various polyphenols and anthocyanidins.[41] In the rainforest of the western Amazon Basin of Peru, Bolivia and Ecuador, the large and woody vine *Uncaria tomentosa* (Willd.) DC. (Cat's Claw, Rubiaceae) grows. Its quinovic acid glycosides and alkaloids act as an anti-inflammatory. In the Andes, *Lepidium meyenii*

Walp. (Brassicaceae, Maca) grows 4,000 meters above sea level. It has stimulant properties due to its isothiocyanates. Another plant of the Andes is the Asteraceae *Smallanthus sonchifolius* (Poepp. & Endl.) H. Rob. (also known as Yacón). It is effective in the treatment of hyperglycemia and kidney problems because it contains oligofructans and phenolic compounds. The tuber also contains high levels of fructo-oligosaccharides.[41] *Paullinia cupana* Kunth. (Sapindaceae, Guarana), a plant of the Amazon rainforest, produces a caffeine-rich fruit with stimulant effects.[41] In Brazil, many medicinal plants have been found.[42] Decoction of roots from *Aleurites molluccanus* (L.) Willd. (Euphorbiaceae) can be applied for the treatment of urinary and ovarian inflammations. The stem bark of *Amburana cearensis* (Allemao) A.C. Sm (Fabaceae) and also the fresh leaves or seeds are especially useful against infected external ulcers and also vaginal and throat infections. The rhizome tissue of *Costus spiralis* Rosc. (Zingiberaceae) acts as an antimicrobial. *Bromelia laciniosa* Mart. ex Schult.f. (Bromeliaceae) is used for the treatment of hepatitis, other liver diseases and diarrhea. *Hancornia speciosa* Gomes (Apocynaceae) is found in north-eastern Brazil. This plant has also been used for the treatment of stomach diseases and diabetes.[42] Mate (*Ilex paraguariensis* A.St.-Hil, Aquifoliaceae) is found in north-eastern Argentina, southern Brazil and Paraguay. It is used as tea, as an ingredient in dietary supplements, or as food. Stimulant and tonic effects are caused by its xanthine alkaloids. Stevia (*Stevia rebaudiana* (Bertoni) Bertoni, Asteraceae) grows in Paraguay and contains stevioside in its leaves. Stevia is used as a low caloric sweetener in some foods.[41]

6.3.2. *Medicinal plants of Africa*

Medicinal plants of Africa are described in "African Herbal Pharmacopoeia."[43] A rich diversity of medicinal plants grows in Africa. One such plant is *Triphyophyllum peltatum* (Hutch, & Dalziel) Airy Shaw (Dioncophyllaceae). It has antimalarial activity against *Plasmodium falciparum.*[44] The alkaloid 5′-O-demethyl-8-O-methyl-7-epidioncophylline A is responsible for this activity. *Dolichos marginata ssp. erecta* (Fabaceae, sphenostylins in root bark) and *Chenopodium procerum* Hochst. ex Moq (Chenopodiaceae, isoflavanoids in roots) show antifungal activity. The South African *Garcinia gerrardii* Harvey ex T.R. Sim(Clusiaceae) contains prenylated xanthones in its root bark that are fungicidal.[44] In Malawi, quinones are found in the root bark of *Diospyrus usambarensis* A. DC. (Ebendaceae) with molluscicdal and fungicidal effects, and Clusiaceae *Hypericum revolutum* Vahl contains chromenes, which are fungicidal as well.[44] In Zimbabwe, *Parinari capensis* Harv. (Chrysobalanaceae) contains diterpene lactones against fungal infections. The antifungal compound tetracycline has also been found in the root bark of *Bauhinia rufescens* Lam. (Fabaceae), growing in Niger. (Vahl) R. Br. (Acanthaceae) grows in Madagascar and has antifungal properties due to the two diterpenes fluricoserpol A and dolabeserpenoic acid A. *Swartzia madagascariensis* Desv. (Fabaceae) is a crop in East Africa. The saponines in the seeds are molluscocidal. The leaves of the endemic tree *Polyscias dichroostachya* Baker (Araliaceae), growing in Mauritius, show molluscocidal activity due to saponines.[44]

6.3.3. *Medicinal plants of Asia*

Two important pharmacopoeias in Asia are the "Indian Pharmacopoeia"[45] and the "Pharmacopoeia of the Peoples' Republic of China."[46] Medicinal plants in Asia are very diverse. This chapter focuses on plants from India, Vietnam, Laos and China.

Many ethnomedicinal plants are found in northeast India. Examples include *Swertia angustifolia* Buch.-Ham ex D. Don (Gentianaceae), used against fever and malaria; *Stemona tuberosa* Lour. (Stemonaceae), used for the treatment of asthma and tuberculosis and *Dillenia indica* L. (Dilleniaceae), which is used against diarrhea and dysentery.[47] An example of a medicinal plant from Vietnam is *Litsea verticillata* Hance (Lauraceae). The leaves and twigs of this plant contain an anti-HIV substance, the litesane sesquiterpene litsea verticillol.[48] Verrucasin L acetate is an active macrocyclic trichothecene sesquiterpernoid, which is found in leaves and stem bark of *Ficus fistulosa* Reinw. Ex Blume (Moraceae). It shows antimalarial activity.[48] In Laos, *Asparagus cochinchiensis* (Lour.) Merr. (Asparagaceae) reveals anti-HIV activity and *Nauclea orientalis* (L.) L. (Rubiaceae) displays antimalarial activity.[48] The Asteraceae *Crassocephalum crepidioides* (Benth.) S. Mooregrows is found in both Thailand and China. The plant contains jacoline and jacobine. Decoctions of the whole plant are applied to treat fever, dysentery, gastroenteritis, urinary tract infection and mastitis.[49] *Senecio integrifolius* var. *fauriri* (L.) Chlairy. (Asteraceae) is a traditional Chinese plant and is used against dysentery, conjunctivitis and tumefaction. It contains the nontoxic alkaloids integrifoline, 7-angeloyltumeforcidine, 1,2-dihydrosenkirkine and 7-angeloylheliotridine.

6.3.4. *Medicinal plants of Australia*

The medicinal plants of Australia are listed in "The Australian and New Zealand Pharmaceutical Formulary."[50] Leaves from *Amyema quandang* (Lindl.) Tiegh. (Loranthaceae), leaves from *Eremophila duttonii* F. Muell. (Myoporaceae) and stem base from *Leidosperima viscidum* R. Br. (Cyperaceae) show activity against the gram-positive bacteria *Bacillus cereus*, *Enterococcus faecalis*, *Staphylococcus aureus* and *Streptococcus pyrogenes*. In traditional medicines, the plants are used against cold (*Leidosperima viscidum*), fever (*Amyema quandang*), respiratory tract infection, sore throat, skin cuts, earache or eye inflamation (*Eremophila duttonii*).[51] *Euphorbia australis* Boiss. (Euphorbiaceae) shows an incomplete inhibition of the gram-negative bacteria *Escherichia coli*, *Klebsiella pneumoniae*, *Pseudomonas aeruginosa* and *Salmonella typhimurium*. *Euphorbia australis* is traditionally used for skin sores and medicinal washes.

Two examples of medicinal plants with antiviral activity against poliovirus are *Dianella longifolia* var. *grandis* (Liliaceae, roots) and *Pterocaulon sphacelatum*. (Asteraceae, green aerial parts). Furthermore, extracts from *Euphorbia australis* Boiss. (Euphorbiaceae) and *Scaevola spinesces* R. Br. (Goodeniaceae) show activity against human cytomegalovirus. Plant extracts against Ross River virus include *Eremophila latrobei* F. Muell. ssp. *glabra* (L.S. Smith) Chinn. (Myoporaceae) and *Pittosporum phylliraeoides* DC. var. *microcarpa* S. Moore (Pittosporaceae).[52]

6.3.5. *Medicinal plants of Europe*

Medicinal plants of Europe are listed in the "European Pharmacopoeia,"[53] "German Pharmacopoeia"[54] and "British Pharmacopoeia."[55] The section below gives examples of medicinal plants from Italy and Spain.

The juice of *Chelidonium majus* L. (Papaveraceae) is used in Italy for the treatment of warts. It is topically applied. The aerial parts of *Echium italicum* L. (Boraginaceae) are applied to treat abdominal pains. The buds and flowers of *Crocus napolitanus* (Liliaceae) are used as an external poultice for the prevention of lice and act as an antiseptic. The bulbs of *Lilium candidum* L. (Liliaceae) are useful for the treatment of shingles (*Herpes zoster*). The whole of *Parmelia* sp. Ach. (Parmeliaceae) is used as a cholagogue.[56]

Phlomis lychnitis L. (Lamioideae) is a Spanish plant. The decoction of this plant is used against hemorrhoids. *Quercus suber* L. (Fagaceae) can be used for the disinfection of wounds and as vulnerary. Furthermore, *Marrubium vulgare* L. (Lamiaceae) can be used for the treatment of asthma. As a final example, *Leuzea conifera* (L.) DC. (Asteraceae) should be mentioned: its decoction is applied to treat gastritis and colitis.[57]

6.4. Medicinal Plants for the Treatment of Cancer

Medicinal plants are of particular interest in the treatment of cancer. Plants with anticancer-potential have been found all over the world.

In Chile, the leaves of Boldo (*Peumus boldus* Molina, Monimiaceae) are hepatoprotective, because they contain aporphine-like alkaloids such as boldine.[41] Boldine has also shown antitumor properties. The South American plant *Annona muricata* L. (Annonaceae), also known as Graviola, contains acetogenine, which is of great interest in cancer chemotherapy because it has been shown to inhibit multi-drug resistance in cancer cells.[41] Pau d'Arco-Lapacho (*Tabebuia impetiginosa* (Mart. Ex DC.) Standl., Bignoniaceae) grows in the Amazon rainforest and other parts of South America and contains lapachol and beta-lapachone I. The latter is assumed to be a potent antitumor compound.[41]

The compound alvaradoin E is found in the leaves of *Alvaradoa haitiensis* Urb. (Picramniaceae), which originates in the Dominican Republic. This compound shows antileukemic activity.[58] The bulb of *Pancratium littorale* Jacq. (Amaryllidaceae) contains pancratistatin 3,4-O-cyclic phosphate sodium salt, which also has anticancer potential. The rocaglate derivate silvestrol was detected in the fruit and twigs of *Aglaia foveolata* Pannell (Meliaceae) collected in Indonesia. Silvestrol reveals cytotoxic effects against human cell lines derived from breast, prostate and lung cancers.[58]

Catharanthus roseus (L.) G. Don (Apocynaceae) is a well-known plant containing *Vinca* alkaloids. The *Vinca* alkaloids bind to two high affinity sites on each tubulin dimer. This binding results in a stabilization of the αβ-heterodimer and consequently in destabilization of microtubules.[59] Paclitaxel was isolated from *Taxus brevifolia* Nutt. (Taxaceae, cortex) and promotes the destabilization of microtubules. It causes mitotic arrest by preventing proper assembly of the mitotic spindle or by reducing chromosome oscillations. There is also a taxol-binding site at the amino terminus of *β*-tubulin.[59] In China, high levels of paclitaxel are found in *Taxus chinensis* (Miq.) de Laub. (Taxaceae).

A herbal mixture of the following plants, all of which grow in India *Claoxylon hassianum*, *Celrodentrum walichii* Merr. (Verbenaceae), *Mussaenda macrophylla* Wall. (Rubiaceae), *Phlogacanthus thyrsiformis* (Hard.) Mabb. (Acanthaceae) and *Thevesia pamata*, has been reported to be effective against cancer.[47]

There are many plants used in TCM that have anticancer potential. Camptothecin is isolated from *Camptotheca acuminata* Decne. (Nyssaceae), which grows in China and inhibits the ligation of DNA after topoisomerase I-mediated DNA strand breaks.[31] *Podophyllum emodi* var. *chinense* (Berberidaceae), a Chinese medicinal plant, contains podophyllotoxin.[31] Other targets to treat cancer include proteins of signal transduction pathways in malignant cancer cells. For example, emodin was found in Chinese *Rheum palmatum* L. (Polygonaceae) and inhibits casein-kinase 2. Verbascoside, isolated from *Garcinia hanburyi* Hook. F. (Clusiaceae) inhibits telomerase. Telomerases are essential for tumor cell immortalization.[31] Another strategy for cancer treatment is inhibition of tumor angiogenesis. Capsaicin, found in various *Capsicum* species L. (Solanaceae) and sinomemine, found in *Sinomenium acutum* (Thunb.) Rehd. et Wils. (Menispermaceae), are natural inhibitors of angiogenesis.[31]

REFERENCES

1. Cassis G. (1998) Biodiversity loss: a human health issue. *The Medical J Australia* **169**: 568–569.
2. Jutro PR. (1991) Biological diversity, ecology, and global climate change. *Environ Health Perspect* **96**: 167–170.

3. Osborn L. (2011) Number of Species Identified on Earth. http://www.currentresults.com/Environment-Facts/Plants-Animals/number-species.php (29.09.2011)
4. GOLD (Genomes OnLine Database). (2011) http://genomesonline.org/cgi-bin/GOLD/bin/gold.cgi?page_requested=Complete+Genome+Projects (07.10.2011)
5. Wang X, Greenfield P, Li D, *et al.* (2011) Complete genome sequence of a nonculturable methanococcus maripaludis strain extracted in a metagenomic survey of petroleum reservoir fluids. *J Bacteriol* **193**: 5595.
6. Krause DO, Little AC, Dowd SE, *et al.* (2011) Complete genome sequence of adherent invasive Escherichia coli UM146 isolated from Ileal Crohn's disease biopsy tissue. *J Bacteriol* **193**: 583.
7. Lander ES, Linton LM, Birren B, *et al.* (2001) International Human Genome Sequencing Consortium. Initial sequencing and analysis of the human genome. *Nature* **409**: 860–921.
8. Leipe DD. (1996) Biodiversity, genomes, and DNA sequence databases. *Curr Opin Genet Dev* **6**: 686–691.
9. Wuketits FM. (1997) The status of biology and the meaning of biodiversity. *Naturwissenschaften* **84**: 473–479.
10. Gaston KJ. (2000) Global patterns in biodiversity. *Nature* **405**: 220–227.
11. Basso LA, da Silva LH, Fett-Neto AG, *et al.* (2005) The use of biodiversity as source of new chemical entities against defined molecular targets for treatment of malaria, tuberculosis, and T-cell mediated diseases — a review. *Memorias do Instituto Oswaldo Cruz* **100**: 475–506.
12. Rollinger JM, Langer T, Stuppner H. (2006) Strategies for efficient lead structure discovery from natural products. *Curr Med Chem* **13**: 1491–1507.
13. Shuler ML. (1994) Bioreactor engineering as an enabling technology to tap biodiversity. The case of taxol. *Ann NY Acad Sci* **745**: 455–461.
14. Tripathi CK, Tripathi D, Praveen V *et al.* (2007) Microbial diversity — biotechnological and industrial perspectives. *Indian J Exp Biol* **45**: 326–332.
15. Langer M, Gabor EM, Liebeton K, *et al.* (2006) Metagenomics: an inexhaustible access to nature's diversity. *Biotechnol J* **1**: 815–821.
16. Bull AT, Goodfellow M, Slater JH. (1992) Biodiversity as a source of innovation in biotechnology. *Annual Rev Microbiol* **46**: 219–252.
17. Torsvik V, Ovreas L, Thingstad TF. (2002) Prokaryotic diversity — magnitude, dynamics, and controlling factors. *Science* **296**: 1064–1066.
18. Hodkinson TR, Waldren S, Parnell JA, *et al.* (2007). DNA banking for plant breeding, biotechnology and biodiversity evaluation. *J Plant Res* **120**: 17–29.

19. Singh A, Nigam PS, Murphy JD. (2010) Renewable fuels from algae: an answer to debatable land based fuels. *Bioresour Technol* **102**: 10–16.
20. Costa JA, de Morais MG. (2010) The role of biochemical engineering in the production of biofuels from microalgae. *Bioresour Technol J* **102**: 2–9.
21. Frusciante L, Barone A, Carputo D, *et al.* (2000) Evaluation and use of plant biodiversity for food and pharmaceuticals. *Filoterapia* **71**: 66–72.
22. Hammer K, Arrowsmith N, Gladis T. (2003) Agrobiodiversity with emphasis on plant genetic resources. *Die Naturwissenschaften* **90**: 241–250.
23. Johns T, Sthapit BR. (2004) Biocultural diversity in the sustainability of developing-country food systems. *Food Nutr Bull* **25**: 143–155.
24. Lewis SL. (2006) Tropical forests and the changing earth system. *Philos Trans R Soc Lond B* **361**: 195–210.
25. Chapin FS, Zavaleta ES, Eviner VT, *et al.* (2000) Consequences of changing biodiversity. *Nature* **405**: 234–242.
26. McMichael AJ, Patz J, Kovats RS. (1998) Impacts of global environmental change on future health and health care in tropical countries. *British Med Bull* **54**: 475–488.
27. Wildt DE. (2000) Genome Resource Banking for Wildlife Research, Management, and Conservation. *ILAR J* **21**: 228–234.
28. Soejarto DD. (1996) Biodiversity prospecting and benefit-sharing: perspectives from the field. *J Ethnopharmacol* **51**: 1–15.
29. Reid WV. (1996) Gene co-ops and the biotrade: translating genetic resource rights into sustainable development. *J Ethnopharmacol* **51**: 75–92.
30. Heubl G. (2010) New aspects of DNA-based authentication of Chinese medicinal plants by molecular biological techniques. *Planta Med* **76**: 1963–1974.
31. Efferth T, Li PCH, Konkimalla VSB, *et al.* (2007) From traditional Chinese medicine to rational cancer therapy. *Trends Mol Med* **13**: 353–361.
32. Zhang YB, Shaw PC, Sze CW, *et al.* (2006) Molecular authentication of Chinese herbal materials. *J Food and Drug Analysis* **15**: 1–9.
33. Sucher NJ, Carles MC. (2008) Genome-based approaches to the authentication of medicinal plants. *Planta Med* **74**: 603–623.
34. Cimino MT. (2010) Successful isolation and PCR amplification of DNA from National Institute of Standards and Technology herbal dietary supplement standard reference material powders and extracts. *Planta Med* **76**: 495–497.
35. Latscha HP, Linti GW, Klein HA. (2003) Analytische chemie: Chemie-Basiswissen III, 4th edn. Berlin Heidelberg: Springer.
36. Mortimer CE, Müller U. (2007) Chemie, 9th edn. Stuttgart: Georg Thieme Verlag.
37. Yao H, Song J, Liu C, *et al.* (2010) Use of ITS2 region as the universal DNA barcode for plants and animals. *PLoS One* **5**: e13102.

38. Lou SK, Wong KL, Li M, *et al.* (2010) An integrated web medicinal materials DNA database: MMDBD (Medicinal Materials DNA Barcode Database). *BMC Genomics* **11**: 402.
39. United States Pharmacopeial Convention (2008) United States Pharmacopeia 2008.
40. Brazilian Pharmacopoeia Committee. *Brazilian Pharmacopoeia.*
41. Desmarchelier C. (2010) Neotropics and natural ingredients for pharmaceuticals: why isn't South American biodiversity on the crest of the wave. *Phytother Res* **24**: 791–799.
42. Benko-Iseppon AM, Crovella S. (2010) Ethnobotanical bioprospection of candidates for potential antimicrobial drugs from Brazilian plants: state of art and perspectives. *Current Protein and Peptide Science* **11**: 189–194.
43. Brendler T, Eloff JN, Gurib-Fakim A, *et al.* (2010) African Herbal Pharmacopoeia, 1st edn. (Association for African Medicinal Plants Standards).
44. Khalid SA. (2009) Decades of phytochemical research on African biodiversity. *Nat Prod Commun* **4**: 1431–1446.
45. Indian Pharmacopoeia Commission (2010) *Indian Pharmacopoeia*, 6th edition.
46. State Pharmacopoeia Commission of the PRC (2011) *Pharmacopoeia of the Peoples' Republic of China 2010*, 9th edn., Stationery Office Books.
47. Rai PK. (2011) Assessment of multifaceted environmental issues and model development of an Indo-Burma hotspot region. *Environ Monit Assess* DOI: 10.1007/s10661-011-1951-8.
48. Soejarto DD, Zhang HJ, Fong HH, *et al.* (2006) Studies on biodiversity of Vietnam and Laos 1998–2005: examining the impact. *J Nat Prod* **69**: 473–481.
49. Roeder E. (2000) Medicinal plants in China containing pyrrolizidine alkaloids. *Pharmazie* **55**: 711–726.
50. The Pharmaceutical Association of Australia and New Zealand (1934) The Australian and New Zealand Pharmaceutical Formulary 1934.
51. Palombo EA, Semple SJ. (2001) Antibacterial activity of traditional Australian medicinal plants. *J Ethnopharmacol* **77**: 151–157.
52. Semple SJ, Reynolds GD, O'Leary MC, *et al.* (1998) Screening of Australian medicinal plants for antiviral activity. *J Ethnopharmacol* **60**: 163–172.
53. European Pharmacopoeia Commission (2008) Europäisches Arzneibuch 6, (Deutscher Apotheker Verlag, Stuttgart).
54. German Pharmacopoeia Commission (2010) Deutsches Arzneibuch 2010, (Deutscher Apotheker Verlag, Stuttgart).
55. British Pharmacopoeia Commission (2010) *British Pharmacopoeia 2010*, 1st edn, (Stationery Office Books (TSO)).

56. Pieroni A. (2000) Medicinal plants and food medicines in the folk traditions of the upper Lucca Province, Italy. *J Ethnopharmacol* **70**: 235–273.
57. Vázquez FM, Suarez MA, Pérez A. (1997) Medicinal plants used in the Barros Area, Badajoz Province (Spain). *J Ethnopharmacol* **55**: 81–85.
58. Kinghorn AD, Chin YW, Swanson SM. (2009) Discovery of natural product anticancer agents from biodiverse organisms. *Curr Opin Drug Discov Devel* **12**: 189–196.
59. Hung DT, Jamison TF, Schreiber SL. (1996) Understanding and controlling the cell cycle with natural products. *Chem Biol* **3**: 623–639.

CHAPTER 2

Biodiversity and Metagenomics

Eva-Maria Surmann and Thomas Efferth*†

ABSTRACT

Metagenomics refers to all the environmental communities and comprises the genomes of hundreds or more organisms from all three domains of life i.e. archaea, bacteria and eukarya, as well as those of viruses in contrast to genomics, which comprise the entire genetic complement of only one organism. The study of genetics and genomics from all the environmental communities without prior isolation and cultivation reveals new insights into the biodiversity, evolution and metabolic potential of the uncultured microbial world. By faster sequencing technologies at reasonable costs, sequencing of entire metagenomes enables the discovery of genes and phylogenetic anchors in reconstructed metagenomes of entire communities. Metagenomics can be applied to marine microbes, soil microbials, microbial communities of the human body, etc. Sophisticated bioinformatic tools are necessary to construct and screen metagenomic libraries. Metagenomics comes more and more into focus in pharmaceutical and biotechnological industries for the discovery of novel biocatalysts and antibiotic drugs.

1. INTRODUCTION

1.1. Evolution and the Phylogenetic Tree of Life

More than three billion years ago, the earth was completely different from what we know of it today, sole living organisms at that time developed

*Institute of Pharmacy and Molecular Biotechnology, University of Heidelberg, Im Neuenheimer feld 364, 69120 Heidelberg, Germany

†Institute of Pharmacy and Biochemistry, Johannes Gutenberg University, Staudinger Weg 5, 55128 Mainz, Germany

mechanisms that enabled the evolution of all complex life forms on earth. They may have started with photosynthesis to generate energy and continued with splitting water molecules, releasing oxygen into the atmosphere that may have been the starting point for the complex, multicellular life.[1] Our earth contains, in addition to plants and microorganisms, a large number of complex animal species, which is estimated between 1 million up to 20 million. Such numbers may sound enormous at first glance, but they might represent only a minor fraction (probably less than 1percent) of animals that have ever existed. During the evolution of life, animals diverged from a common ancestor more than 540 million years ago through modifications of DNA that led to morphological diversity. Yet, they all share common specific gene families that are responsible for major patterns of body composition. Animal diversity increased dramatically in Cambrium (545–490 million years ago) resulting in the development of some of the modern phyla known today such as arthropods or molluscs, and to the appearance of large, complex animals in the Early-Middle Cambrian — a period called the "Cambrian explosion." So far, 35 animal phyla have been identified that are grouped according to common shared characteristics.[2] Although those complex life forms seem to be more attractive, the quasi invisible microbes are by far more abundant and diverse than multicellular taxa. They are "the engine of life" as they maintain the essential atmospheric and chemical conditions through the conversion of carbon, nitrogen, oxygen and sulfur into forms accessible to other life forms.[1]

To capture the origin and evolution of life, it is necessary to study phylogenetic relationships among individuals. Based on anatomical and embryological comparisons many diverse phylogenetic schemes and trees have been suggested and put into question during the last decades. To circumvent critical and inconsistent morphological comparisons, genetic characters have been used successfully to determine phylogenies of organisms.[2]

First, there are only few genes and proteins that are responsible and even essential for the regulation of the development of organisms. These regulation factors are termed "genetic toolkit," and seem to be present in any given animal. This toolkit mainly consists of transcription factors regulating the expression of genes involved in signalling pathways, which mediate cell-to-cell interactions involved in formation and composition of the body. Toolkit genes are conserved throughout evolution. An example is

Hox genes, first identified in *Drosophila* and later also in frogs, mice and humans (vertebrates). The discovery of the same genetic toolkit in totally different and even long-diverged phyla raises many developmental and evolutionary questions, but also allows some insights into the mysteries of life. How do totally different structures that are controlled by the same genetic toolkit develop and how does this finding influence understanding about phylogenetics and evolution.[2]

Possible answers may be: Genetic differences may arise from distinct gene expression patterns during different developmental phases; expressed gene products may interact with distinct interaction networks dependent on the time of expression. Moreover, genetic differences result from different coding contents, including the number and biochemical functions of toolkit genes. The most striking events are mutation, duplication and divergence based on DNA replication errors or unequal cross-over events during recombination that lead to sequence development and the formation of gene families. Interestingly, the extent of gene duplication was positively correlated with the complexity of the animal phylum.[2]

However, in evolution, sequence homology was conserved within and among phlya, as identified for toolkit genes and also for "housekeeping" genes. Therefore, these sequences must have been present in the ancestor of this group. Consequently, initial gene sequence diverged, leading to many distinct phlya and species.[2]

In summary, changes in developmental gene regulation are dominant mechanisms leading to different morphological evolution. The degree of divergence was shown to be correlated with the relationship of phyla. Thus, evolutionary conserved gene families enable the construction of phylogenetic trees based on genetic characteristics (e.g. the presence or absence of particular genes or the linkage of a group of genes).[2] Genetic analysis based on small sub-unit rRNA (SSU rRNA) comparisons lead to the construction of the popular tripartite tree of life created by C.R. Woese and co-workers in the late 1970s. According to this phylogenetic tree, life on earth is divided into three kingdoms,[3] of Archaea, Bacteria and Eukaryota instead of a main division between plants and animals as proposed by Linnaeus at the beginning of the twentieth century or the five-kingdom system suggested by R.H. Whittaker in 1969.[4] Since then, microbiology has followed a constant revolution, and the invention of molecular methods such as amplification, cloning and sequencing of small subunit ribosomal RNA genes from the environment facilitates access to microbial diversity.[3]

The greatest phylogenetic diversity can be found in the microbial world, which had about three billion years more to evolve and diversify to reach their recent variety. The huge number of microbes play a key role in maintaining atmospheric and chemical conditions necessary for survival of all life forms on earth. Thus, especially bacterial phyla show an explosive radiation. Therefore, they represent the maximal diversity on earth, since they are able to occupy every imaginable niche with the exception of a few extreme environments, such as salt crystallizers or hydrothermal vents that are dominated by archaea. Classical DNA reassociation data together with new analytical approaches estimates several million bacterial species solely present in soil.[5] By contrast, archaea split into two (or perhaps three) kingdoms with much more restricted diversity and scaled differentiation, mostly living under extreme energy-challenging conditions. Several attempts were made to explain the evolution of eukaryota, the most famous of which was the archaezoa hypothesis. However, most of the deepest nodes of the eukaryotic tree remain a mystery.[2] It is important to keep in mind that these ratings are based on findings by conventional laboratorial techniques. The real biodiversity may still not be adequately estimated, and new developments may lead to totally different distributions among kingdoms.

In the context of phylogenetics, it is necessary to introduce the term "biodiversity" that covers many aspects of biological variation. Biodiversity was defined in 1992 by the Convention on Biological Diversity:[6]

> "*Biological diversity means the variability among living organisms from all sources including, inter alia (i.e. among other things), terrestrial, marine and other aquatic ecosystems and the ecological complexes of which they are part; this includes diversity within species, between species and ecosystems.*"

Thus, biodiversity comprises all forms, levels and combinations of natural variation. A common concept puts forward three categories of biodiversity: ecological diversity that includes the diversity of biomes, bioregions, landscapes, ecosystems, habitats, niches and populations, organismal diversity (domains (or kingdoms), phyla, families, genera, species, subspecies, populations and individuals), and genetic diversity that comprises variations in populations, individuals, chromosomes, genes and nucleotides.

These groups can be considered separately, but share common elements as well, such as "populations" that appears in all three categories. Therefore, it is difficult to find exact definitions with precise demarcations of all levels of biodiversity and going further into detail is beyond the scope of this overview.[6]

Most genetic studies focus on microorganisms because they show the greatest diversity on earth. Genetic analyses are necessary to study taxonomy, evolution, physiology and ecology of microbes.[7] One can get an insight into the huge microbial biodiversity, by the *in vitro* cultivation of microbes. The most challenging problem is that most microorganisms are difficult to capture for cultivation due to specialized growth requirements. Thus, we are not aware of their existence.[8,9] Accordingly, only a minority of microorganisms living in a given habitat are cultivable. For example, only 0.001–0.1 percent of the microorganisms in seawater, 0.25 percent in fresh water, 0.25 percent in sediments and 0.3 percent in soil can be grown in culture. Conversely, 99 percent of the microbial diversity are not accessible to classical cultivation methods.[9,10]

The first analysis of uncultured bacteria in the environment focused on the construction of a 5S rRNA cDNA library derived from the symbiotic microbial community within the tubeworm *Riftia pachyptila*.[11,12] Direct cloning of DNA from an environment was initially suggested by Pace *et al.*[13] and first applied by Schmidt and co-workers,[14] who constructed a λ phage library from a seawater sample to screen for 16S rRNA genes. Since then, viral communities had been the subject of several metagenomic investigations and were among the earliest to be studied, since they play a key role in microbial evolution and ecology.[12]

The increased interest and activity in metagenomic research during the past decade, accompanied by the rapid progress in technology paved the way for the study of single genes, pathways, organisms and even whole communities of many habitats and improved our knowledge of global diversity.[12]

1.2. What is Metagenomics?

In contrast to genomics, which comprises the entire genetic complement of one organism, metagenomics refers to entire environmental communities and comprises the genomes of hundreds or more organisms from all three domains of life (archaea, bacteria and eukarya), as well as those of viruses. The study of genetics and genomics from whole environmental communities without prior isolation and cultivation reveals new insights into the biodiversity, evolution and metabolic potential of the uncultured microbial world.[8,15] Besides, it should be noted that metagenomic approaches do not replace studies on individual species using classical genomics and microbiology.[1] However, in addition to metagenomics, in RNA transcripts and proteins can also be isolated from environmental samples, giving rise to the new fields of metatranscriptomics and metaproteomics.[8]

The term "metagenomics" was first introduced by Handelsman in 1998[16] and defined as "the habitat-based investigation of mixed microbial populations at the DNA level" and specified in 2004[17] as "the genomic analysis of microorganisms by direct extraction and cloning of DNA from an assemblage of microorganisms." Further definitions were formed by Riesenfeld,[12] who described the metagenomic approach as the functional and sequence-based analysis of the collective microbial genomes contained in an environmental sample." Recently, the analysis of metagenomics has been more broadly defined as "sampling the genome sequences of a community of organisms inhabiting a common environment" or as "any type of analysis of DNA obtained directly from the environment."[8]

The analysis of metagenomics is based on a "gene-centric approach" that characterizes environments by determining the function of the proteins encoded in the sequenced genes.[18] Genes that are more frequently present in one community in comparison to another one may have a beneficial function for these organisms, e.g. genes encoding for proteorhodopsin are found widely distributed in marine surface waters to generate energy, whereas other genes associated with different mechanisms of energy production are used in soil. In addition, genes of unknown function harbor the possibility to discover new proteins or enzymes, which maybe important for the pharmaceutical or chemical industry.[8]

However, the difficulties in cultivation have influenced our view of microbial diversity and limited our appreciation of the microbial world. Therefore, metagenomics provides a relatively unbiased view concerning the community composition (species richness and distribution), as well as the metabolic properties of a community. In principle, any environment can be analyzed for metagenomics as far as nucleic acids can be extracted from the sample including soil, aquatic habitats, the oral cavity and feces, as well as the hospital metagenome.[8,12] Most interest is focused on the marine environment, in which large metagenomic studies have been done, including the Global Ocean Sampling Expedition, which has generated a data set of 6.12 million predicted proteins from 7.7 million shotgun sequences. Another interesting and international initiative is the Human Microbiome Project that analyzes human-inhabiting microbial populations.[8]

2. METAGENOMIC INSIGHTS INTO ENVIRONMENTAL MICROBIAL COMMUNITIES

2.1. Metagenomic Combination of Phylogenetics and Function within Species

In the past, either single genes or entire genomes of an organism have been studied to reveal the diversity and evolution of microbial populations in a given habitat.[7] This method neither reveals the complexity of the microbial diversity, nor the role of the different members of the population within the community. It also does not provide information about the physiology of organisms, since microorganisms with identical 16S rDNA sequences can have totally different genomes and completely different physiologies.[19–21] Nonetheless, 16S rRNA/DNA studies remain important for microbial ecology. Nowadays, it is important to link the identity of different microorganisms within a habitat to the processes they perform in their environment. The first metagenomic studies linked a metabolic function with its phylogenetic source, revealing information and the potential role of one particular species in its community.[12,21] Different approaches are possible to link a phenotype or a specific metabolic function with the identity of the original host:

One approach is to screen metagenomic libraries related to a specific activity to search for one particular phenotype. The phylogenetic origin can then be identified by PCR of 16S rDNA or by sequencing genes flanking the region of interest. The flanking genes can be compared to homologous genes in GenBank or diverse other databases.[12] However, one has to be aware that horizontal gene transfer between microbes may falsify the results.[12,22] Nevertheless, this screening strategy has already been successfully applied and new antibiotics have been identified, such as terragine,[23] turbomycin A and B,[24] or acyl tyrosines,[25] as well as hydrolytic and degradable enzymes, biosynthetic pathways, antibiotic resistance enzymes and membrane proteins.[26]

Conversely, metagenomic libraries can be screened for specific phylogenetic anchors, such as rRNA genes (e.g. 16S rRNA), or diverse other genes with phylogenetic potential such as *radA/recA* homologues. This strategy followed by sequencing of genes flanking the anchor.[12,27] In many cases, ORFs can be identified with hypothetical or unknown function. Nonetheless, these sequences provide useful information about the G+C content, codon usage or promoter sites that might be helpful to understand the expression strategy of uncultured and unknown phyla. The function of these genes might be unravelled as soon as it will be identified for a homologue of the same gene family. However, the huge number of conserved hypothetical proteins did not change in many previous sequencing projects. Therefore, an important task for the future is the identification of about 70 percent of the predicted proteins.[12,28]

These two metagenomic approaches reveal important information about phylogeny and function within species as illustrated by examples:

The sequencing of 6 fosmid clones and 12 BAC clones from uncultured soil *Acidobacterium* for example demonstrated the huge genetic diversity of Acidobacteria and revealed many genes with homology to housekeeping genes involved in DNA transport, repair, cell division, etc. Genes, encoding cyclic β 1′-2′ glucan synthetase or polyhydroxybutyrate depolymerase, provided insights into the ecological roles of Acidobacteria. In addition, genes homologous to genes from members of the α-Proteobacteria phylum (e.g. penicillin-binding protein, zinc metalloprotease, etc.) were also found, which might be the result of horizontal gene transfer between species.[26,29]

Another example is the discovery of *Cenarchaeum symbiosum*, a dominant archaeal symbiont in *Axinella* sp. sea sponge. This was found in a culture-independent 16S rRNA gene survey.[30] Genome sequencing of *Cenarchaeum symbiosum* revealed the genetic variation within the population and symbiotic relationship and, thus, the role of population biology in this sponge.[12,31,32]

An important finding was the discovery of rhodopsin-like photoreceptors in marine bacteria by Béjà and co-workers in 2000.[33] These photoreceptors have only been found in Archaea so far. These bacteriorhodopsins couple light energy with carbon-cycling. They were found in many marine Proteobacteria optimized for various wavelengths at distinct ocean depths.[26]

2.2. Metagenomic Determination of Phylogenetics and Function of Entire Microbial Communities

Sequencing of entire metagenomes, by faster sequencing technologies at reasonable costs, enables the discovery of genes and phylogenetic anchors in reconstructed metagenomes of entire communities. In most cases, the metagenome of only simple communities has been sequenced and even these communities showed enormous microheterogeneity making reconstruction of genomes a difficult task.[26]

The first metagenome reconstructions were performed for viral communities in the ocean[34] and human feces.[35] Bacteriophages have a profound effect on the diversity and population structure of bacterial communities by mediating genetic exchange between bacteria.[26,36] Since simple phylogenic studies such as targeting the 16S rRNA of microbial communities are not possible in viruses, complete viral genome sequencing is necessary for viral phylogeny. Breitbart and co-workers found 1,200 viral genotypes in human faeces, demonstrating a huge diversity of bacterial species in the human intestine.[35] The same group generated 2,000 viral sequences collected at two different marine locations, showing huge differences between each other. In relation to the faecal samples, a predominance of gram-negative bacteria in seawater and gram-positive bacteria in the gut was observed.[34]

Further attempts were made by sequencing the metagenome of an acid mine drainage (AMD) biofilm, containing a relatively simple and highly interacting community with three bacterial and three archaeal species. Acid mine drainages emerge as a consequence of bacterial iron oxidation in abandoned mines, resulting in acidification based on the insolubility of pyrite.[12,37,38] Tyson and co-workers extracted the DNA directly from the biofilm and constructed a small insert library. Differences in G+C content associated with 16S rRNA analysis revealed almost complete genome sequences for *Leptospirillum* group II and *Ferroplasma* type II and partial genome sequences for *Leptospirillum* group III, *Ferroplasma* type I, and G-plasma. Each genome harbored genes that were responsible for the fixation of carbon via the reductive acetyl coenzyme A pathway. In addition, they found genes in the *Ferroplasma* type I and II genomes that were similar to sugar and amino acid transporters, supporting a heterotrophic lifestyle of these organisms. In contrast, the only genes responsible for N_2 fixation were found in *Leptospirillum* group III.[12,37]

2.3. Biodiversity and Global Impact of Marine Microbes

The marine environment is most important for metagenomic research, where many exciting discoveries have been made (Fig. 1). Surprisingly, most marine bacteria are responsible for half of the photosynthesis rate on earth, sustaining essential chemical balances of the atmosphere.[1]

The ocean is one of the most important ecosystems on earth that is mainly driven by marine microbes and that needs to be further explored to understand the complex interactions between organisms. Modern marine microbial ecology started in the 1970s when it was discovered that most respiration in the oceans results from microorganisms.[39–41] The fact that the ocean occupies about two-thirds of the planet's surface with an average depth of 3,800 m clearly demonstrates the

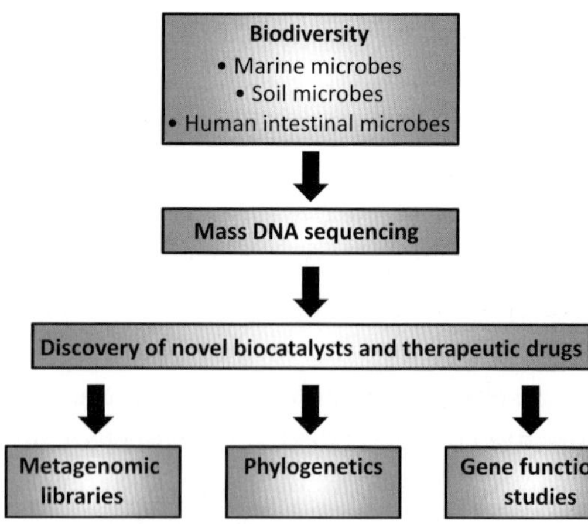

Figure 1 The potential of biodiversity by means of metagenomics.

role of the marine microbial community in global biogeochemical cycling.[15]

The marine environment can be separated in many different habitats such as the sea surface microlayer (SML), the bulk water column or the habitats around the seafloor that contain totally different community structures. Thus, the marine environment with its manifold niches shows enormous microbial biodiversity, since marine microbes are found to be extremely abundant and diverse.[42,43] Pelagic bacteria achieve densities up to 10^6/ml of seawater, being responsible for most oceanic biomass and half of the total primary production on the planet, as well as 95 percent of the total respiration.[43–46]

Most microbial studies were done in the photic zone of the ocean, where high cell densities and activities can be found and where most primary production occurs. The deeper oceanic layers are also interesting because marine microbes have to cope with extremes in pressure, temperature, salinity and nutrient availability. Therefore, they show unique biochemistry. Enzymes produced by these microbes show particular biocatalytic characteristics and may be interesting for biotechnological applications.[15,43] Besides, microbes are mainly involved in symbiotic relationships with various marine invertebrates, including sponges, corals, squids, sea squirts, and tunicates. In the case of marine sponges, bacterial symbionts were reported to produce many of the products finally released by these invertebrate, such as biologically active and pharmacologically valuable natural products that primarily serve as defense mechanisms for the sponge, but also show a huge potential as novel therapeutical drugs.[43,47,48]

To date, diverse enzymes have been isolated from cultivable marine microbes, such as the non-specific nuclease isolated from the bacteriophage of the thermophile *Geobacillus* sp. 6K51[49] or the cold-adapted lipase from the γ-proteobacterium *Pseudoalteromonas haloplanctis* TAC125.[43,50] Furthermore, several novel hydrolytic enzymes have been found in bacterial metagenomic DNA of the Antarctic sea water,[51] as well as an alkane hydroxylase gene in a metagenomic library derived from samples of the deep-sea sediment in the Pacific.[43,52] These examples represent only a small fraction of the ocean's capacity, demonstrating the huge potential of the marine environment as source of new biocatalysts that can be discovered by metagenomic research.[43]

Actinobacteria were found to be important contributors to the marine environment.[42] New cultivation methods as well as the metagenomic approach revealed novel insights in biodiversity and biogeography of actinobacteria, as well as new sources for natural products, although actinomycetes were not linked to major ecological processes in the marine environment in the past.[42,53] 16S rRNA studies of SML of oligotrophic waters of the Bay of Banyuls-sur-Mer in France and of the Olympic Harbour in Barcelona, both environments with unique properties and a high microbial diversity and density, showed only a minor fraction of the presence of Actinobacteria.[54] A similar discovery was made for the oligotrophic bulk water column of the North Atlantic.[55] Cyanobacteria were found as a dominant phylum in the water column, whereas actinobacteria were less present and were in a range similar to the β-proteobacteria and other bacterial groups. The maximum diversity of actinomycetes was found in marine sediments, where more novel actinobacteria have been isolated earlier due to new technical developments.[42] Actinomycetes were also identified in many marine vertebrates and invertebrates, including sessile ones, which produce mostly bioactive metabolites that might be required as chemical defense. One of these active metabolites is the neurotoxin tetrodotoxin, which was believed to be solely produced by the pufferfish, but now many diverse prokaryotic taxa have been identified to produce this toxin as well, including actinomycetes.[42,56] Actinomycetes antagonize the growth of other bacterial strains, thereby revealing competitive interactions.[57–59] In summary, actinobacteria occupy many diverse niches of the marine environment. Despite the fact that they are few in number, they show high diversity and different ecological roles that have to be further clarified to determine their exact role in the marine environment.[42]

Moreover, metagenomic studies of the microbial community in the Sargasso Sea revealed new insights into the biogeochemistry of marine ecosystems. In the past, oceanic nitrification was thought to be solely dependent on bacteria. In 2004 Venter[55] and co-workers found an archaeal ammonium monooxygenase. Furthermore, genes involved in the uptake of various forms of phosphorus were identified, revealing insights into the phosphorus cycle of organisms living in phosphorus-limited environments.[12]

Another example is that of marine cyanobacteria, which are grouped by 16S rRNA analysis into the two genera, *Prochlorococcus* and *Synechococcus*.[60]

These widely distributed cyanobacteria account for two-thirds of the photosynthesis in the oceans and, therefore, might be the most abundant photosynthetic organisms on earth. Recently, three new strains of *Prochlorococcus* were found to belong to the same species according to conventional 16S rRNA analyses.[61,62] However the genome of one strain had 2.4 Mb, while those of the other two strains only had 1.7 Mb, which was related to the stability of the particular strain's environment. Strains with a smaller genome size live on the ocean's surface and in depths where stable nutrient conditions and light intensities exist. Larger genomes reflect a more variable environment in intermediate depths. The additional 0.7 Mb in the genome encode genes responsible for the use of different nitrogen sources that are required to have enough versatility to survive in changing environments. Non-essential genes may have been discarded by the two strains with smaller genomes as adaptation to constant environments. This mechanism is referred to as "genome streamlining."[7,60,63]

In conclusion, marine metagenomics offers information about the role of marine microbes in marine food chains and biogeochemical cycling and presents the possibility to discover new enzymes and compounds with potential biotechnological applications.[43] Largescale metagenome sequencing projects include the sampling of oligotrophic seawater from the Sargasso Sea[55] and the Global Ocean Sampling (GOS) expedition.[64]

2.4. The Complexity of Analyzing Soil Metagenomics

Soil is a very complex and challenging environment containing huge microbial communities and a high diversity of species. The versatility of soil microorganisms is especially interesting for the industry in order to discover new natural products such as antibiotics or novel enzymes.[65,66]

The high biodiversity of soil is due to its extreme spatial heterogeneity, multiphase nature including gases, water and solid material, and complex chemical and biological characteristics. For example, one gram of forest soil harbors approx. 4×10^7 prokaryotic cells[67] and one gram of cultivated soils and grasslands contains 2×10^9 prokaryotes.[65,68] Even more important in the context of metagenomes is the fact that the number of distinct prokaryotic genomes range from 2,000 to 18,000 per gram of soil, excluding rare and unrecovered species. This demonstrates the enormous abundance of this environment.[65,69–72]

Soil consists of distinct mineral particles that differ in size, shape and chemical characteristics, and harbor the soil biota as well as organic compounds. The most abundant organisms present in soil are prokaryotes that mostly adhere to or absorb soil particles, live on the surface of soil aggregates, or in complex pore spaces between and inside them, producing the largest part of the soil biomass.[73,74] Soil microorganisms are mainly influenced by the availability of nutrients and water, since soil undergoes dramatic cyclic changes (such as water saturation or extreme aridity) that may cause death to many members of the microbial population. This results in considerable fluctuation of the composition of soil communities and community structure. In addition, further environmental factors, such as pH, oxygen or temperature, have a profound effect on microbial populations that need to be further investigated.[75]

The problem of isolating DNA out of soil is that many compounds, such as humic acids bind to DNA and, therefore, interfere with enzymatic reactions that renders cloning or PCR, as well as transformation and DNA hybridization to a difficult task. However, some phylogenetic studies were already done by the amplification of 16S rRNA genes with universal primers for bacteria and archaea.[76,77] Other marker genes analyzing the microbial biodiversity in different soil habitats include *dnak* (Hsp-70-type molecular chaperone)[78] and *amoA* (ammonia monooxygenase).[65,79]

The construction of soil libraries starts with fragmentation of soil DNA by restriction-enzyme digestion or mechanical shearing, insertion of DNA fragments into vectors and ends with the transformation of recombinant plasmids into the host. This sounds quite simple, but in fact, due to the large size of the soil metagenome, large numbers of clones are required for full coverage. The quality of metagenomic soil libraries is dependent on many factors, such as the composition of the soil sample itself, the DNA isolation method, the vector used for cloning or the host, replicating the plasmid and expressing the protein. To compare samples of different soil habitats, the soil samples should be representative of that habitat. Information about the physical, chemical and biotic properties (such as particle size, soil type, water content, pH and further) is absolutely necessary for analysis and comparison.[65,68]

DNA extraction soil can be done by two different approaches: One is to lyse cells directly in the sample matrix by a combination of enzymes, high temperature and detergents or mechanical disruption followed by the separation of DNA from the matrix and cell debris.[80] Conversely, the other way to extract DNA is to firstly separate cells from the soil matrix by mild mechanical forces or chemical procedures and then lyse them.[81] DNA obtained by the second approach is almost entirely prokaryotic, whereas the first DNA extraction method also reveals extracellular and eukaryotic DNA.[76,82,83] In addition, DNA obtained by firstly separating cells from the matrix was shown to be less contaminated with matrix compounds. Furthermore, the average size of the isolated DNA is larger in the case of firstly separating cells and, therefore, more suitable for the generation of large insert libraries.[84] However, the amount of isolated and purified DNA ranges from less than 1 μg to 500 μg per gram of soil. More DNA can be recovered by using the direct lysis approach.[82] It is also presumed that the direct lysis method better represents the microbial diversity of the soil sample, since microorganisms that adhere to particles are also included.[65,85,86]

2.5. Microbial Communities in the Human Body

In fact, humans harbor trillions of microbes that sustain many basic bodily processes, such as the digestion of food, the degradation of toxins and the destruction of harmful, disease-causing microbes. In general, there are ten times more prokaryotic cells in the human body than there are human cells, whose genomes contain at least 100 times as many genes as our own genome. These symbionts are capable of regulating our metabolism in a way that benefits both themselves and us.[1,28]

The sequencing of the metagenome of the human gastrointestinal tract (GI) revealed major differences among individuals, as well as microheterogeneity within individual GI populations.[88] Previously, it was thought that the most dominant microorganisms of the human GI are Firmicutes and Bacteroides, but metagenomic analysis revealed Actinobacteria and Archaea as the most abundant ones.[28,87] Using 16S rDNA and metabolic function analyses, Gill and co-workers showed enriched metabolism of glycans, amino acids, and xenobiotics; methanogenesis; and 2-methyl-D-erythritol

4-phosphate pathway–mediated biosynthesis of vitamins and isoprenoids in the human gut.[87]

To analyze the complex microbial world of the human skin, Grice and colleagues sequenced 16S rRNAs from the inner elbow of five healthy human subjects. They found 113 so-called "operational taxonomic units" (OTUs) at the level of 97 percent similarity that belong to six bacterial divisions. In all layers of the human skin, *Proteobacteria* were shown to be a dominant group and both inter- and intrapersonal variations in community composition and structure were observed. Surprisingly, similarities in community structures were identified for the mouse and the human skin. However, analyzing human skin may be a novel approach to determine the role of skin microbiota in health and disease and to reveal the complex interactions between the skin and its inhabiting microorganisms.[89]

In summary, the human microbiome is an interesting field of metagenomics that may result in the development of novel applications and guidelines in human and animal nutrition, drug discovery, and medicine. In addition, it may also greatly increase our knowledge of complex diseases, such as obesity or cancer.[1]

2.6. Comparative Metagenomics

Another interesting area of metagenomics represents the comparison of species abundance, diversity, as well as metabolic capabilities of a community among different habitats to capture the characteristics and demands of specific environments.

Accordingly, Tringe and co-workers characterized and compared the metabolic potential of terrestrial and marine microbial communities.[90] They investigated samples from agricultural soil and from deep-sea whale fall carcasses — environments that are both rich in nutrients, but offer totally different nutrient sources. The analysis of the microbial diversity by PCR of small rRNA in soil revealed a huge diversity of bacteria, only few archaeal species, some fungi, as well as unicellular eukaryotes. The whale fall samples were less diverse due to the specialized properties of this ecological niche. Because of limitations in the assembly of sequences, genes were analyzed independent of the origin. Individual genes only found in particular environments were termed "environmental gene tags" (EGTs). They can be considered as hallmarks of specific habitats. However, the predicted protein complement of a community was similar to that of other communities that live in related environments with similar metabolic demands. Conversely, differences were apparent at the individual gene level, e.g. operons, and higher order processes of marine and terrestrial microbes. In soil, for example, 73 putative orthologs of cellobiose phosphorylase that are involved in degradation of plant material were found, but none were detected in genomic sequences of the Sargasso Sea. In contrast, 466 distinct homologs of bacteriorhodopsin could solely be identified in the Sargasso Sea. In addition, photosynthesis was the primary energy source in the oligotrophic waters, whereas starch and sucrose metabolism were identified as the main energy production pathway in soil. Finally, most conjugation systems, plasmids and antibiotics were mainly discovered in terrestrial microbes.[55,90]

Moreover, the comparison of metagenomic libraries from the deep Mediterranean (3,000 m) and the Pacific ALOHA (North-Pacific subtropical Gyre) water column revealed that bathypelagic Mediterranean communities were similar to mesopelagic communities in the Pacific. 16S rRNA analysis and sequencing identified genomes from Rhizobiales within the Alphaproteobacteria, *Cenarchaeum symbiosum*, Planctomycetes, Acidobacteria, Chlorofelxi and Gammaproteobacteria — a community structure similar to that found in the aphotic zone of the mesopelagic populations living in the Pacific. Metabolic genes were mostly linked to catabolism, transport (including transporters for amino acids and carboxylic acids), and degradation of complex organic molecules such as xenobiotics. This provides evidence for a heterotrophic lifestyle of deep-sea microbes. The discovery of genes encoding for dehydrogenases revealed the aerobic carbon monoxide oxidation as an additional source of energy production. Finally, a comparison between the deep Mediterranean and the Pacific ALOHA at identical depths showed that in the absence of light, temperature was the most critical parameter shaping community structures, whereas pressure had no effect in depths of 4,000 m.[15]

These examples illustrate the identification of "fingerprints" that provide insights into environments of microbial communities. This information may be useful to predict particular characteristics of habitats, such as energy sources or pollution levels, and to understand the interactions and adaptation mechanisms of microbial populations within their specialized environments.[90]

3. REFORMULATION OF THE NICHE AND CONCEPT OF SPECIES IN THE CONTEXT OF METAGENOMICS

Microbial communities contain millions of different individual organisms that are dependent on each other and that share lots of common genes. Closely related organisms might be considered as one species, but genetic elements are also transferred from one microbe to another due to changing environmental conditions. These facts pose some questions such as: how can a species be defined, how do microbial communities react to changes in their environment and what determines the niche of a microbial species.[91] The new concept of metagenomics clearly results in a reformulation of the niche and species concept, answering some of these fundamental questions. In this section, different definitions of the term "niche" are explained and considered in the context of metagenomics. In addition, it is demonstrated that the species itself will not necessarily be in the focus in the future anymore. Instead, genome composition and isolated genes will come more into consideration.[92]

In the past decades, phenotypic capabilities in substrate utilisation, environmental tolerances or pathogenic traits were used to determine the niche of a microbial species. Earlier, niche was defined very vaguely as "the place in the environment that has the potential to support a species."[93,94] Later, Hutchinson reformed the concept of niche by taking into account independent environmental variables, scaled along orthogonal coordinates.[95] They determine a region, where each point is a combination of different values associated with a particular fitness allowing species to exist. In addition, events such as interspecific competition prevent the species from exploiting its entire fundamental niche. This results in the occupancy of only a part of the niche, where the species is competitively dominant. This situation is termed realized niche.[95] Therefore, species and not environments have niches: thus, no unoccupied niches exist.[96]

However, species actively change their inhabiting environments or even establish new ones instead of solely being passive occupants of them — a process called "ecosystem engineering"[97] or "niche construction."[98] Species may alter their habitat in a way that an occupancy by other species is facilitated and coexistence of other species is promoted.[99,100] In contrast, allelopathy results in a chemical modification of the habitat in a way that the existence of other species is prevented.[101] For example, bacteria produce bacteriocin that inhibits growth of different species.[102]

Species can also react to changing environmental conditions by generating alternative phenotypes, a phenomenon called "phenotypic plasticity."[103] In this context, phenotypic plasticity is associated with a specific trait and not to the entire genotype. It can be heritable or adaptive.[104] In summary, modification of an environment by species and phenotypic plasticity leads to the growth of biodiversity.

Nonetheless, it may be tricky to link genetic and functional characteristics of a specific microbial niche. For example, plasmid horizontal transfer is widely distributed among different bacterial species and even genera, in order to facilitate the colonization of new habitats. Hence, specific functions conferred by these plasmids cannot be attributed to a specific bacterial species.[92] However, horizontal gene transfer in bacteria is more important to diversification and speciation than mutation and adaptation.[105,106] Thus, genes obtained through horizontal transfer result in a new metabolic potential of the recipient, which is able to colonize new environments with different conditions. This new genetic trait leads to an alteration in the fundamental and realized niches of the recipient and subsequently also of its lifestyle. Consequently, the genetic background has to be defined as another variable of the microbial niche, since genetic changes modify or expand the realized niche. This is an illustrative example, that organisms play an active role in their own niche construction.[92]

With respect to metagenomics, an alternative view of the niche concept was recently suggested.[92] Clapham defined the niche as "a property of the community [...], involving both abiotic (pH, temperature, etc.) and biotic conditions" such as competition or predation.[107] However, operational genetic functional units (OGFU), a combination of genetic and functional traits form another variable of the niche concept, including the microbial genetic background. Thus, OGFU's properties as well as the community's traits, such as the pool of horizontally transmitted genes would be integrated into the niche concept. For example, an OGFU's fitness response to a particular environmental parameter can change after receiving a gene with the property to cope with toxic soil compounds.[92]

Metagenomics also gives reason to reformulate the concept of species, as genome structures and genes become the central point of interest instead of the species as a subject of study.[28]

Attempts to find an appropriate universal definition of "species" can be traced back to Darwin and have not ended until today. Over the years, many definitions have been proposed and opinions have widely differed on what constitutes a microbial species. Nonetheless, it is obvious that all organisms fall into groups of phenotypic similarity, sequence identity, resemblance in genomic contents and ecology. In the past, prokaryotic species were clustered according to phenotypic similarities, but with the progress of biotechnological applications sequence analysis have come more into consideration.[108]

As suggested by Fraser,[28] a microbial species can be defined by its pan-genome i.e. the set of both core genes that are essentially present in every strain and non-essential genes. The latter include all genes present in a species' gene pool, but do not have to be present in each individual strain. In addition, a certain number of genomes are necessary to determine a pan-genome. In his study, Hervé Tettelin[109] has clearly illustrated this concept: He found in two different *Streptococcus* groups that every genome brings an average of 33 novel, strain-specific genes to the group's pan-genome, leading an open pan-genome that increases as more strains are considered. In contrast, the five genomes of *Bacillus anthracis* show a more closed pan-genome, which remains constant as more strains are included. This may be due to the fact that *Streptococcus* strains live in highly variable environments, whereas strains of *B. anthracis* do not. However, the number of genomes that are essential to capture 95 percent of the genes of a particular species is between 13 and 15, which varies among species. Besides, recombination and LGT events among prokaryotes generally make it difficult to define the microbial species.[28,110]

4. METAGENOMIC LIBRARIES

4.1. Construction of Metagenomic Libraries

Metagenomics is based on a wide variety of novel methods and approaches. First, samples from environments, such as soil or seawater, are collected and DNA mass extraction from the whole community of the sample is performed. Then, DNA fragments are cloned into vectors to generate metagenomic libraries that consist of millions of DNA fragments from all the microbes present in the sampled community.[1] Depending on the intention of a study, two different library types that vary in size of inserts can be distinguished: Small-insert and large-insert metagenomic libraries.

Large-insert metagenomic libraries harbor inserts of up to 100 kb and are constructed either in lambda phage, cosmid, fosmid, or BAC vectors. They are mainly used to unravel the function of a microorganism in its environment. This type of library normally contains fewer than 100,000 clones, numbers that are actually several orders too small to capture the complete microbial diversity of a sample. However, existing libraries have already given insight into the microbial ecology of many ecosystems despite incomplete metagenome coverage. In general, large-insert libraries are beneficial to capture complex pathways requiring many genes that are often arranged in operons.[12] It also provides the opportunity to identify the origin of a particular metabolic function by sequencing 16S rRNA genes in close proximity to genes encoding metabolically active proteins or vice versa.[7,65]

Small-insert libraries, constructed in conventional plasmid vectors (less than 15 kb), are required to discover new metabolic functions encoded by single genes or small operons and to reconstruct metagenomes by linking overlapping sequences to assemblies called "contigs."[12,65] Besides, harsh methods, such as bead beating can be applied to generate small pieces of DNA that are more favorable for creating small-insert rather than large-insert libraries. Unfortunately, massive sequencing is necessary for analyzing thousands of clones. Even the reassembly of genomes from many small fragments proved to be very difficult.[7]

A frequently used host organism for the construction of DNA libraries is *Escherichia coli.* Libraries can also be transferred into other hosts such as *Streptomyces* or *Pseudomonas* species by specific shuttle cosmid or BAC vectors.[111,112] The choice of the appropriate vector system depends on many parameters, for example the quality of the DNA, the desired average insert size of the library, the desired vector copy number, the host or the screening strategy.[65]

4.2. Screening of Metagenomic Libraries

The high-throughput and sensitive screening of DNA libraries, especially in the case of small-insert libraries is a burdensome and challenging task. Screening essays are based either on a particular metabolic activity (function-driven approach) or on the analysis and subsequent comparison of nucleotide sequences (sequence-driven approach).

Sequence-based screening bears the potential to identify sequences that are probably not expressed in the host species. It mainly provides information about the different functions of a community, linkage of traits, genomic arrangements, as well as the extent of horizontal gene transfer. The main aim is to link a particular gene with its origin. It is even possible to assemble the complete genome of individual species to determine their specific role in the microbial community. In addition, the genome of the community can be considered in its entirety, which reveals insights into population ecology and evolution.[1] The screening for sequences is based on sequencing of primers and probes that are derived from known genes and therefore target phylogenetic anchors, such as 16S rRNA or highly conserved sequences to discover homologous ones[113] such as genes encoding for polyketide synthases,[84,114,115] gluconic acid reductases[116] or nitrole hydratases.[117] Thus, PCR and hybridization are limited to the identification of new members of already known gene families. In the case of analyzing phylogenetic anchors, it is crucial to differentiate between the anchor of the cloned DNA and the host homologue present in the chromosome. This problem can be solved by using a vector with high or inducible copy number,[118] or by using the terminator PCR that blocks amplification of the host cell's homologue[119,120] or by using taxon-specific oligonucleotide probes and primers for 16S rRNA genes that do not identify the *E. coli* genes.[121]

Apart from the sequence-driven approach, *functional screening* can be used to identify genes that are not recognized by their sequence due to restrictions in the existing databases and the still poor availability of sequencing capacity.[12] Functional screening of clone libraries is useful to discover novel enzymes, such as degradative enzymes,[122–124] antibiotics[23,25] or antibiotic resistances.[12] The strategies to detect a specific activity should be simple, since a large number of clones have to be tested in order to receive positive ones.[12,65] New selection methods facilitate the discovery of active clones, such as the creation of reporter fusions that are expressed simultaneously with the gene of interest. Positive functional clones can also be identified by the use of host strains or mutants of host strains that need heterologous complementation for growth under selective conditions. Although functional screening has the potential to identify full-length genes and, therefore, functional gene products, expression and functioning of the gene product only works in an appropriate host, which explains the low gene detection frequencies during functional screening.[65] In addition to *E. coli*, *Streptomyces*, *Pseudomonas* and eukaryotic host strains have been used to expand the range of host, in order to increase the discovery of new gene products.[23,111,112] Furthermore, enrichment steps for microorganisms harboring the desired traits have been used before library construction, such as carbon or nitrogen sources as growth substrates or antibiotic and metal resistance. Nonetheless, selective growth of microorganisms leads to selection and, therefore, negatively influences microbial diversity.[82,124]

Fluorescent *in situ* hybridization (FISH) is commonly used in metagenomic analysis by using a fluorescent labeled probe that hybridizes with 16S rRNA genes. The limitation of FISH is its low sensitivity, and only highly expressed genes can be detected. To overcome this problem, improvements in sensitivity have been accomplished, such as RING-FISH (recognition of individual genes-FISH)[125] or FISH-MAR, which is linked to microautoradiography to detect taxons that utilize particular substrates.[12]

Another application of functional analysis is the use of stable isotopes to study the cycling of elements in microbial populations. Substrates labeled with stable carbon or nitrogen isotopes can be utilized by community members that incorporate the isotopes into their DNA. The labeled DNA can be separated from unlabeled DNA by density centrifugation. Constructing metagenomic libraries with labeled DNA is useful to study genes associated with the active species of a community and reveals insights into the metabolism of populations.

Recently, a new high-throughput screening strategy has been developed, which is called "substrate-induced gene expression cloning" (SIGEX) and is used for the discovery of genes encoding catabolic pathways.[126] Usually, genes encoding catabolic pathways are arranged in operons and regulated by specific elements near the catabolic genes. Their expression can be induced by the addition of the relevant substrates. Cloning of metagenomic DNA upstream of a *gfp*-gene of an operon-trap *gfp* vector places the expression of *gfp* under the control of promoters present in the metagenomic DNA. Clones expressing *gfp* upon substrate addition can be isolated by fluorescence-activated cell sorting.[126]

In summary, a combination of both sequence-based and functional-based screening of different types of libraries is eligible. Although the construction of metagenomic libraries sounds theoretically simple, it bears some difficulties in creating and analyzing this information. First, a large amount of DNA has to be extracted and cloned from an environmental sample to obtain more than single sequence coverage and to represent rare members of the community. Then, many clones and sequences have to be analyzed to obtain convincing data, and finally, it is difficult to capture the minor species from a given sample in log-growing populations. High-throughput methods are absolutely necessary to capture functionally active genes, phylogenetic anchors and novel genes.[12,65]

4.3. From Sequencing to Sequence Analysis

The switch from single genomes to metagenomes requires both a large sequencing capacity and very powerful bioinformatic tools to organize and analyze the huge amounts of data. Massive random shotgun and bi-directional sequencing provides access to diverse environments and enables the discovery of the most frequent functional genes and, thereby, metabolic functions of an ecosystem. In addition, shotgun sequencing of the Sargasso Sea revealed the diversity and importance of the proton pump rhodopsin in marine environments.[3,55,127] Shotgun sequencing of low-diversity environments such as acid mines, actually led to the assembly of complete genomes as recently shown for *Leptospirillum* and *Ferroplasma* spp.[37] New sequencing techniques, such as pyrosequencing, have dramatically reduced time and cost constraints of DNA sequencing.[7,128,129] However, the analysis of complex communities needs enormous sequencing efforts that may just lead to genome assemblies of the most predominant members.[55,90]

Having overcome the problem of sequencing limitations, sequences generated from environmental or clinical samples need to be analyzed with existing databases. These database are mostly rich in sequences from common, cultivable and easily achievable organisms.[18] Thus, there are some biases in the composition of databases that may limit the view of global diversity.[18] The real extent of biodiversity may be underestimated.[3]

In addition to functional assignment of proteins, taxonomic assignments (binnings) are essential to study community structure, dynamics and alterations in composition, and to analyze the evolution of the community, i.e. the extent of horizontal gene transfer (HGT) or the barriers forming the species.[18,130] To deal with the following problems i.e. binning is burdensome; metagenomics is short; in-frame stop codons; frame shifts; and lack of information, several groups have tried to assign ORFs to the taxon of their closest relatives in a homology search.[55,131,132] This strategy reveals some difficulties, since there are poorly known taxa and misleading HGT events.[133] Pignatelli and co-workers have previously shown that diversity is limited in the existing databases, and that even the best results for taxonomic classification are risky.[18] Thus, it is necessary to sequence more genes and genomes from poorly known or unknown taxa to update data in the current databases.[18]

5. METAGENOMICS IN INDUSTRY: DISCOVERY OF NOVEL BIOCATALYSTS AND ANTIBIOTICS

The metagenomic analysis led to the identification and characterization of many novel biocatalysts and drugs.[21] The analysis of soil metagenomes revealed many more antibiotics and mechanisms of antibiotic resistance than found in cultivation-based techniques thus far.[24]

The screening for bleomycin resistance genes of a metagenomic DNA library of activated sludge revealed two different genes that showed no similarity to the genes of bleomycin-producing Actinomycetes or clinical isolates.[134] The antibiotic bleomycin was firstly identified in *Streptomyces verticillus* and used as an antitumor agent.[135,136] However, bleomycin-producing Actinomycetes harbor resistance genes (*ble*) that code for bleomycin resistance proteins (BRPs), sequestering the antibiotic to prevent DNA cleavage.[136] It was proved that this resistance mechanism spread over to bacteria of clinical isolates that normally do not produce bleomycin.[137–139] Recently, a novel antibiotic — palmitoylputrescine — was found in bromeliad tank water and a new mechanism of tetracycline inactivation was found in the oral metagenome.[21,25,140] Thus, non-clinical environments provide a huge potential to discover new antibiotics, as well as resistance mechanisms that may lead to new developments in drug research, extending the range of effective pharmaceuticals against infectious diseases.[21,134]

Nonetheless, most pharmaceutical companies focus on new drugs against long-term chronic diseases such as obesity or high cholesterol, in comparison to the economically unfavorable discovery of novel antibiotics that are essential to overcome the "global antibiotic resistance problem."[142] However, it should not be overseen that natural products from bacteria or fungi constituted of 63 percent of all newly used anti-infectives between 1983 and 1994.[143]

Biosynthetic pathways accessed by metagenomic studies are mostly involved in vitamin biosynthesis. Clones harboring genes that encode 2,5-diketo-D-gluconic acid reductase, which produces vitamin C out of glucose, have been detected in metagenomic libraries derived from enrichment cultures.[116,144]

Prokaryotes also play a key role in environmental bioremediation, since they are able to degrade all naturally occurring compounds and most xenobiotics.[21] However, genes that are necessary for remediation are different in culture from those found in the natural occurring environmental bioremediation. Therefore, metagenomics reveals insights into diverse xenobiotic degradation pathways of natural environments that may be important in cleaning up oil spills, groundwater, sewage or nuclear waste in order to restore healthy ecosystems.[1,145,146]

The industry primarily focuses on prokaryotic metagenomes, since prokaryotes show maximal biodiversity and their genomes can easily be targeted by conventional screening applications.[9] In contrast, eukaryotes have only infrequently been used for metagenomic research. This is due to their large genomes that are associated with high sequencing costs and their high proportion of DNA that does not code for proteins. With the development of higher-throughput sequencing technologies and decreasing costs, eukaryotes should come more into the focus of metagenomic research.[8] The chemical and pharmaceutical industries mainly switched established conventional chemical processes to biotechnological areas that require metagenomic research in order to discover new enzymes, biocatalysts and applications.[9] In addition to the discovery of new enzyme classes, variations in functional properties within the classes are also important for biotechnology. Consequently, enzymes that function under specific conditions such as higher temperature, pH extremes or elevated salinity levels at the center of interest.[21] As an example, a novel thermostable esterase was identified with a pH range of 5.5–7.5 and an extremely wide temperature range with an optimum at 47°C.[147] Further key features of an ideal biocatalyst include activity, stability, specificity and efficiency. In addition, the novel enzyme should contain a single backbone with superior functionality and an entirely new sequence to prevent competing property rights. Finally, the heterologous expression of new enzymes should generate adequate pure protein in appropriate amounts at reasonable costs.[9]

6. CONCLUSION

The discovery of microbes in the seventeenth century marked the beginning of modern microbiology. Since then, humankind has benefited enormously from the study of cultivated microorganisms that represent only a miniscule fraction of microbes actually present on earth. In the past decades, we have realized the importance of microbial communities. We have expanded our knowledge of evolution, biodiversity, and ecology that has led to major advances in medicine, agriculture, energy production and bioremediation. The study of genetic and genomic information from environmental communities all over the world without prior cultivation as provided by metagenomics currently reveals new insights into the microbial community, reshaping the landscape of microbiology and revealing the secrets of the uncultured world.[1]

REFERENCES

1. National Academy of Science, 2007
2. Carroll SB, Weatherbee SD, Langeland JA. (1995) Homeotic genes and the regulation and evolution of insect wing number. *Nature* 375: 58–61.
3. López-García P, Moreira D. (2008) Tracking microbial biodiversity through molecular and genomic ecology. *Research in Microbiology* 159: 67–73.

4. Offner S. (2001) A universal phylogenetic tree. *The American Biology Teacher* **63**: 164–170.
5. Gans J, Wolinsky M, Dunbar J. (2005) Computational improvements reveal great bacterial diversity and high metal toxicity in soil. *Science* **309**: 1387–1390.
6. Gaston KJ. (1996) What is biodiversity? In KJ Gaston (ed.), *Biodiversity: a Biology of Numbers and Difference.* Oxford, UK: Blackwell Science Ltd, pp. 1–9.
7. Pedrós-Alió C. (2006) Genomics and marine microbial ecology. *International Microbiology* **9**: 191–197.
8. Hugenholtz P, Tyson GW. (2008) Metagenomics. *Nature* **455**: 481–483.
9. Langer M, Gabor EM, Liebeton K, *et al.* (2006) Metagenomics: An inexhaustible access to nature's diversity. *Biotechnol J* **1**: 815–821.
10. Amann RI, Ludwig W, Schleifer KH. (1995) Phylogenetic identification and in situ detection of individual microbial cells without cultivation. *Microbiol Rev* **59**: 143–169.
11. Stahl DA, Lane DJ, Olsen GJ, Pace NR. (1984) Analysis of hydrothermal ventassociated symbionts by ribosomal RNA sequences. *Science* **224**: 409–411.
12. Riesenfeld CS, Schloss PD, Handelsman J. (2004) Metagenomics: genomics analysis of microbial communities. *Annu Rev Genet* **38**: 525–552.
13. Pace NR, Stahl DA, Lane DJ, Olsen GJ. (1985) Analyzing natural microbial populations by rRNA sequences. *ASM News* **51**: 4–12.
14. Schmidt TM, DeLong EF, Pace NR. (1991) Analysis of a marine picoplankton community by 16S rRNA gene cloning and sequencing. *J Bacteriol* **173**: 4371–4378.
15. Martín-Cuadrado AB, López-García P, Alba JC, *et al.* (2007) Metagenomics of the deep Mediterranean, a warm bathypelagic habitat. *PLoS ONE* **19**: e914.
16. Handelsman J, Rondon MR, Brady SF, *et al.* (1998) Molecular biological access to the chemistry of unknown soil microbes: a new frontier for natural products. *Chem Biol* **5**: R245–R249.
17. Handelsman J. (2004) Metagenomics: application of genomics to uncultured microorganisms. *Microbiol Mol Biol Rev* **68**: 669–685.
18. Pignatelli M, Aparicio G, Blanquer I, *et al.* (2008) Metagenomics reveals our incomplete knowledge of global diversity. *Bioinformatics* **24**: 2124–2125.
19. Hahn MW, Pöckl M. (2005) Ecotypes of planktonic actinobacteria with identical 16S rRNA genes adapted to thermal niches in temperate, subtropical and tropical freshwater habitats. *Appl Environ Microbiol* **71**: 766–773.
20. Jaspers E, Overmann J. (2004) Ecological significance of microdiversity: identical 16S rRNA gene sequences can be found in bacteria with highly divergent genomes and ecophysiologies. *Appl Environ Microbiol* **70**: 4831–4839.

21. Steele HL, Streit WR. (2005) Metagenomics: Advances in ecology and biotechnology. *FEMS Microbiol Lett* **247**: 105–111.
22. Healy FG, Ray RM, Aldrich HC, *et al.* (1995) Direct isolation of functional genes encoding cellulases from the microbial consortia in a thermophilic, anaerobic digester maintained on lignocellulose. *Appl Microbiol Biotechnol* **43**: 667–674.
23. Wang GY, Graziani E, Waters B, *et al.* (2000) Novel natural products from soil DNA libraries in a streptomycete host. *Org Lett* **2**: 2401–2404.
24. Gillespie DE, Brady SF, Bettermann AD, *et al.* (2002) Isolation of antibiotics turbomycin A and B from a metagenomic library of soil microbial DNA. *Appl Environ Microbiol* **68**: 4301–4306.
25. Brady SF, Clardy J. (2000) Long-chain N-acyl amino acid antibiotics isolated from heterologously expressed environmental DNA. *J Am Chem Soc* **122**: 12903–12904.
26. Riesenfeld CS, Goodman RM, Handelsman J. (2004) Uncultured soil bacteria are a reservoir of new antibiotic resistance genes. *Environ Microbiol* **6**: 981–989.
27. Sandler SJ, Hugenholtz P, Schleper C, *et al.* (1999) Diversity of rad A genes from cultured and uncultured archaea: comparative analysis of putative RadA proteins and their use as a phylogenetic marker. *J Bacteriol* **181**: 907–915.
28. Mongodin EF, Emerson JB, Nelson KE. (2005) Microbial metagenomics. *Genome Biol* **6**: 347.
29. Quaiser A, Ochsenreiter T, Lanz C, *et al.* (2003) Acidobacteria form a coherent but highly diverse group within the bacterial domain: evidence from environmental genomics. *Mol Microbiol* **50**: 563–575.
30. Preston CM, Wu KY, Molinski TF, De-Long EF. (1996) A psychrophilic crenarchaeon inhabits a marine sponge: Cenarchaeum symbiosum gen. nov., sp. nov. *Proc Natl Acad Sci USA* **93**: 6241–6246.
31. Schleper C, Holben W, Klenk HP. (1997) Recovery of Crenarchaeotal ribosomal DNA sequences from freshwaterlake sediments. *Appl Environ Microbiol* **63**: 321–323.
32. Schleper C, DeLong EF, Preston CM, *et al.* (1998) Genomic analysis reveals chromosomal variation in natural populations of the uncultured psychrophilic archaeon Cenarchaeum symbiosum. *J Bacteriol* **180**: 5003–5009.
33. Béjà O, Aravind L, Koonin EV, *et al.* (2000) Bacterial rhodopsin: evidence for a new type of phototrophy in the sea. *Science* **289**: 1902–1906.
34. Breitbart M, Salamon P, Andresen B, *et al.* (2002) Genomic analysis of uncultured marine viral communities. *Proc Natl Acad Sci USA* **99**: 14250–14255.
35. Breitbart M, Hewson I, Felts B, *et al.* (2003) Metagenomic analyzes of an uncultured viral community from human feces. *J Bacteriol* **185**: 6220–6223.

36. Wommack KE, Colwell RR. (2000) Virioplankton: viruses in aquatic ecosystems. *Microbiol Mol Biol Rev* **64**: 69–114.
37. Tyson GW, Chapman J, Hugenholtz P, *et al.* (2004) Community structure and metabolism through reconstruction of microbial genomes from the environment. *Nature* **428**: 37–43.
38. Edwards KJ, Bond PL, Druschel GK, *et al.* (2000) Geochemical and biological aspects of sulfide mineral dissolution: lessons from Iron Mountain, California. *Chem Geol* **169**: 383–397.
39. Pomeroy LR. (1974) The ocean's food-web, a changing paradigm. *BioScience* **24**: 499–504.
40. Hobbie JE, Daley RJ, Jasper S. (1977) Use of Nuclepore filtres for counting bacteria by fluorescent microscopy. *Appl Environ Microbiol* **33**: 1225–1228.
41. Zimmerman R, Meyer-Reil LA. (1974) A new method for fluorescence staining of bacterial populations on membrane filters. *Kiel Meeresforsch* **30**: 24–27.
42. Ward AC, Bora N. (2006) Diversity and biogeography of marine actinobacteria. *Curr Opin Microbiol* **9**: 279–286.
43. Kennedy J, Marchesi JR, Dobson ADW. (2008) Marine metagenomics: strategies for the discovery of novel enzymes with biotechnological applications from marine environments. *Microbial Cell Factories* **7**: 1–8.
44. Field CB, Behrenfeld MJ, Randerson JT, Falkowski P. (1998) Primary production of the biosphere: integrating terrestrial and oceanic components. *Science* **281**: 237–240.
45. Azam F. (1998) Oceanography: Microbial Control of Oceanic Carbon Flux: The Plot Thickens. *Science* **280**: 694–696.
46. del Giorgio PA, Duarte CM. (2002) Respiration in the open ocean. *Nature* **420**: 379–384.
47. Newman DJ, Hill RT. (2006) New drugs from marine microbes: the tide is turning. *J Ind Microbiol Biotechnol* **33**: 539–544.
48. Sipkema D, Franssen MC, Osinga R, *et al.* (2005) Marine sponges as pharmacy. *Mar Biotechnol (NY)* **7**: 142–162.
49. Song Q, Zhang X. (2008) Characterization of a novel non-specific nuclease from thermophilic bacteriophage GBSV1. B.M.C. *Biotechnol* **8**: 43.
50. de Pascale D, Cusano AM, Autore F, *et al.* (2008) The cold-active Lip1 lipase from the Antarctic bacterium Pseudoalteromonas haloplanktis TAC125 is a member of a new bacterial lipolytic enzyme family. *Extremophiles* **12**: 311–323.
51. Acevedo JP, Reyes F, Parra LP, *et al.* (2008) Cloning of complete genes for novel hydrolytic enzymes from Antarctic sea water bacteria by use of an improved genome walking technique. *J Biotechnol* **133**: 277–286.

52. Xu M, Xiao X, Wang F. (2008) Isolation and characterization of alkane hydroxylases from a metagenomic library of Pacific deep-sea sediment. *Extremophiles* **12**: 255–262.
53. DeLong EF, Karl DM. (2005) Genomic perspectives in microbial oceanography. *Nature* **437**: 336–342.
54. Agogue H, Casamayor EO, Bourrain M, *et al.* (2005) A survey on bacteria inhabiting the sea surface microlayer of coastal ecosystems. *FEMS Microbiol Ecol* **54**: 269–280.
55. Venter JC, Remington K, Heidelberg JF, *et al.* (2004) Environmental genome shotgun sequencing of the sargasso sea. *Science* **304**: 66–74.
56. Wu Z, Xie L, Xia G, *et al.* (2005) A new tetrodotoxin-producing actinomycete, Nocardiopsis dassonvillei, isolated from the ovaries of pufferfish Fugu rubripes. *Toxicon* **45**: 851–859.
57. DeLong EF, Franks EG, Alldredge AL. (1993) Phylogenetic diversity of aggregate-attached vs. free-living marine bacterial assemblages. *Limnol Oceanogr* **38**: 924–934.
58. Simon M, Grossart HP, Schweitzer B, Ploug H. (2002) Microbial ecology of organic aggregates. *Aquat Microbiol Ecol* **26**: 175–211.
59. Grossart HP, Schlingoff A, Bernhard M, Simon M, Brinkhoff T. (2004) Antagonistic activity of bacteria isolated from organic aggregates of the German Wadden Sea. *FEMS Microbiol Ecol* **47**: 387–396.
60. Partensky F, Hess WR, Vaulot D. (1999) Prochlorococcus, a marine photosynthetic prokaryote of global significance. *Microbiol Mol Biol Rev* **63**: 106–127.
61. Dufresne A, Salanoubat M, Partensky F, *et al.* (2003) Genome sequence of the cyanobacterium Prochlorococcus marinus SS120, a nearly minimal oxyphototrophic genome. *Proc Natl Acad Sci USA* **100**: 10020–10025.
62. Rocap G, Larimer FW, Lamerdin J, *et al.* (2003) Genome divergence in two Prochlorococcus ecotypes reflects oceanic niche differentiation. *Nature* **424**: 1042–1047.
63. Giovannoni SJ, Tripp HJ, Givan S. (2005) Genome streamlining in a cosmopolitan oceanic bacterium. *Science* **309**: 1242–1245.
64. Yooseph S, Sutton G, Rusch DB, *et al.* (2007) The Sorcerer II Global Ocean Sampling expedition: expanding the universe of protein families. *PLoS Biol* **5**: e16.
65. Daniel R. (2005) The metagenomics of soil. *Nat Rev* **3**: 470–478.
66. Osburne MS, Grossmann TH, August PR, MacNeil IA. (2000) Tapping into microbial diversity for natural products drug discovery. *ASM News* **66**: 411–417.
67. Richter DD, Markewitz D. (1995) How deep is soil? *Bioscience* **45**: 600–609.

68. Paul EA, Clark FE. (1989) Soil microbiology and biochemistry. Academic Press, San Diego.
69. Torsvik V, Sorheim R, Goksoy J. (1996) Total bacterial diversity in soil and sediment communities. *J Ind Microbiol* **17**: 170–178.
70. Torsvik V, Daae FL, Sandaa RA, Øvreås L. (1998) Novel techniques for analyzing microbial diversity in natural and perturbed environments. *J Biotech* **64**: 53–62.
71. Torsvik V, Daae FL, Sandaa RA, Øvreås L. (2002) Microbial diversity and function in soil: from genes to ecosystems. *Curr Opin Microbiol* **5**: 240–245.
72. Doolittle W. (1999) Phylogenic classification and the universal tree. *Science* **284**: 2124–2128.
73. Foster RC. (1988) Microenvironments of soil microorganisms. *Bio Ferti Soils* **6**: 189–203.
74. Hassink J, Bouwman LA, Zwart KB, Brussaard L. (1993) Relationship between habitable pore space, soil biota, and mineralization rates in grassland soils. *Soil Bio Biochem* **25**: 47–55.
75. Kieft TL, Soroker E, Firestone MR. (1987) Microbial biomass response to a rapid change increase in water potential when dry soil is wetted. *Soil Bio Biochem* **19**: 119–126.
76. Lloyd-Jones G, Hunter DWF. (2001) Comparison of rapid DNA extraction methods applied to contrasting New Zealand soils. *Soil Bio Biochem* **33**: 2053–2059.
77. Dunbar J, Takala S, Barns SM, *et al.* (1999) Levels of bacterial community diversity in four arid soils compared by cultivation and 16S rRNA gene cloning. *Appl Environ Microbiol* **65**: 1662–1669.
78. Yap WH, Li X, Soong TW, Davies JE. (1996) Genetic diversity of soil microorganisms assessed by analysis of hsp70 (dnaK) sequences. *J Industr Microbiol* **17**: 179–184.
79. Webster G, Embley TM, Posser JL. (2002) Grassland management regimens reduce small-scale heterogeneity and species diversity of β-proteobacterial ammonia oxidisers. *App Environ Microbiol* **68**: 20–30.
80. Ogram A, Sayler GS, Barkay T. (1987) The extraction and purification of microbial DNA from sediments. *J Microbiol Methods* **7**: 57–66.
81. Holben WE, Jansson JK, Chelm BK, Tiedje JM. (1988) DNA probe method for the detection of specific microorganisms in the soil bacterial community. *Appl Environ Microbiol* **54**: 703–711.
82. Gabor EM, de Vries EJ, Janssen DB. (2003) Efficient recovery of environmental DNA for expression cloning by indirect methods. *FEMS Microbiol Ecol* **44**: 153–163.

83. Treusch AH, Kletzin A, Raddatz G, *et al.* (2004) Characterization of large-insert DNA libraries from soil for environmental genomic studies of Archaea. *Environ Microbiol* **6**: 970–980.
84. Courtois S, Frostegård A, Göransson P, *et al.* (2001) Quantification of bacterial subgroups in soil: comparison of DNA extracted directly from soil or from cells previously released by density gradient centrifugation. *Environ Microbiol* **3**: 431–439.
85. Steffan RJ, Goksoyr J, Bej AK, Atlas RM. (1988) Recovery of DNA from soils and sediments. *Appl Environ Microbiol* **54**: 2908–2915.
86. Leff LG, Dana JR, McArthur JV, Shimkets LJ. (1995) Comparison of methods of DNA extraction from stream sediments. *Appl Environ Microbiol* **61**: 1141–1143.
87. Gill SR, Pop M, DeBoy RT, *et al.* (2006) Metagenomic Analysis of the Human Distal Gut Microbiome. *Science* **312**: 1355–1359.
88. Blaser M, Bork P, Fraser C, Knight R, Wang J. (2013) The microbiome explored: recent insights and future challenges. *Nat Rev Microbiol* **11**: 213–217.
89. Grice EA, Kong HH, Renaud GG, Young AC. (2008) NISC Comparative Sequencing Program. Bouffard GG, Blakesley RW, Wolfsberg TG, Turner ML, Segre JA. A diversity profile of the human skin microbiota. *Genome Res* doi:10.1101/gr.075549.107.
90. Tringe SG, von Mering C, Kobayashi A, *et al.* (2005) Comparative metagenomics of microbial communities. *Science* **308**: 554–557.
91. Jurkowski A, Handelsman J. (2007) The new sience of Metagenomics: revealing the secret of our microbial planet. *National Research Council report.*
92. Marco D. (2008) Metagenomics and the niche concept. *Theory Biosci* **127**: 241–247.
93. Grinnell J. (1917) The niche-relationships of the California Thrasher. *Auk* **34**: 427–433.
94. Elton C. (1927) Animal ecology, Sidgwick and Jackson, London.
95. Hutchinson GE. (1957) Concluding remarks. *Cold Spring Harb Symp Quant Biol* **22**: 415–427.
96. Herbold B, Moyle PB. (1986) Introduced species and vacant niches. *Am Nat* **128**: 751–760.
97. Jones CG, Lawton JH, Shachak M. (1994) Organisms as ecosystem engineers. *Oikos* **69**: 373–386.
98. Day RL, Laland KN, Odling-Smee FJ. (2003) Rethinking adaptation: the niche-construction perspective. *Perspect Biol Med* **46**: 80–95.

99. Li J, Helmerhorst EJ, Leone CW, *et al.* (2004) Identification of early microbial colonizers in human dental biofilm. *J Appl Microbiol* **97**: 1311–1318.
100. Callaway RM, Walker LR. (1997) Competition and facilitation: a synthetic approach to interactions in plant communities. *Spec Feature Ecol* **78**: 1958–1965.
101. Weidenhamer J, Romeo J. (1989) Allelopathic properties of Polygonella myriophylla: field evidence and bioassays. *J Chem Ecol* **15**: 1957–1970.
102. Jamuna M, Jeevaratnam K. (2004) Isolation and partial characterization of bacteriocins from Pediococcus species. *Appl Microbiol Biotechnol* **65**: 433–439.
103. Bradshaw AD. (1965) The evolutionary significance of phenotypic plasticity in plants. *Adv Genet* **13**: 115–155.
104. Scheiner SM. (1993) Genetics and evolution of phenotypic plasticity. *Annu Rev Ecol Syst* **24**: 35–68.
105. Lawrence J. (2002) Gene transfer in bacteria: speciation without species? *Theor Popul Biol* **61**: 449–460.
106. Ochman H, Lawrence JG, Groisman E. (2000) Lateral gene transfer and the nature of bacterial innovation. *Nature* **405**: 299–304.
107. Clapham WB. (1973) *Natural Ecosystems.* New York: MacMillan.
108. Ward DM, Cohan FM, Bhaya D, *et al.* (2008) Genomics, environmental genomics and the issue of microbial species. *Heredity* **100**: 207–219.
109. Tettelin H, Riley D, Cattuto C, Medini D. (2008) Comparative genomics: the bacterial pan-genome. *Curr Opin Microbiol* **11**: 472–477.
110. Medini D, Donati C, Tettelin H, *et al.* (2005) The microbial pan-genome. *Curr Opin Genet Dev* **15**: 589–594.
111. Courtois SCM, Cappellano M, Ball FX, *et al.* (2003) Recombinant environmental libraries provide access to microbial diversity for drug discovery from natural products. *Appl Environ Microbiol* **69**: 49–55.
112. Martinez A, Kolvek SJ, Yip CL, *et al.* (2004) Genetically modified bacterial strains and novel bacterial artificial chromosome shuttle vectors for constructing environmental libraries and detecting heterologous natural products in multiple expression hosts. *Appl Environ Microbiol* **70**: 2452–2463.
113. Piel J. (2002) A polyketide synthasepeptide synthetase gene cluster from an uncultured bacterial symbiont of Paederus beetles. *Proc Natl Acad Sci USA* **99**: 14002–14007.
114. Ginolhac A, Jarrin C, Gillet B, *et al.* (2004) Phylogenetic analysis of polyketide synthase I domains from soil metagenomic libraries allows selection of promising clones. *Appl Environ Microbiol* **70**: 5522–5527.

115. Seow KT, Meurer G, Gerlitz M, *et al.* (1997) A study of iterative type II polyketide synthases, using bacterial genes cloned from soil DNA: a means to access and use genes from uncultured microorganisms. *J Bacteriol* **179**: 7360–7368.
116. Eschenfeldt WH, Stols L, Rosenbaum H, *et al.* (2001) DNA from uncultured organisms as a source of 2,5-diketo-D-gluconic acid reductases. *Appl Environ Microbiol* **67**: 4206–4214.
117. Precigou S, Goulas P, Duran R. (2001) Rapid and specific identification of nitrile hydratase (Nhase)-encoding genes. *FEMS Microbiol Lett* **204**: 155–161.
118. Handelsman J, Liles MR, Mann DA, *et al.* (2003) Cloning the metagenome: culture-independent access to the diversity and functions of the uncultivated microbial world, In: Functional Microbial Genomics, ed. B. Wren, N. Dorrell, pp. 241–55. New York: Academic.
119. Liles MR, Manske BF, Bintrim SB, *et al.* (2003) A census of rRNA genes and linked genomic sequences within a soil metagenomic library. *Appl Environ Microbiol* **69**: 2684–2691.
120. Rondon MR, August PR, Bettermann AD, *et al.* (2000) Cloning the soil metagenome: a strategy for accessing the genetic and functional diversity of uncultured microorganisms. *Appl Environ Microbiol* **66**: 2541–2547.
121. Vergin KL, Urbach E, Stein JL, *et al.* (1998) Screening of a fosmid library of marine environmental genomic DNA fragments reveals four clones related to members of the order Planctomycetales. *Appl Environ Microbiol* **64**: 3075–3078.
122. Henne A, Daniel R, Schmitz RA, Gottschalk G. (1999) Construction of environmental DNA libraries in Escherichia coli and screening for the presence of genes conferring utilization of 4-hydroxybutyrate. *Appl Environ Microbiol* **65**: 3901–3907.
123. Gupta R, Berg QK, Lorenz P. (2002) Bacterial alkaline proteases: molecular approaches and industrial applications. *Appl Microbiol Biotechnol* **59**: 15–32.
124. Knietsch A, Waschkowitz T, Bowien S, *et al.* (2003) Metagenomes of complex microbial consortia derived from different soils as sources for novel genes conferring formation of carbonyls from short-chain polyols on *Escherichia* coli. *J Mol Microbiol Biotechnol* **5**: 46–56.
125. Zwirglmaier K, Ludwig W, Schleifer KH. (2004) Recognition of individual genes in a single bacterial cell by fluorescence in situ hybridization—RING-FISH. *Mol Microbiol* **51**: 89–96.
126. Uchiyama T, Abe T, Ikemura T, Watanabe K. (2005) Substrate-induced gene-expression screening of environmental metagenome libraries for isolation of catabolic genes. *Nature Biotechnol* **23**: 88–93.

127. Rusch DB, Halpern AL, Sutton G, *et al.* (2007) The Sorcerer II Global Ocean Sampling expedition: Northwest Atlantic through Eastern Tropical Pacific. *PLoS Biol* **5**: e77.
128. Margulies M, Egholm M, Altman WE, *et al.* (2005) Genome sequencing in microfabricated high-density picolitre reactors. *Nature* **437**: 376–380.
129. Fuhrman JA. (2003) Gene sequences from the sea. *Nature* **424**: 1001–1002.
130. Tamames J, Moya A. (2008) Estimating the extent of horizontal gene transfer in metagenomic sequences. *BMC Genomics* **9**: 136.
131. Tringe SG, Zhang T, Liu X, *et al.* (2008) The airborne metagenome in an indoor urban environment. *PLoS One* **3**: e1862.
132. Krause L, Diaz NN, Bartels D, *et al.* (2006) Finding novel genes in bacterial communities isolated from the environment. *Bioinformatics* **22**: e281-e289.
133. Koski LB, Golding GB. (2001) The closest BLAST hit is often not the nearest neighbour. *J Mol Evol* **52**: 540–542.
134. Mori T, Mizuta S, Suenaga H, Miyazaki K. (2008) Metagenomic screening for bleomycin resistance genes. *Appl Environ Microbiol* **74**: 6803–6805.
135. Suzuki H, Nagai K, Yamaki H, *et al.* (1969) On the mechanism of action of bleomycin: scission of DNA strands in vitro and in vivo. *J Antibiot* **22**: 446–448.
136. Umezawa H, Maeda K, Takeuchi T, Okami Y. (1966) New antibiotics, bleomycin A and B. *J Antibiot* **19**: 200–209.
137. Gatignol A, Durand H, Tiraby G. (1988) Bleomycin resistance conferred by a drug-binding protein. *FEBS Lett* **230**: 171–175.
138. Shen B, Du LC, Sanchez C, *et al.* (2002) Cloning and characterization of the bleomycin biosynthetic gene cluster from Streptomyces verticillus ATCC15003. *J Nat Prod* **65**: 422–431.
139. Tao M, Wang L, Wendt-Pienkowski E, *et al.* (2007) The tallysomycin biosynthetic gene cluster from Streptoalloteichus hindustanus E465–94 ATCC 31158 unveiling new insights into the biosynthesis of the bleomycin family of antitumor antibiotics. *Mol Biosyst* **3**: 60–74.
140. Diaz-Torres ML, McNab R, Spratt DA, *et al.* (2003) Novel tetracycline resistance determinant from the oral metagenome. *Antimicrob Agents Chemother* **47**: 1430–1432.
141. Brady SF, Clardy J. (2004) Palmitoylputrescine, an antibiotic isolated from the heterologous expression of DNA extracted from Bromeliad Tank water. *J Nat Prod* **67**: 1283–1286.
142. Leeb M. (2004) Antibiotics: a shot in the arm. *Nature* **431**: 892–893.
143. Newman DJ, Cragg GM, Snader KM. (2003) Natural products as sources of new drugs over the period 1981–2002. *J Nat Prod* **66**: 1022–1037.

144. Entcheva P, Liebl W, Johann A, *et al.* (2001) Direct cloning from enrichment cultures, a reliable strategy for isolation of complete operons and genes from microbial consortia. *Appl Environ Microbiol* **67**: 89–99.
145. Marchesi JR, Weightman AJ. (2003) Comparing the dehalogenase gene pool in cultivated alpha-halocarboxylic aciddegrading bacteria with the environmental metagene pool. *Appl Environ Microbiol* **69**: 4375–4382.
146. Eyers L, George I, Schuler L, *et al.* (2004) Environmental genomics: exploring the unmined richness of microbes to degrade xenobiotics. *Appl Microbiol Biotechnol* **66**: 123–130.
147. Rhee JK, Ahn DG, Kim YG, Oh JW. (2005) New thermophilic and thermostable esterase with sequence similarity to the hormone-sensitive lipase family, cloned from a metagenomic library. *Appl Environ Microbiol* **71**: 817–825.

CHAPTER 3

Chemical Ecology of Medicinal Plants

Christian Kersten, Stephanie Lenz and Janina Wich

ABSTRACT

Chemical ecology is an interdisciplinary field that investigates the incidence, function and synthesis of plant-based secondary metabolites. Plant metabolism can be separated into primary and secondary metabolism. Primary metabolism comprises all processes crucial for growth and plant development, while secondary metabolism serves other specific functions. The diversity of secondary metabolites has its origin in the evolution of life itself; these compounds represent adaptations plants have developed to gain a competitive edge in their environment. Pathways for the biosynthesis of secondary metabolites are derived from primary metabolic pathways. Effective chemical defense strategies protect plants e.g. from herbivores, insects or microorganisms, which in turn develop detoxification and adaptation strategies urging the plant to further evolve novel defense systems. This co-evolution leads to reciprocal adaptation. Plants have evolved many different strategies to defend themselves against attackers such as herbivores. These strategies are categorized as direct, indirect, constitutive and induced mechanisms as well as tolerance. The combination of various mechanisms allows plants to efficiently defend themselves against enemies. In particular, the induced indirect defense mechanisms were more thoroughly investigated due to the availability of necessary techniques, for example the determination of up-regulated genes. There was much interest in the correlation and interaction of plants with organisms from higher trophic levels. In this chapter, we discuss plant-plant, plant-fungi, and plant-animal interactions.

1. INTRODUCTION

Chemical ecology is an interdisciplinary field that investigates the incidence, function and synthesis of plant-based secondary metabolites. In

Institute of Pharmacy and Biochemistry, Johannes Gutenberg University, Staudinger Weg 5, 55128 Mainz, Germany.

particular, chemical knowledge of secondary metabolites may facilitate the discovery of medical applications for these secondary metabolites. Hence, the advancement of chemical ecology is closely related with progress in analytic chemistry.

Secondary metabolites are a huge group of very diverse chemicals encompassing volatile and non-volatile compounds, small molecules such as alkaloids and terpenoids, and macromolecules such as peptides or polysaccharides. Secondary metabolites are not directly involved in basic cell functions, such as growth, development or reproduction.[1] During the past two decades, the significance of these chemicals as defense agents has stimulated renewed interest. A role in defense has been experimentally confirmed for many of these compounds.[2] Secondary metabolites are also important in attracting pollinators and competing with neighboring plants.

The diversity of secondary metabolites has its origin in the evolution of life itself; these compounds represent adaptations that plants have developed to gain a competitive edge in their environment (Fig. 1).

Pathways for the biosynthesis of secondary metabolites are derived from primary metabolic pathways by incidental duplication and subsequent mutation of the copied genes during evolution. One example of the

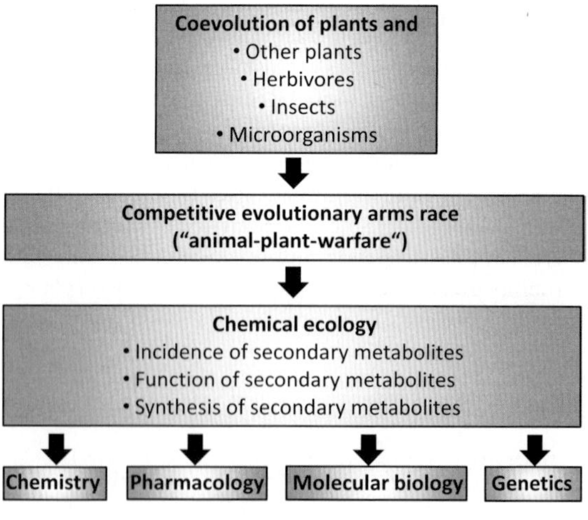

Figure 1 Chemical ecology as a result of co-evolution of plants and other organisms.

similarity between pathways is the polyketide synthases and the essential fatty acid synthases, both of which display the same type of enzymatic activity. This type of analogy has been demonstrated for many secondary metabolites. However, there are some exceptions, where the evolutionary links are missing.[1] Some biosynthesis pathways may be derived by lateral gene transfer from symbiotic fungi.[3] Due to rapid diversification and selection processes, some substances originated for specific ecological niches. In contrast to these highly specific structures, there also exist many products generated in more classical pathway-type reactions such as the terpenoids or alkaloids. Secondary metabolism is much more variable than primary metabolism, even in the same species. This is because of the independent evolutionary pressures acting on the enzymes in a plant. There are entire biosynthetic pathways that exclusively exist in the secondary metabolic network. Diverse secondary metabolic pathways produce a large diversity of chemical structures with limited genetic resources.[1] The enormous diversity of secondary metabolites can be in large part attributed to the highly variable chemistry of plants.

Effective chemical defense strategies protect plants e.g. from herbivores, insects or microorganisms, which in turn develop detoxification and adaptation strategies urging the plant to further evolve novel defense systems. This co-evolution leads to reciprocal adaptation, which may be specific, pairwise or diffuse. Pairwise co-evolution is the mutual evolution of two species in which each adapts to the other's characteristics. Diffuse co-evolution represents the adaptation of one species driven by interaction with several other species. This interplay can lead to a competitive escalation or an evolutionary arms race. An example of a defensive evolutionary invention is the sulfur-based glucosinolate-myrosinase defense system in Brassicaceae, which is characteristic of this plant family. Furthermore, some genera of Brassicaceae have evolved additional novel structures typical of other plant families, e.g. tropane alkaloids. However, the correlation of chemical similarity with phylogenetic propinquity does not completely match. In fact, host shifts of related herbivores better fit with plant chemistry than does the phylogeny of the plants. On the whole, evolutionary and phylogenetic history contains valuable information on plant-animal interactions.[3]

Genetic variation results in differential reproductive successes, in which the most advantageous phenotypes will have the largest genetic contributions to the next generation. To succeed and reproduce, a plant must make certain tradeoffs, e.g. investment in growth versus defense. It is thought that large, long-living species make greater investments than rare, small species do.[4]

2. A GLANCE AT HISTORICAL METHODS AND THEORIES

Plants have been used for medicinal purposes as well as for the production of poisons, pigments and flavors since ancient times, even without knowledge of their chemical background. During the twentieth century, the progress in methodology resulted in the discovery of an enormous amount of natural compounds. Interest grew as to why these compounds appear in plants and other organisms and what their role may be.[5] The question was, what evolutionary advantage does a plant gain by producing these secondary metabolites? In the year 1888, Stahl,[6] who is frequently quoted but heavily criticized, described experiments on the subject of plants' chemical protection against slugs and snails. He concluded that plant secondary compounds must have evolved under the selection pressures of herbivores. It was not until the 1980s that the functional aspects of secondary metabolism in plants were heralded as the "playground of biochemical evolution."[7] However, at this time it was still thought that secondary compounds were only ecologically important due to coincidence.[8] Mueller and Börger[9,10] published the theory of phytoalexins in 1940,[9] declaring phytoalexins to be "plant derived antibiotics that are synthesized *de novo* in living plant tissue as a response to pathogenic attack,"[11] thereby forming an immunologic system of defense. As a plant's sensitivity to or resistance against disease is an evolutionarily advantage, and the secondary metabolic pathways that bring about such resistance are heritable, theories on the evolutionary role of secondary metabolites are supported. Around the same time, the theory of allelopathy (from Greek: allelo = respective; pathos = damage) was developed. Its claims were and still are controversial, stating that plants can release compounds, be it volatile compounds through their leaves or liquid compounds through their roots, which are incorporated by other plants to cause harm or protection.

Phytochemical studies started at the end of the eighteenth century, when plants began to be analyzed for their major bioactive compounds. In the first half of the nineteenth century, crude plant extracts were purified by processes of crystallization and re-crystallization, liquid-liquid differential extraction, distillation and other separation techniques. The generation of secondary metabolites in plants was explained by analogous reactions in organic chemistry or by comparisons to known structures.[5] Tracer techniques with radioactively labeled nuclides were developed in the 1950s.[8] However, these techniques required large amounts of compounds and also great ability and imagination on the part of the chemists to solve the puzzle of chemical structures and biosynthesis routes.[5] Studies were much simplified, when improved mass spectrometry (MS) and nuclear magnetic resonance (NMR) techniques allowed analysis without chemical degradation. Later, MS combination systems with gas chromatography (GC-MS) or liquid chromatography (LC-MS) with advanced sensitivity were developed.[8] Their widespread use in the 1970s led to great successes in understanding the chemistry of natural products.[5] Difficulties in isolating active enzymes were overcome in the early 1970s with the introduction of column chromatography and the use of plant cell and organ cultures.[8] Box 1 gives more information about the development of analytic methods. In the 1960s and 1970s, phytochemical studies expanded to include microorganisms, insects, lichens, algae and marine organisms.[5] In the 1970s, a new discipline named (Phyto) Chemical Ecology came into being.[8] Over time, findings led to a new chemotaxonomic approach, using natural products as tools for the classification of species and taxa, which nowadays can be confirmed by DNA studies.[5]

The latter is regulated by continuous modification and new functionalization of duplicated genes. Because the duplicate copy is free from the demands on the parent gene, new traits can fuel the evolution of metabolic diversity.[8]

What accounts for the environmental pressure that necessitates the energetic expenses of secondary metabolism? A closer look on the history of life on earth may bear the answer to this question: The earth was

Box 1: Development of analytic methods.[5]

X-Ray diffraction analysis

1895	Discovery of X-rays by W. C. Röntgen
1925	First use of X-rays with organic compounds (e.g. benzene)
1945	Elucidation of the structure of penicillin and Vitamin B12
1953	Revelation of the double helical structure of DNA
Today	Routine technique
Drawback	Only usable for crystalline compounds (not cells, which contain aqueous medium); individual stereoisomers only measurable with the presence of heavy atoms (e.g. bromine)

NMR analysis

1957	First experiments; resonance of the N-atom was the most commonly used signal
1960	First H-atom resonance frequency for isobutyl and tert-butyl-groups
1960s	Reliable results for chemical elucidation of natural products (e.g. In 1961 both alpha-thujaplicinol and pisatin — confirming the phytoalexin-theory of 1940)
1963	Elucidation of the stereochemistry of the methyl groups in betulin through long range spin-spin coupling and decoupling proton experiments
Since the 1960s	Increasing power of magnets has allowed higher resolution spectra to be obtained with smaller amounts of the compound, resulting in the ability to examine minor metabolites.
Development of new techniques	e.g. TOCSY-1D, COSY, TOCSY, ROESY, NOESY, INADEQUATE, DOSY, HETCOR, HSQC and HMBC
1994–1996	Elucidation of the complete structure of marine polyethers, the causative agents leading to poisonous effects of the red tide

(*Continued*)

Box 1: *(Continued)*

Chromatography methods

Important because NMR, as well as X-ray diffraction analysis, require high purity of the examined compound

Development from TLC on paper or silica gel to GC, HPLC and CCC

Recently Combinations of chromatographic separation with spectroscopic data collection and detection forms hyphenated tandem techniques, e.g. GC-MS, LC-UV-MS or LC-MS-UV-NMR

formed 4.6 billion years ago. Bacteria and primitive unicellular prokaryotic cells appeared 800 million years (MY) later. Autotrophic photosynthetic cyanobacteria emerged 3.1 billion years ago, which lead to the production of oxygen and its accumulation in the atmosphere, causing the first massive extinction of anaerobic organisms 2 billion years ago. Terrestrial bryophytes developed 550 MY ago, with fungi and lichen arising first, followed by the first terrestrial vascular plants 440 MY ago. These vascular plants formed the first forests. Only 40 MY later, i.e. about 400 MY ago, insects appeared on the earth. Since then, insects and plants have been living together, each influencing the course of the other's evolution.[5] This persistent co-evolution between plants and their enemies (animal herbivores and parasites as well as other plant pathogens) is thought to be a major reason for the emergence of the great biodiversity found today on earth. Current theory states that plants synthesized and accumulated toxins as a response to the selection pressure of phytophagous insects. By adapting to and surviving the exposure to these toxins, a few insect species learn to tolerate these otherwise toxic plants. Thus, the plants evolved new or altered toxins to combat their assailants. This back-and-forth process resulted in highly specific relationships between insect and plant species that are closely linked ecologically, as was described in 1964 by Ehrlich and Raven.[12] Evidence of sequestration of toxic compounds by phytophagous organisms further supports the idea of co-evolution. Toxic compounds in the diets of these insects serve in their own defense or as building blocks for insect venoms or pheromones. Certain organisms have formed a close, specialized plant-predator-relationship, such as the monarch butterfly (*Danaus plexippus*), which uses cardenolides from *Asclepia scurassavica* to make itself poisonous.[5]

Recently, the so-called "-omics" technologies (genomics, transcriptomics, proteomics, metabolomics) have found an entrance into chemical ecology. In the past, medicinal plant compounds were discovered by bioassay-guided fractionation, examining the bioactivity of a first crude plant extract, which was then repeatedly fractionized and examined for bioactivity. Now, metabolomics allows for the identification and quantification of all primary and secondary metabolites in an organism.[13] In ecogenetics and ecogenomics, the interactions of plants and herbivores are investigated, but not without several restraints. For one, there is necessarily reductionism in trying to understand the interactions of plants with their environments; combining potted plants in greenhouses with random herbivores can only lead to very theoretical models.[14] Ideally, one should investigate the interactions of plants and herbivores that share the same habitat under natural selection pressures of that habitat, that is to say, in their natural environment. However, this is impractical, and techniques such as ecogenetics can provide practical useful information. Plant micrometabolomics is a specialization of the metabolomics techniques to analyze the metabolites present in a certain plant cell or plant tissue.[15]

From time to time, compounds from the rich chemical pool of secondary metabolites are adopted into primary metabolic pathways, gaining essential functions as phytohormones and signaling compounds. For instance, some flavonoids, known as floral color compounds, may affect auxin transport and developmental processes in *Arabidopsis.* Interestingly, these compounds predate the evolution of flowering plants. Some metabolites, such as canavanine, have acquired both primary and secondary functions. Canavanine is a toxic aginine-like antimetabolite that accumulates in the seeds of certain legumes. Its primary function is the remobilization of nitrogen during germination,[8] but it also serves as a protection against predators.

As new technologies are emerging, well established results are leading to new applications. Ethnobotanical and ecological approaches have given rise to the concept of ethnopharmacy, culminating in the discovery of new bioactive compounds to address diverse human afflictions. The issue of sustainability has been addressed as well. For example, sustainability concerns have arisen surrounding the Pacific yew tree (*Taxus brefivolia*), from which the anticancer-drug paclitaxel is obtained. Extensive harvesting endangered the survival of the species until a related analogue, bacattin III, was found in a more readily available source, *Taxus baccata*, allowing semisynthesis of the drug.[5]

3. PLANT DEFENSE MECHANISMS

Interactions between plants and other species can be beneficial, i.e. in the case of pollination. On the other hand there exist many heterotrophs, such as herbivores, fungi and pathogens, which attack plants to obtain specific compounds or to fulfill their life cycles. To defend themselves against these attackers, plants have evolved a myriad of different strategies.[16–18]

These defense strategies can be classified as either direct or indirect and can involve either induced or constitutive mechanisms. Another type of defense is known as tolerance. Of course, these strategies can also be combined within a single plant species (Fig. 2).

Direct defenses cause immediate harm to the attackers by physical barriers (e.g. thorns, trichomes or waxes) or by plant metabolites, for instance

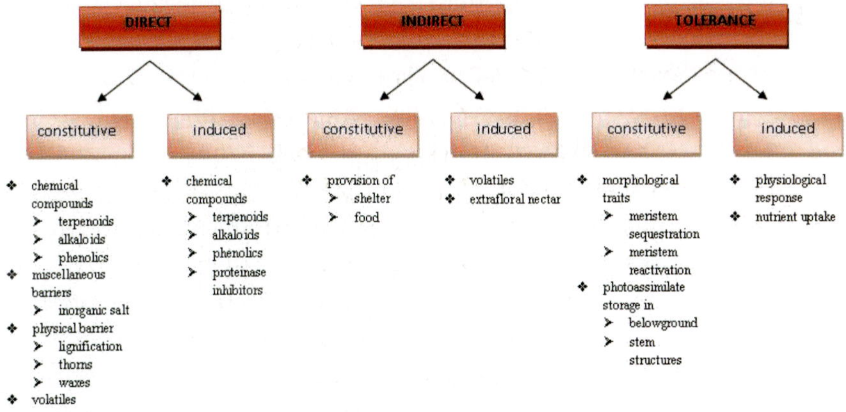

Figure 2 Overview of plant defense mechanism.

terpenoids, alkaloids or fatty acids as common defense compounds or proteinase inhibitors as specialized defense proteins.[19] These metabolites are commonly toxic, repellant or antidigestive.[20] Volatile organic compounds, especially those used in indirect defense, can also act as repellants against oviposition.[18] Such compounds even play a role in protection against abiotic stresses, such as oxidative stress.[21]

In contrast to direct defense mechanisms, indirect defenses act by activation of higher trophic levels. Plants attract the enemies of herbivores, e.g. parasitoids and predators, to get rid of the attackers by emitting volatiles or producing extrafloral nectar.[6,19] The provision of shelter, which can be obtained by modifying existing morphological structures, and food, e.g. pollen, to such enemies are further indirect strategies.[18]

Direct and indirect defenses can be either constitutive or induced. Important for the determination is the timing of the defense deployment. A defense is constitutive or "static," if the mechanism is always present. It is induced or "active," when the mechanism appears only in response to an attack. If a plant's tissue is damaged by mechanical or chemical interference, signal transduction pathways can induce local and/or systemic defenses. The metabolites activated by such pathways are often more specialized than the metabolites in "static" mechanisms.[17,19,20] As a constitutive defense mechanism, for example, plants may continuously emit volatiles that not only attract enemies of herbivores, but are also important for communication with other plants. Among other constitutive defense possibilities are chemical compounds such as alkaloids, which, for example, are continually present in the ripe fruits of the *Solanum* species for protection against frugivores. One difficulty in producing chemical defenses is the storage of toxic compounds without poisoning the plant itself. To this end, many plants have evolved strategies to store the toxins in their inactive forms.[4,22,23]

Proteinase inhibitors, for example, act as an induced defense in the potato plant, while they form a constitutive defense in potato tubers.[17] The choice between constitutive or induced defense mechanisms is a question of energy trade-offs, because maintaining a balance between energy directed towards growth and energy directed towards defense is important for a plant's overall fitness. Induced defense requires less of an energy investment than constitutive defense, because it avoids the production of idle resources.

This is favorable under natural selection pressures.[18,20] On the other hand, it may be more favorable for plants or parts of plants at high risk of attack to express constitutive defenses.[23]

Plants are called tolerant, if they can sustain tissue loss with little or no decrease in fitness relative to their undamaged state.[18] This kind of defense is expressed in morphological traits, storage of photo-assimilated materials in belowground and stem structures, physiological responses and nutrient uptake mechanisms.[18]

3.1. Plant-Plant Interactions

Allelopathy, chemical communication between plants, has long been seen merely as a tactic for the chemical-producing plant to harm the exposed plant.[24] Certainly, there are many examples of this type of allelopathic interaction. Milfoil (*Myriophyllum spicatum*) releases the hydrolysable tannin, tellimagrandin II, as an allelochemical from its shoots. Milfoil is an aquatic plant, and the tannin inhibits the growth of algae on the plant by complexing and inactivating the extracellular enzymes of the algae. The phenolic allelochemicals 2-chlorophenol and salicyl aldehyde present in aqueous extracts of *Typhadomingensis* roots, stems and leaves, are inhibitory when added to the growth medium of the water fern, *Salvinia minima.*[22] When *Centaurea maculosa* experiences root herbivory, it increasingly exudes the flavanol (±)-catechin, which has negative effects on the surrounding plant species. Although floral scent is generally recognized as an attractant for pollinators, there is evidence that floral volatile compounds were originally involved in defense as well. Methyl benzoate is the main compound in volatiles of the snapdragon flower *Antirrhinum majus* responsible for inhibiting the growth of *Arabidopsis* roots. Sagebrush (*Artemisia tridentana*) emits volatiles that inhibit seed germination in neighboring plants, but at the same time its presence reduces herbivory in nearby plants. However, it is not yet known whether the same compounds are responsible for both effects.

The sagebrush example leads us to a second frequently-observed type of allelopathy, in which allelopathy results in a positive effect on the exposed plants. Exposing plants of one barley cultivar to volatiles of a plant of a different cultivar changes the behavior of aphids and their natural enemies towards the barley plants. Additionally, many examples of compounds that directly repel or deter herbivores such as caffeine have been described. A solution of four previously-identified chemicals derived from the root exudates of couch grass (*Elytrigia repens*) makes barley plants less palatable to the bird cherry-oat aphid, *Rhopalosiphum padi*, when applied to the soil around the plant. This aphid is a serious pest of many cereal grains. However, the compounds themselves did not repel the aphids; instead, they appear to alter the composition of the receiving barley plants, making them less attractive to the aphids.[25] When barley plants were exposed to volatiles of the thistle *Cirsium vulgare*, aphids were less likely to attack the barley plants.[26] Likewise, volatiles from the weed *Chenopodium album*, reduced the amount of aphids settling on barley, when the two species were sown side by side in the laboratory and in the field.[27] While allelopathy between different plant species has been intensively investigated, communication between different genotypes of the same plant species has not stimulated similar consideration, but may also be ecologically relevant. However, it has been demonstrated that sowing different genotypes of the same crop species results in a reduction in the occurrence of damage-causing organisms that normally use the plants as hosts.[28–30] For instance, growing certain genotypes of barley plants in a 1:1 mixture reduced aphid attraction in the field,[31] and mixed cultivar farming in cereal crops reduced the incidence of fungal pathogens and aphid-borne plant viruses.[29,30] Furthermore, integration of multiple plant genotypes can reduce biotic damage and increase yields.[32–34] In contrast to the examples of direct defense by allelopathy, indirect defense mechanisms have been observed as well. A plot planted with two cultivars of spring barley crop in a 1:1 ratio resulted in significantly more reported ladybirds (*C. septem punctata*), an enemy of barley-infesting herbivores, than plots with either of the cultivars alone.[35] Self-recognition and kin-recognition help the plant to detect and distinguish potential resource competitors from beneficial neighboring plants and to respond in an appropriate way.

3.2. Defense Mechanisms of Plants against Herbivores

Because plants are rooted and immobile, they have evolved unique mechanisms of defense against herbivore attack. A combination of diverse mechanisms leads to an effective defense.

The indirect defense follows the principle, "the enemy of my enemy is my friend," under which an attacked plant releases volatile compounds attracting carnivorous enemies (predators or parasitoids) of the attacking herbivore.[36] In particular, the induced indirect defense has recently been in the focus of interest, because of its effectiveness, the great variety of substances involved, and span of interactions across different trophic levels. The chemical mediators of these defense systems, called herbivore-induced plant volatiles (HIPVs), can interact with neighboring plants or undamaged parts of the same plant as well as microorganisms or arthropods, in addition to attracting carnivore enemies of the herbivore. The specificity of the interaction lies in the variety of HIPV mixtures. The lima bean (*Phaseolus lunatus*) represents an illustrative example of a tri-trophic system involving the two-spotted spider mite (*Tetranychus urticae*) and the carnivorous mite *Phytoseiulus persimilis.* Lima beans infested with *T. urticae* emit HIPVs that attract *P. persimilis* to remove (feed on) *T. urticae.*[37] Infochemical is a natural chemical involved in the interaction between two individuals that conveys a response to the receiver. In this case, both the emitter and the receiver gain an advantage from the infochemical. This type of mutually-beneficial interaction is called synomone according to the allele chemical terminology (Table 1). Another example is the interaction between plants, caterpillars and parasitic wasps. Maize plants (*Zea mays*) infected by the African cotton leafworm (*Spodoptera littoralis*) use

Table 1 Allelochemical Terminology (modified according to Arimura *et al.*, 2009[36])

Allelochemical	An infochemical that mediates an interaction between two individuals of different species
Allomone	An allelochemical of organism 1 evoking a behavorial or physiological response in organism 2, favorable only to organism 1
Antimone	An allelochemical of organism 1 evoking a behavorial or physiological response in organism 2, favorable neither to organism 1 nor to organism 2
Kairomone	An allelochemical of organism 1 evoking a behavorial or physiological response in organism 2, favorable only to organism 2
Synomone	An allelochemical of organism 1 evoking a behavorial or physiological response in organism 2, favorable to both organism 1 and organism 2

volatiles to attract *Cotesia marginiventris* and *Microplitis rufiventris* (two parasitic wasp species), which not only attack caterpillars, but also the eggs of herbivorous insects after induction by egg deposition.[38] Additionally, it is important to mention that HIPV-mediated interaction can occur below ground as well. Weevil larvae (*Otiorhynchus sulcatus*) attack the roots of *Thuja occidentalis* and this induces the release of chemicals attracting the parasitic nematode, *Heterorhabditis megidis.* This phenomenon was first observed with the root-feeding beetle (*Diabrotica virgifera virgifera*) and maize plants.[39]

The most important groups of HIPVs are the volatile terpenoids (see Chapter on Plant defense agents) and the so-called green leaf volatiles (GLVs). GLVs are produced by the lipoxygenase pathway (LOX) after leaf injury or in response to biotic or abiotic stress. The GLVs include C_6-alcohols, aldehydes and esters. While most other volatiles are produced after up-regulation of biosynthesis genes and are released after several hours, GLVs are released within seconds (as shown in *Arabidopsis* leafs).[36] (Z)-3-hexenal is immediately formed after herbivore damage, followed by (Z)-3-hexenol and its acetate, which are built near the wounded site supplied by the hexenal from the damaged tissue (Fig. 3). This rapid reaction could be explained by the sequestration of substrate and enzyme in separate tissue compartments followed by combination of the substrate and enzyme after wounding. However, this hypothesis has yet to be thoroughly investigated. GLVs are mainly emitted under stress situations, as observed in tobacco plants under ozone exposure and reed leaves under elevated temperature or harsh lights.[40,41] It has been suggested that oxidative stress is an outcome of herbivore damage inducing GLV emission.[36]

While GLVs undergo oxidation in the atmosphere (e.g. by O_3), they have a short lifetime ((Z)-3-hexenal has a half-life of around 2 h). Hence, its ecological function is an acute attraction, for example of parasitic wasps, which are not attracted by the O_3-degraded GLVs. (Z)-3-hexenol attracts generalist predators, while (E)-3-hexenal reveals considerable effect on the sesquiterpene synthesis in plants.[36]

In addition to volatile emission after herbivore damage, some plants have developed another way to attract predatory arthropods: extrafloral nectar (EFN). EFN's main components are sucrose, glucose, fructose and other sugars, amino acids and other organic compounds. They are

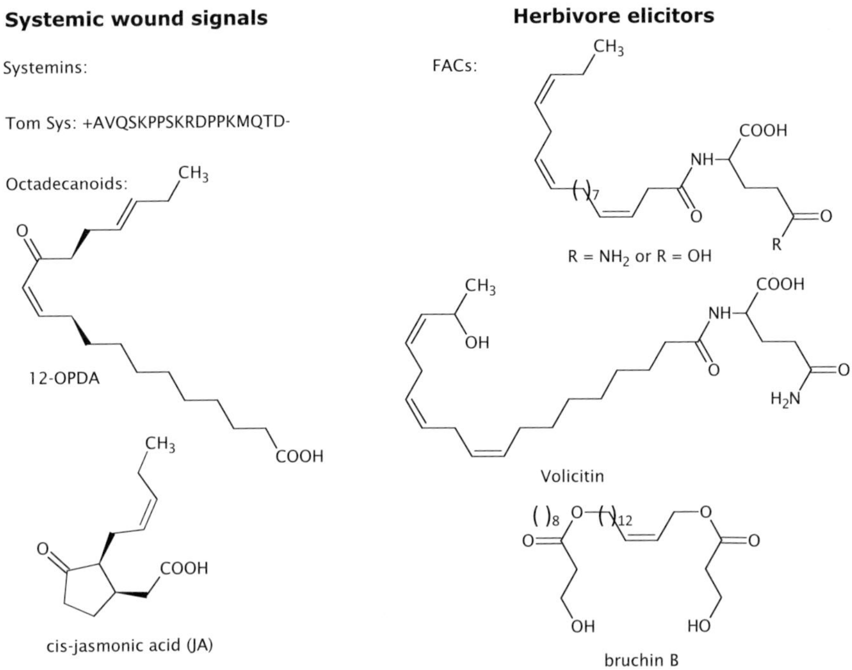

Figure 3 Examples of elicitors and systemic wound signals (according to Kessler & Baldwin, 2002).[18] Polypeptides of the Solanaceae family activating the octadecanoid-pathway are called systemins. Systemins are involved in signaling triggered by systemic wounds and are named after the plant in which they occur (e.g. Tom Sys in tomato, Tob Sys I&II in tobacco). Octadecanoids such as 12-oxophytodienoic acid (12-OPDA), jasmonic acid (JA) and methyl-jasmonate (MeJA) are called Oxylipins. They play a central role in eliciting gene expression for direct and indirect secondary defense metabolites. Herbivore specific elicitors, such as the fatty-acid-amino-acid-compounds (FACs) of *Manduca* spp. or *Spodopteraexigua* (Volicitin) can be found in the oral secretion or saliva of the herbivore. β-glucosidase induces HIPV emission via ROS generation. Bruchins are elicitors found in Bruchideae oviposition fluid.

generated in specialized secreting organs and are found in over 93 non-gymnosperm plant families.[42] The production of EFN is obligate and can be either dependent or independent of circadian rhythms. Production can be increased by treatment with jasmonic acid or by leaf damage, and the composition of the nectar may change (for example, resulting in an increased concentration of amino acids). In cotton plants (*Gossypium herbaceum*), root damage induced an EFN release in above-ground plant structures as well as in roots.[43] The general function of EFN is to attract plant defenders like arthropods such as *Araneae* (spiders), *Diptera* (flies), *Coleoptera* (bugs) and *Hymenoptera* (insect order containing ants, wasps, etc.). Many of these predators and parasites reduce the count of herbivores on the plant. For better detection, some EFN are colored and are often combined with HIPVs emission.[19]

3.3. Induction of Plant Defenses and Signal Transduction

As many plant defense mechanisms are induced, there must be certain triggers for release of directly toxic or indirectly harmful volatile compounds. The two main activators are wounding of the plant and so-called elicitors (also referred to as herbivory-associated molecular patterns "HAMPs"), either in the form of oral secretions or oviposition fluid. It has been suggested that microbes found in the digestive organs of herbivores are involved in elicitor production.[20]

Two classes of elicitors in oral secretions inducing indirect defenses have been found in lepidopteran larvae. The first class comprise lytic enzymes such as β-glucosidase of *Pierisbrassicae*, glucose oxidase of *Helicoverpazea saliva* and whitefly's alkaline phosphatase.[44] The second class consists of fatty-acid-amino-acid conjugates (FACs) such as volicitin (Fig. 4) of *Spodopteraexigua* (the beet armyworm), which induces the release of HIPVs in maize plants.[45] Furthermore, bruchins (long chain diols diesterfied with 3-hydroxypropanoic acid) are excreted with the oviposition fluid of pea and cowpea weevils (Bruchideae). Brucins attract parasitic wasps that attack the eggs.[46] In many cases, early steps of the elicitation process are unknown because few elicitor receptors have been identified. One case in which the pathway has been elucidated is that of glucose oxidase. In this case, the glucose oxidase is expected to increase H_2O_2 production, which forms reactive oxygen species (ROS).[18,47]

The wound response induced after herbivore attack or mechanical damage mainly follows the octadecanoid pathway (C_{18} fatty acids). It is a systemic response triggered by ROS production in the wounded tissue, which leads to calcium influx, which further leads to enhanced activity of NADPH-oxidase (located in cell membranes). In the following section, we take a closer look at the signal transduction pathway.[19,20]

The first step in response to mechanical damage and/or contact with elicitors is the generation of reactive oxygen species (ROS) such as H_2O_2, which serves as a second messenger.[36] This leads to a depolarization of the cell membrane and an influx of cytosolic calcium. One striking way in which the responses of different plant types differ is that some plants react after a single herbivore attack or mechanical damage, whereas others emit HIPVs only after a continuous period of wounding. It has been shown that lima beans and cotton (*Gossypium hirsutum*) do not respond to contact with a single elicitor. The fully induced defense response can only be reached by continuous wounding.[36,48] Furthermore, individual elicitors can each provoke different responses in the same plant.

Immediately after a single incidence of leaf wounding, the mitogen-activated protein kinase (MAPK) pathway called the WIPK (wound-induced protein kinase) pathway is activated. WIPK pathway activation is probably essential for the formation of the phytohormone, jasmonic acid (JA) and its derivate methyl-jasmonate (MeJA).[49] WIPK activates the transcription of a $\omega-3$ fatty acid desaturase (FAD), which catalyses the reaction of linoleic acid to linolenic acid, an important JA precursor[50] of the octadecanoid pathway mentioned above. One of the key enzymes in JA synthesis is lipoxygenase 2 (LOX2), which is up-regulated by phospholipase D (PL) after wounding stimuli.[51] In addition, phospholipase A mediates the release of lineolic acids across the cell membrane.[19] It has been shown that not only free JA, but also many conjugates (especially those with isoleucine) are also effective.[19]

Another enzyme of the MAPK-pathway, the salicylic acid-induced protein kinase (SIPK), reaches its maximum activity approximately 5–10 minutes after damage occurs. Salicylic acid (SA) is the main antagonist of JA.[19,20,36] It interferes with the octadecanoid pathway at the level of 12-oxophytodienoic acid (OPDA). It is important to know that the balance of SA and JA leads to the production of a unique blend of volatiles, which finally attracts the predator. Different mixtures defend against different herbivores.[52–54] For example, chewing caterpillars induce higher levels of JA relative to SA, while phloem-sucking arthropods induce higher SA levels.[36] In the final stage of HIPV emission, transcription factors (TF) up-regulate the synthesis of different volatiles.[19]

In contrast to SA, which always acts as an antagonist to JA, the phytohormone ethylene can act as a synergist or antagonist depending on the plant and the situation. Like SA, ethylene synthesis is triggered by wounding. Plant defense genes and JA are induced synergistically in *Arabidopsis thaliana*. JA-induced emission of volatiles is promoted as well as JA-precursor synthesis of OPDA. In contrast, ethylene has an antagonistic effect on JA-induced transcript accumulation in wild tobacco plants (*Nicotiana attenuata*).[18]

It is well known that plants respond to herbivorous attack not only at the site of damage, but also in a systemic way. One group of important intercellular signaling molecules are the 18 amino-acid peptides called systemins mentioned above. These peptides were first discovered in the Solanaceae family.[55] Similar to many other signals, systemins activate the octadecanoid pathway. However, systemins have been observed to effect plants other than Solanaceae as well. Poplar trees, for example, are able to trigger the activation of gene expression systematically via systemins so that induced defenses are expressed in undamaged leaves far away from those that had been damaged by the forest tent caterpillar (*Malacosoma disstria*).[56] In tobacco plants, JA is translocated to unwounded sides and induces nicotine expression.[57,58] Finally, plants exhibit different pathways for long distance signaling.[19]

In conclusion, signal transduction begins with herbivore attack and triggers (continuous) wounding and/or elicitors. The subsequent depolarization via Ca^{2+} influx causes the formation of ROS. Consequently, the MAPK pathway is activated, linolenic acid is released from the cells, and the octadecanoid pathway syntheses jasmonic acid and other oxylipins. Crosstalk such as that between SA and ethylene may result in complicated feedback mechanisms and regulation of metabolic pathways finally leading to specific defense reactions, especially the emission of volatiles.

3.4. Interactions between Fungi and Plants

Pathogenic fungi must attach to the cuticular surface of a plant in order to establish themselves on a host plant. In pea plants, the reduced wax surface of the cuticle leads to an increased likelihood of infestation by the fungal aphid pathogen, *Pandora neoaphidis.*[59]

Crucifers and Brassicaceae, families containing many economically important species, have been thoroughly examined for infection by a wide range of fungi. Both families form phytoalexins, often containing dithiocarbamate or thiolcarbamate groups, derived mostly from tryptophan. However, a susceptibility to fungus infection prevails, suggesting that the fungi possess an effective detoxification mechanism. Some crucifer fungi use specific enzymes for detoxification. Therefore, attempts have been made to design inhibitors specific to these detoxification enzymes.[60]

Fungi synthesize a number of notable defense-related compounds. In 1928, the penicillins were found in the microfungus *Penicillium notatum*. Closer inspection has revealed that this fungus synthesizes a rich array of bioactive compounds for defense against other microbes. The strobilurins are antifungal compounds first isolated from mycelial cultures of the basidiomycete, *Strobilurus tenacellus*. They are able to reversibly bind to the ubihydrochinone oxidation centre of the cytochrome bc_1 complex of fungi, thereby inhibiting electron transfer in the fungal respiratory chain. Because of their potent antifungal activity, these compounds served as lead structures for several agricultural fungicides. A phytotoxin named hydroxycornestin holds potential to lead to new herbicides for cornfields. Cultures of *Paecilomyces varietii* have exhibited high levels of phytotoxicity in broadleaf weeds, but little activity in *Zea mays*.[61] Truffles (*Tuber* spp.), too, bear a potency in agricultural techniques. Truffle species require symbiosis with other organisms to fulfill their life cycle. Fruiting bodies can only be formed when the fungus is in contact with the roots of a host plants. Once established, some truffles such as the black species, *T. melanosporum*, causes a unique "burnt" zone surrounding the host plant, where no herbaceous plants grow. Ethylene and indole-3-acetic acid (IAA) are volatile hormones released by truffles that act as potent herbicides and cause root hair elongation in the non-host plant *Arabidopsis thaliana*. This is one of many examples in which fungal secondary metabolites affect plants.[62]

3.5. Pollinator Correlation with Plants

Pollinators are most often attracted by volatiles or floral color. Ethylene, one of the three plant hormones emitted into the air in biologically active quantities, regulates the growth and development of flowers, which further controls that way a plant uses volatiles to attract pollinators.

Since communication through volatiles represents transmission via an open channel, fragrances emitted to allure pollinators can easily attract nectar-robbing bees or florivores. Therefore, some plants have developed highly specialized attractants. Some orchids, for instance, fool social hunting wasps by emitting green leaf volatiles from their flowers. The wasps recognize these volatiles as those emitted by plants with leaf damage from feeding caterpillars, one of their prey. The flowers of other orchids emit volatiles simulating the smell of virgin female wasps, tricking the male wasps into pollinating the flowers. Whether it is by smelling like rotting carcasses, dung or wet soil, many plants have become specialized in mimicking the exact scent its pollinator is looking for. The more specific a signal is in the cacophonous volatile cloud, the more effective it is in causing pollination.[21]

4. DEFENSIVE AGENTS

Many plant defense agents are toxic to herbivores and pathogenic microbes. Some, such as poison hemlock and aconite, even have toxic effects on larger mammals and humans. The mechanisms behind plant defense agents are as variable as the chemical structures. The following section provides an introduction to this great chemical variety. The most thoroughly investigated and common types of chemical defense agents are alkaloids, terpenoids, phenolics and glucosinolates, which are described below along with several others.[22]

While some chemical defense mechanisms are well understood, many others remain unknown. Saponins such as avenacoside B (Fig. 4A) from *Avena sativa* disrupt cellular membranes.[63] Cyanogenic glycosides (Fig. 4B) release hydrogen cyanide, which inhibits mitochondrial respiration.[64] Cardenolides (Fig. 4C) are inhibitors of the Na^+/ K^+ATPase.[65,66] Polyacetylenecicutoxin (Fig. 4D) of the water hemlock (*Cicutavirosa*) blocks voltage-dependent potassium channels and thereby prolongs the repolarisation process.[67] *Papaver somnifera* alkaloids, morphine (Fig. 4E) and bismorphine, which is formed in an H_2O_2 and peroxidase dependent reaction, affect the central nervous system and produce analgesia.[23]

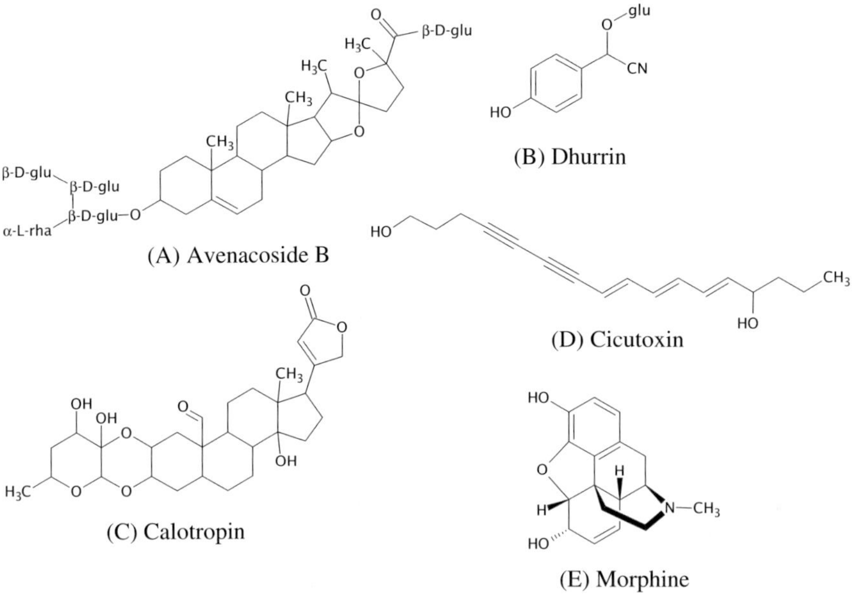

Figure 4 Various notable herbal defense chemicals.[23]

In addition to these species or genus specific compounds, there are groups of secondary metabolites for defense that are found across a wide variety of plant families. Firstly, the terpenoids must be mentioned. Terpenoids prevent slugs and snails from feeding on plants. Examples of terpenoids include avenacoside B and calotropin as well aspetasin (Fig. 5A) and furanopetasin of *Petasites hybridus*. The diterpenoid sacutilol A and B (Fig. 5B) stop fish from feeding on the brown alga *Dictyota acutiloba*. Saponins, mentioned above, are triterpenes (see Fig. 4A).[68]

Many terpenes have potent antibacterial and antifungal activity and are toxic to or inhibit the growth of insects, nematodes and fishes. Another example of terpene use is the administration of the sesquiterpene artemisinin of *Artemisia annua* for the treatment of malaria (Fig. 6A). It is effective against the malaria parasite *Plasmodium falciparum*, and acts by killing the parasite in its asexual state. The blockbuster anticancer drug paclitaxel (Fig. 6B), which binds to tubulin, is a diterpene as well.[9]

The terpenoids are derived from common five-carbon isoprene units. Their synthesis follows either the mevalonate pathway (MVA) in the

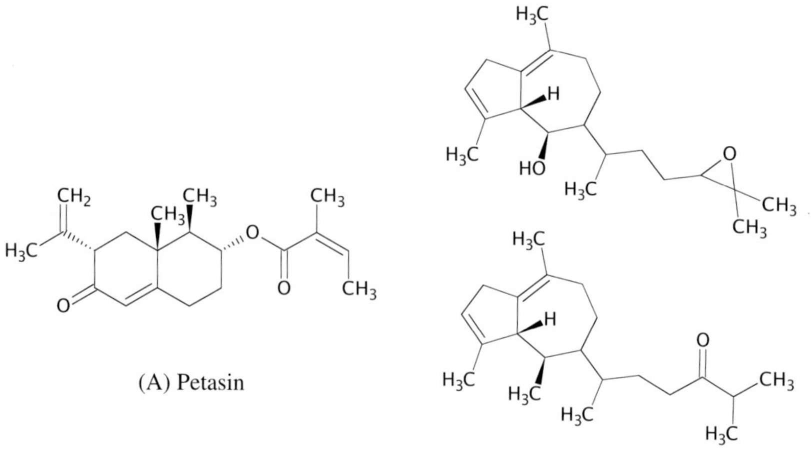

(A) Petasin

(B) Acutilol A (epoxide) & B

Figure 5 Terpenoids.[22]

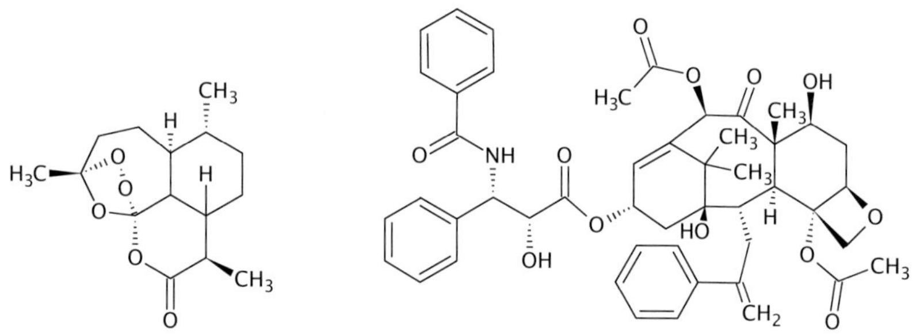

(A) Artemisinin

(B) Paclitaxel

Figure 6 Medically used terpenes artemisinin and paclitaxel.[69]

cytoplasm or the 2-C-methyl-D-erythritol 4-phosphate pathway (MEP) in plastids. The sesquiterpenes (C_{15}) are formed via the MVA pathway, the mono- (C_{10}) and diterpenes (C_{20}) via the MEP pathway. The structural diversity of terpenes is brought about by the diversity of terpene synthases (TPSs), which synthesize monoterpenes from geranyldiphosphate, sesquiterpenes from fernesyldiphosphate, and diterpenes from geranylgeranyldiphosphate.[4]

Many terpenes are volatile and play a role in indirect defense. Chemoattractant terpenoids are shown in Fig. 7.

Alkaloids are thought to be "multipurpose" defense agents.[70] A large variety of alkaloid targets resemble the large diversity of alkaloids. In general, they are synthesized continuously in vulnerable tissue. Some alkaloids are species-specific, such as the pyrrolizidine alkaloids (PA) of *Senecio* sp.[71] Although no increase in alkaloid concentrations was found in damaged *Delphinium* sp.,[72] the best-studied system for induced defense is *Nicotiana sylvestris*, which demonstrates a significant increase in alkaloid concentrations (Fig. 8) after herbivore feeding.[22]

The toxic bluebell (*Hyacinthoides nonscripta*), synthesizes alkaloids as well. It owes its toxicity to glycosidase inhibition by the PAs DMDP, homoDMDP, (Fig. 8), homoDMDP-7-apioside and

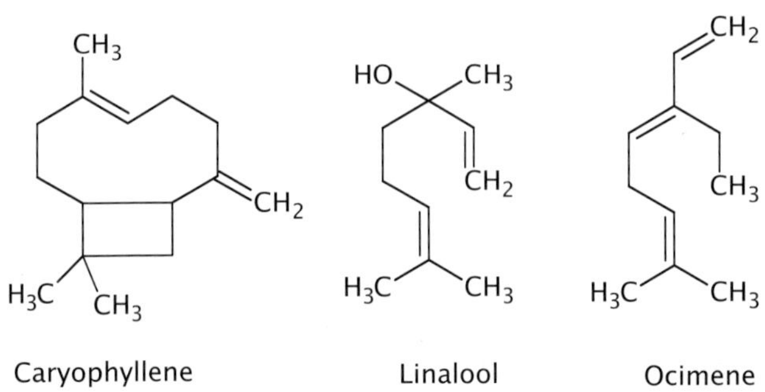

Figure 7 Volatile terpenes: the sesquiterpene caryophyllene and the monoterpenes linalool and ocimene.[36]

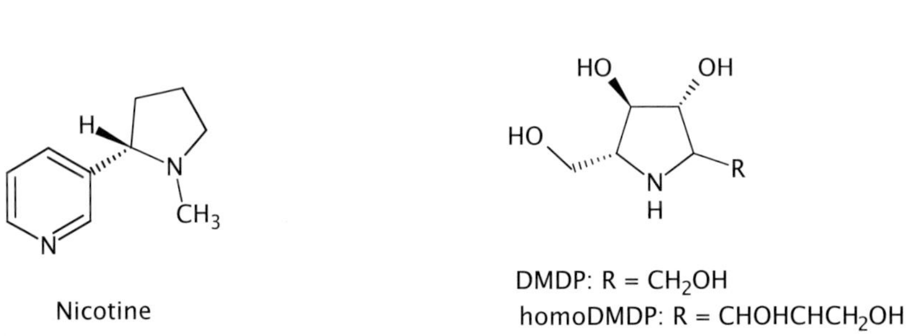

Figure 8 *Nicotiana sylvestris* alkaloid, nicotine and *Hyacinthoides nonscripta* alkaloids, DMDP and homoDMDP.[22]

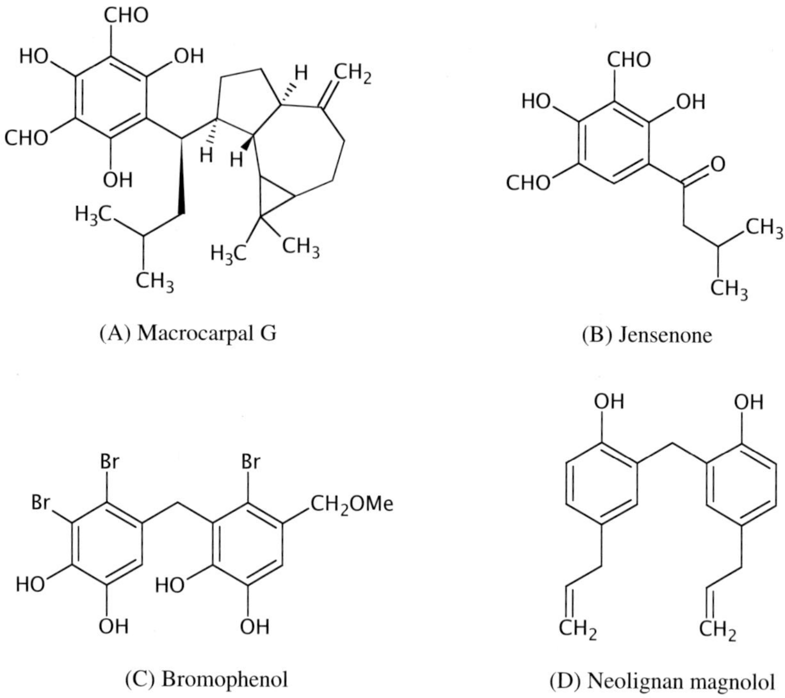

Figure 9 Phenolic defense agents.[22]

1,4,-dideoxy-1,4-imino-D-arabinitol. Of course, many other examples of plant defense involving alkaloids have been investigated.[73]

Another important group to mention is the phenolics. *Eucalyptus* species produce the defensive compounds macrocarpal G and jensenone to protect against possum and koala (Fig. 9A and B).[74] The red alga *Odonthalia corymbifera,* a potent bromophenol, prevents *Heliotis discus nannai* from feeding on the alga (Fig. 9C).[11] The defensive role of the neolignan magnolol (Fig. 9D) has been investigated in *Magnolia virginiana.* The synthesis of magnolol interrupts the growth of specialist moth larvae of *Callosamia securifera* that develop on leaves of *C. angulifera* and *C. promathea.*[75]

Finally, it must be mentioned that the composition of a plant's chemical defense rarely consists of a single substance. Usually, complex mixtures of many compounds interact synergistically to drive away attackers. This synergism results in a defense much more potent than the sum of the effects produced by each single compound. One possible mechanism of synergistic interaction is inhibitory activity of one compound towards the detoxification of the other(s). Feeding studies with essential oils on *Spodoptera litura* larvae have shown that the synergistic toxicity can be up to nine times higher than that of any single ingredient in the mixture.[76,23]

5. THE CHALLENGE OF DEFENSE

As co-evolution favors neither plants nor herbivores, many herbivores have developed mechanisms of resistance against phytochemicals.

One simple mechanism of resistance is avoidance of toxic parts of a plant, for example by spatial shift or niche restriction. Alternatively, insects may feed on the plant at a stage during which the toxic compounds have not yet been synthesized (phonological shift), or may completely avoid the plant. Insects detect plants visually, by contact, or by olfaction.[77] Detection mechanisms may be genetically conserved or they can be part of a learning process. For instance, many insects avoid bitter tasting food sources. As these bitter compounds have diverse chemical structures, this behavior necessitates a lack of specificity on the part of the insects. For example, the tobacco hornworm (*Manduca sexta*) is not able to differentiate between the toxic alkaloid caffeine and the non-toxic salicin, and therefore avoids both.[78] Grasshoppers and weevils keep away from bitter cyanogenic glycosides even in non-toxic concentrations.[79] Thus bitterness may not always be an "honest" signal.[80]

Another mechanism that herbivores have developed against plant defenses is manipulation of the chemical defense. Deactivation of the chemical before herbivores even begin feeding is one possible tactic. One very clever example is the larva of *Erinnyis alope*, which feed on *Carica papaya* leaves. Toxic agents of this papaya plant are stored in laticifers and oil ducts and are transported to the feeding site after damage occurs. However, the larvae first sever a part of the leaf and then start feeding, so that there can be no influx of toxic oil and latex into the damaged tissues.[23,81] Some chewing insects are able to reduce the potency of a plant's defenses with their oral secretions. For example, the tobacco earworm excretes glucose oxidase, decreasing the level of nicotine in *Nicotioana tabacum*. Gall-inducing insects such as the sawfly are able to decrease the toxic level of phenolics in the gall, where larvae develop.[80,82]

Accumulated toxic agents can also be excreted or discarded via exuvia or even used by the insect for defense against its own predators (sequestration).[83] This strategy requires selective transport and storage of toxins, but can be energetically advantageous as in the case of various leaf beetles species. While some basal species develop their chemical defenses *de novo*, other species obtain these agents from their host plant and modify them only minimally for evolutionary innovation via certain enzymes.[84,85]

Furthermore, resistance can occur metabolically by means of overproduction of detoxification enzymes. This so-called biotransformation of toxic chemicals represents the main resistance developed by insects over millions of years of co-evolution.[86] There are two ways to develop metabolic resistance. Firstly, toxins may resemble certain substances already present in an herbivore's diet and thus be chemically detoxified by existing enzymes. Secondly, there are specific mutations in genes encoding for metabolic enzymes that enhance detoxification.[87]

The most important families of metabolic enzymes in detoxification are the cytochrome P450 monooxygenases (P450, CYP), the glutathione-S-transferases (GSTs) and the carboxylesterases (COEs).[88] Related enzymes that catalyze detoxification reactions have been found in *Depressaria pastinacella*,[4] *M. sexta*[89] and *Helicoverpa* earthworm species.[90] Another resistance mechanism found in *Helicover pazea* is the ability to detect jasmonate and salicylate, which results in an overproduction of P450s.[80] GSTs usually link glutathione to toxic electrophilic compounds to increase their water solubility and thus their excretion. Activity of GSTs can also be induced by allochemicals. The great diversification pattern of GSTs in generalistic herbivores hints at adaption to a wide range of plant chemicals.[68,92,80] Some insects such as the polyphagous arctiid moth *Estigmenea crea* have developed highly specific detoxification mechanisms. These moths sequester spyrrolizidine alkaloids (PAs) from Asteraceae species via N-oxidation by a specific flavin-dependent monooxygenase. The diamondback moth (*Plutella xylostella*) that copes with the "mustard oil bomb" of its host plant, are lease of glucosinolates and a myrosinase, which hydrolyzes the glucosinolate into a toxic agent, from separate compartments. The moth has developed a sulfatase gut enzyme preventing the glucosinolate hydrolysis and thereby reducing its toxicity.[80,93]

Finally, evolution of mutations at the toxins' target sites is a possible resistance mechanism. One well-documented example is the resistance of the monarch butterfly and two leaf beetles (*Chrysochus*) to ouabain. A single amino acid modification in the ouabain target site, the Na^{+}/K^{+}ATPase, leads to complete resistance.[94] The rarity of this kind of resistance may be due to the high evolutionary cost of the modification, as the mutation is probably accompanied by a reduction in the efficiency of the enzyme's normal function. Insect populations treated with insecticides for prolonged periods have been shown to develop other resistance mechanisms.[95,80]

In addition to herbivore resistance mechanisms, plants have further challenges to cope with. In open communities, the emitter of volatiles has little control over who the receiver is.[21] Hence, the volatiles produce a combination of intended and unintended effects. The risk of attracting nectar-robbing bees or inviting some lepidopteran species that uses volatiles as a feeding stimulant is always present. Some larvae even postpone feeding until receiving the "bon appetit" signal of their host's volatile emission.[21]

All resistance traits are costly to produce, especially when fitness-limiting resources such as nitrogen must be invested. Furthermore, many resistance mechanisms are maintained under enemy-free conditions too.[19] Hence, there must be a balance between obligate and induced defenses. On the one hand, continuous production of defense compounds is very costly and requires the plant to cope with the problem of storing the toxic chemicals. This problem is usually solved by storing precursors, such as glucosides, separated from their activating enzymes, thioglucosidase myrosinase.[23] On the other hand, expressing only induced defense can be dangerous. If the initial damage occurs very rapidly and there is a lag-time in up-regulating the needed substances, induced defenses may be ineffective.[19] Therefore, many plants reveal non-uniform distributions of defense chemicals. Frequently attacked parts maintain high concentrations of defense chemicals, while less frequently attacked parts have low concentrations.[22] Additionally, in the case of indirect defense, obligate production can be counterproductive, because if the attracted predators find no herbivores to eat, they might start to associate the volatile with hunger and avoid the plant. As mentioned above, a balance of different herbal defense mechanisms is necessary.

Plants usually use obligate as well as induced defenses in both direct and indirect ways.

6. CONCLUSIONS AND PERSPECTIVES

In conclusion, plants have evolved many different strategies to defend themselves against attackers such as herbivores. These strategies are categorized as direct, indirect, constitutive and induced mechanisms as well as tolerance. The combination of various mechanisms allows plants to efficiently defend themselves against enemies. In particular, the induced indirect defense mechanisms were more thoroughly investigated due to the availability of necessary techniques, for example the determination of up-regulated genes. There was much interest in the correlation and interaction of plants with organisms from higher trophic levels. Through discovery of the important secondary metabolites such as jasmonate and ethylene, related compounds were identified and pathways elucidated. Without doubt, the major role of plant secondary metabolites is defense. However, secondary metabolites can also have other functions such as pollination or allelopathy.

Nonetheless, more research in this field is needed. Some signal transduction pathways remain unclear and others have yet to be established. The mechanism of tolerance, which is largely unknown, demands elucidation as well. Beyond communication within a plant and between plants, plant interactions with organisms from other trophic levels should be analyzed. It stands to be clarified why GLVs, a subgroup of the HIPV mediators, react so quickly during plant interactions with organisms from higher trophic levels. How are certain enzymes and substrates brought together after being separated in undamaged tissue? Most metabolites have been investigated in leaves, but few have been investigated in roots. Further research, especially in roots, is necessary to identify novel metabolites with potential value for medicinal usage.

Thus far, research on plant interactions has been possible due to the development of analytic methods, which are essential in identifying interacting phytochemicals. With further improvement of these methods, better screenings and identifications of plant compounds may be possible. Evolution is a permanent and ongoing process that will capture researchers' attention for years to come.

REFERENCES

1. Jenke-Kodama H, Müller R, Dittmann E. (2008) Evolutionary mechanisms underlying secondary metabolite diversity. *Prog Drug Res* **65**: 121–140.
2. Dicke M. (2000) Chemical ecology of host-plant selection by herbivorous arthropods: a multitrophic perspective. *Biochem Syst Ecol* **28**: 601–617.
3. Futuyama DJ, Agrawal AA. (2009) Macroevolution and the biological diversity of plants and herbivores. *Proc Natl Acad Sci USA* **106**: 18054–18061.
4. Cianfrogna JA, Zangerl AR, Berenbaum MR. (2002) Dietary and developmental influences on induced detoxification in an oligophage. *J Chem Ecol* **28**: 1349–1364.
5. Macías FA, Galindo JL, Galindo JC. (2007) Evolution and current status of ecological phytochemistry. *Phytochemistry* **68**: 2917–2936.
6. Stahl E. (1888) Pflanzen und Schnecken Biologische Studien über Schutzmittel der Pflanzen gegen Schneckenfrass. *Jenaer Zeitschrift für Medizin und Naturwissenschaften* **22**: 557–684.
7. Luckner M. (1990) *Secondary metabolism in microorganisms, plants and animals.* VEB G Fischer Verlag, Jena.
8. Hartmann T. (2007) From waste products to ecochemicals: fifty years research of plant secondary metabolism. *Phytochemistry* **68**: 2831–2846.
9. Mueller KO, Börger H. (1940) Berlin-Dahlem. *Arb. Biol. Reichsanstalt. Land-u. Forstwirtsch* **23**: 189.
10. Mueller KO. (1966) Phytoalexin theory of K.O. Mueller. Abhandlungen der Sächsischen Akademie der Wissenschaften zu Leipzig. *Mathematisch-Naturwissenschaftliche Klasse* **49**: 23.
11. Kurata K, Taniguchii K, Takashima K, *et al.* (1997) Feeding-deterrent bromophenols from Odonthalia corymbifera. *Phytochemistry* **45**: 485–487.
12. Ehrlich PR, Raven PH. (1964) Butterflies and plants: a study in coevolution. *Evolution* **18**: 586–608.
13. Prince EK, Pohnert G. (2010). Searching for signals in the noise: metabolomics in chemical ecology, *Anal Bioanal Chem* **396**: 193–197.
14. Kant MR, Baldwin IT. (2007) The ecogenetics and ecogenomics of plant–herbivore interactions: rapid progress on a slippery road. *Curr Opin Genet Dev* **17**: 519–524.
15. Moco S, Schneider B, Vervoort J. (2009) Plant micrometabolomics: the analysis of endogenous metabolites present in a plant cell or tissue. *J Proteome Res* **8**: 1694–1703.
16. Erb M, Lenk C, Degenhardt J, Turlings TC. (2009) The underestimated role of roots in defense against leaf attackers. *Trends Plant Sci* **14**: 653–659.

17. Gatehouse JA. (2002) Plant resistance towards insect herbivores: a dynamic interaction. *New Phytol* **156**: 145–169.
18. Kessler A, Baldwin IT. (2002) Plant responses to insect herbivory: the emerging molecular analysis. *Annu Rev Plant Biol* **53**: 299–328.
19. Arimura G, Kost C, Boland W. (2005) Herbivore-induced, indirect plant defences. *Biochim Biophys Acta* **1734**: 91–111.
20. Wu J, Baldwin IT. (2009) Herbivory-induced signalling in plants: perception and action. *Plant Cell Environ* **32**: 1161–1174.
21. Baldwin IT. (2010) Plant volatiles. *Curr Biol* **20**: R392–R397.
22. Harborne JB. (1999) Recent advances in chemical ecology. *Nat Prod Rep* **16**: 509–523.
23. Wittstock U, Gershenzon J. (2002) Constitutive plant toxins and their role in defense against herbivores and pathogens. *Curr Opin Plant Biol* **5**: 300–377.
24. Glinwood R, Ninkovic V, Pettersson J. (2011) Chemical interaction between undamaged plants — Effects on herbivores and natural enemies. *Phytochemistry* **72**: 1683–1689.
25. Glinwood R, Pettersson J, Ahmed E, *et al.* (2003) Change in acceptability of barley plants to aphids after exposure to allelochemicals from couch-grass (*Elytrigia repens*). *J Chem Echol* **29**: 261–274.
26. Glinwood R, Ninkovic V, Ahmed E, Pettersson J. (2004) Barley exposed to aerial allelopathy from thistles (Cirsium spp.) becomes less acceptable to aphids. *Ecol Entomol* **29**: 188–195.
27. Ninkovic V, Glinwood R, Dahlin I. (2009) Weed-barley interactions affect plant acceptance by aphids in laboratory and field experiments. *Entomol Exp Appl* **133**: 38–45.
28. Cadet P. (2007) Management of nematodes and a stalk borer by increasing within-field sugarcane cultivar diversity. *Plant Pathol* **56**: 526–535.
29. Mundt CC. (2002) Use of multiline cultivars and cultivar mixtures for disease management. *Annu Rev Phytopathol* **40**: 381–410.
30. Power AG. (1991) Virus spread and vector dynamics in genetically diverse plant-populations. *Ecology* **72**: 232–241.
31. Ninkovic V. (2002) Mixing barley cultivars affects aphid host plant acceptance in field experiments. *Entomol Exp Appl* **102**: 177–182.
32. Karban R, Shiojiri K, Huntzinger M, McCall AC. (2006) Damage-induced resistance in sagebrush: volatiles are key to intra- and interplant communication. *Ecology* **87**: 922–930.
33. Thelen GC, Vivanco JM, Beth Newingham B, *et al.* (2005) Insect herbivory stimulates allelopathic exudation by an invasive plant and the suppression of natives. *Ecol Lett* **8**: 209–217.

34. Tratwal A. (2007) The possibilities of reduction of winter barley chemical protection by growing variety mixtures Part II. effect on yield. *J Plant Prot Res* **47**: 79–86.
35. Ninkovic V. (2011) Effect of within-species plant genotype mixing on habitat preference of a polyphagous insect predator. *Oecologia* **166**: 391–400.
36. Arimura G, Matsui K, Takabayashi J. (2009) Chemical and molecular ecology of herbivore-induced plant volatiles: proximate factors and their ultimate functions. *Plant Cell Physiol* **50**: 911–923.
37. Sabelis MW, Takabayashi J, Janssen A, *et al.* (2007) Ecology meets plant physiology: Herbivore-induced plant responses and their indirect effects on arthropod communities. In: Ohgushi T, Craig TP, Price PW (eds.), *Ecological Communities: Plant Mediation in Indirect Interaction Webs.* Cambridge: Cambridge University Press, pp. 188–217.
38. D'Alessandro M, Held M, Triponez Y, Turlings TC. (2006) The role of indole and other shikimic acid derived maize volatiles in the attraction of two parasitic wasps. *J Chem Ecol* **32**: 2733–2748.
39. Rasmann S, Köllner TG, Degenhardt J, *et al.* (2005) Recruitment of entomopathogenic nematodes by insect-damaged maize roots. *Nature* **434**: 732–737.
40. Beauchamp J, Hansel A, Kleist E, *et al.* (2005) Ozone induced emissions of biogenic VOC from tobacco: relationships between ozone uptake and emission of LOX products. *Plant Cell Environ* **28**: 1334–1343.
41. Loreto F, Barta C, Brilli F, Nogues I. (2006) On the induction of volatile organic compound emissions by plants as consequence of wounding or fluctuations of light and temperature. *Plant Cell Environ* **29**: 1820–1828.
42. Koptur S. (1992) Extrafloral nectary-mediated interactions between insects and plants. In: Bernays E (ed.), *Insect–Plant Interactions Vol. IV*. Boca Raton: CRC Press, pp. 81–129.
43. Wäckers FL, Bezemer TM. (2003) Root herbivory induces an above-ground indirect defence. *Ecol Lett* **6**: 9–12.
44. Mattiacci L, Dicke M, Posthumus MA. (1995) Beta-glucosidase — an elicitor of herbivore-induced plant odor that attracts host-searching parasitic wasps. *Proc Natl Acad Sci USA* **92**: 2036–2040.
45. Alborn HT, Turlings TCJ, Jones TH, *et al.* (1997) An elicitor of plant volatiles from beet armyworm oral secretion. *Science* **276**: 945–949.
46. Doss RP, Oliver JE, Proebsting WM, *et al.* (2000) Bruchins: insect-derived plant regulators that stimulate neoplasm formation. *Proc Natl Acad Sci USA* **97**: 6218–6223.

47. Felton GW, Eichenseer H. (1999) Herbivore saliva and its effects on plant defense against herbivores and pathogens. In: *Induced Plant Defenses Against Pathogens and Herbivores: Ecology and Agriculture.* St. Paul, MN: Am Phytopathol Soc Press, pp. 19–36.
48. Spiteller D, Pohnert G, Boland W. (2001) Absolute configuration of volicitin, an elicitor of plant volatile biosynthesis from lepidopteran larvae. *Tetrahedron Lett* **42**: 1483–1485.
49. Seo S, Okamoto M, Seto H, *et al.* (1995). Tobacco MAP kinase: a possible mediator in wound signal transduction pathways. *Science* **270**: 1988–1992.
50. Kodama H, Nishiuchi T, Seo S, *et al.* (2000) Possible involvement of protein phosphorylation in the wound-responsive expression of Arabidopsis plastid N-3 fatty acid desaturase gene. *Plant Sci* **155**: 153–160.
51. Bell E, Creelman RA, Mullet JE. (1995) A chloroplast lipoxygenase is required for wound-induced jasmonic acid accumulation in Arabidopsis. *Proc Natl Acad Sci USA* **92**: 8675–9679.
52. Reymond P, Farmer EE. (1998) Jasmonate and salicylate as global signals for defense gene expression. *Curr Opin Plant Biol* **1**: 404–411.
53. Maleck K, Dietrich RA. (1999) Defense on multiple fronts: how do plants cope with diverse enemies? *Trends Plant Sci* **4**: 215–219.
54. Turner JG, Ellis C, Devoto A. (2002) The jasmonate signal pathway. *Plant Cell* **14**: S153–S164.
55. Pearce G, Moura DS, Stratmann J, Ryan CA. (2001) Production of multiple plant hormones from a single polyprotein precursor. *Nature* **411**: 817–820.
56. Arimura G, Huber DP, Bohlmann J. (2004) Forest tent caterpillars (Malacosoma disstria) induce local and systemic diurnal emissions of terpenoid volatiles in hybrid poplar (Populus trichocarpa × deltoides): cDNA cloning, functional characterization, and patterns of gene expression of (−)-germacrene D synthase, PtdTPS1. *Plant J* **37**: 603–616.
57. Li L, Li C, Lee GI, Howe GA. (2002) Distinct roles for jasmonate synthesis and action in the systemic wound response of tomato. *Proc Natl Acad Sci USA* **99**: 6416–6421.
58. Baldwin IT, Schmelz EA, Ohnmeiss TE. (1994) Wound-induced changes in root and shoot jasmonic acid pools correlate with induced nicotine synthesis in Nicotiana sylvestris Spegazzini and Comes. *J Chem Ecol* **20**: 2139–2157.
59. Müller C, Riederer M. (2005) Plant surface properties in chemical ecology. *J Chem Ecol* **31**: 2621–2651.
60. Pedras MS. (2008) The chemical ecology of crucifers and their fungal pathogens: boosting plant defenses and inhibiting pathogen invasion. *Chem Rec* **8**: 109–115.

61. Spiteller P. (2008) Chemical defence strategies of higher fungi. *Chemistry* **14**: 9100–9110.
62. Splivallo R, Ottonello S, Mello A, Karlovsky P. (2011) Truffle volatiles: from chemical ecology to aroma biosynthesis. *New Phytol* **189**: 688–699.
63. Osbourn A. (1996) Saponins and plant defence — a soap story. *Trends Plant Sci* **1**: 4–9.
64. Jones PR, Andersen MD, Nielsen JS, *et al.* (2000) The biosynthesis, degradation, transport and possible function of cyanogenic glucosides. In: Romeo JT, Ibrahim R, Varin L, DeLuca V (eds.), *Evolution of Metabolic Pathways. Recent Advances in Phytochemistry*. Amsterdam: Pergamon **34**: 191–247.
65. Schatzmann H. (1953) Herzglykoside als Hemmstoffe für den aktiven Kalium- und Natriumtransport durch die Erythrocytenmembran. *Helv Phys Acta* **11**: 346–354. [Title translation: Cardenolides as inhibitors of active potassium and sodium transport through the erythrocyte membrane].
66. Repke KRH, Portius HJ. (1963) Über die Identität der Ionenpumpen-ATPase in der Zellmembran des Herzmuskels mit einem Digitalis-Rezeptorsystem. *Experientia* **19**: 452–458. [Title translation: Concerning the identity of the ion pump ATPase in the cell membrane of the heart muscle with a Digitalis receptor system].
67. Wittstock U, Lichtnow KH, Teuscher E. (1997) Effects of cicutoxin and related polyacetylenes from Cicutavirosa on neuronal action potentials: a comparative study on the mechanism of the convulsive action. *Planta Med* **63**: 120–124.
68. Francis F, Vanhaelen N, Haubruge E. (2005) Glutathione S-transferases in the adaptation to plant secondary metabolites in the Myzuspersicae aphid. *Arch Insect Biochem Physiol* **58**: 166–174.
69. Gershenzon J, Dudareva N. (2007) The function of terpene natural products in the natural world. *Nat Chem Biol* **3**: 408–414.
70. Wink M, Schmeller T, Latz-Brüning B. (1998) Modes of action of allelochemical alkaloids: interaction with neuroreceptors, DNA and other molecular targets. *J Chemical Ecology* **24**: 1881–1937.
71. Hartmann T, Dierich B. (1998) Chemical diversity and variation of pyrrolizidine alkaloids of the senecionine type: biological need or coincidence? *Planta* **206**: 443–451.
72. Ralphs MH, Stegelmeier BL. (1997) Ability of apomorphine and lithium chloride to create food aversions in cattle. *J Range Manage* **50**: 371–373.
73. Watson AA, Nash RJ, Wormald MR, *et al.* (1997) Glycosidase-inhibiting pyrrolidine alkaloids from Hyacinthoides non-scripta. *Phytochemistry* **46**: 255–259.

74. Lawler IR, Foley WJ, Eschler BM, *et al.* (1998) Intraspecific variation in Eucalyptus secondary metabolites determines food intake by folivorous marsupials. *Oecologia* (Berlin) **116**: 160–169.
75. Johnson KS, Scriber JM, Nair M. (1996) Phenylpropenoid phenolics in sweetbay magnolia as chemical determinants of host use in saturniid silkmoths (Callosamia). *J Chem Ecol* **22**: 1955–1969.
76. Hummelbrunner LA, Isman MB. (2001) Acute, sublethal, antifeedant, and synergistic effects of monoterpenoid essential oil compounds on the tobacco cutworm, Spodoptera litura (Lep., Noctuidae). *J Agric Food Chem* **49**: 715–720.
77. Chapman RF. (2003) Contact chemoreception in feeding by phytophagous insects. *Annu Rev Entomol* **48**: 455–484.
78. Glendinning JI, Davis A, Ramaswamy S. (2002) Contribution of different taste cells and signaling pathways to the discrimination of "bitter" taste stimuli by an insect. *J Neurosci* **22**: 7281–7287.
79. Zagrobelny M, Bak S, Rasmussen AV, *et al.* (2004) Cyanogenic glucosides and plant–insect interactions. *Phytochemistry* **65**: 293–306.
80. Després L, David JP, Gallet C. (2007) The evolutionary ecology of insect resistance to plant chemicals. *Trends Ecol Evol* **22**: 298–307.
81. Tune R, Dussourd DE. (2000) Specialized generalists: constraints on host range in some plusiine caterpillars. *Oecologia* **123**: 543–549.
82. Nyman T, Julkunen-Tiitto R. (2000) Manipulation of the phenolic chemistry of willows by gall-inducing sawflies. *Proc Natl Acad Sci USA* **97**: 13184–13187.
83. Willinger G, Dobler S. (2001) Selective sequestration of iridoid glycosides from their host plants in Longitarsus flea beetles. *Biochem Syst Ecol* **29**: 335–346.
84. Kuhn J, Pettersson EM, Feld BK, *et al.* (2004) Selective transport systems mediate sequestration of plant glucosides in leaf beetles: a molecular basis for adaptation and evolution. *Proc Natl Acad Sci USA* **101**: 13808–13813.
85. Termonia A, Hsiao TH, Pasteels JM, Milinkovitch MC. (2001) Feeding specialization and host-derived chemical defense in Chrysomeline leaf beetles did not lead to an evolutionary dead end. *Proc Natl Acad Sci USA* **98**: 3909–3914.
86. Berenbaum MR. (2002) Postgenomic chemical ecology: from genetic code to ecological interactions. *J Chem Ecol* **28**: 873–895.
87. Wen Z, Baudry J, Berenbaum MR, Schuler MA. (2005) Ile115Leu mutation in the SRS1 region of an insect cytochrome P450 (CYP6B1) compromises substrate turnover via changes in a predicted product release channel. *Protein Eng Des Sel* **18**: 191–199.
88. Feyereisen R. (2005) Insect cytochrome P450. In: Gilbert LI *et al.* (eds.), *Comprehensive Molecular Insect Science.* Elsevier, pp. 1–77.

89. Stevens JL, Snyder MJ, Koener JF, Feyereisen R. (2000) Inducible P450s of the CYP9 family from larval Manduca sexta midgut. *Insect Biochem Mol Biol* **30**: 559–568.
90. Li X, Berenbaum MR, Schuler MA. (2002) Plant allelochemicals differentially regulate Helicoverpazea cytochrome P450 genes. *Insect Mol Biol* **11**: 343–351.
91. Li X, Schuler MA, Berenbaum MR. (2002) Jasmonate and salicylate induce expression of herbivore cytochrome P450 genes. *Nature* **419**: 712–715.
92. Francis F, Lognay G, Wathelet JP, Haubruge E. (2001) Effects of allelochemicals from first (Brassicaceae) and second (Myzuspersicae and Brevicorynebrassicae) trophic levels on Adalia bipunctata. *J Chem Ecol* **27**: 243–256.
93. Ratzka A, Vogel H, Kliebenstein DJ, *et al.* (2002) Disarming the mustard oil bomb. *Proc Natl Acad Sci USA* **99**: 11223–11228.
94. Holzinger F, Wink M. (1996) Mediation of cardiac glycoside insensitivity in the Monarch butterfly (*Danaus plexippus*): Role of an amino acid substitution in the ouabain binding site of Na^+, K^+-ATPase. *J Chem Ecol* **22**: 1921–1937.
95. Raymond M, Berticat C, Weill M, *et al.* (2001) Insecticide resistance in the mosquito *Culex pipiens*: what have we learned about adaptation? *Genetica* **112–113**: 287–296.

CHAPTER 4

Chemical Ecology of Marine Organisms

André Antunes and Thomas Efferth*[†]

ABSTRACT

Research on marine chemical ecology has provided a great deal of basic information that has advanced the knowledge in the fields of organic chemistry, biochemistry, ecology, behavior, and evolution. The understanding of how chemical signals from the environment, chemical defenses, and other chemical agents intervene in life processes can shed light on the forces driving the ecology and evolution of organisms. So far, several thousand marine natural products have also been chemically defined; many of which are biologically active compounds possessing novel functional groups and molecular structures. Interest in biotechnological applications for marine natural products has increased over the past decade, as the knowledge of the chemistry of marine organisms has developed. Applied studies have identified applications of these compounds as pharmaceuticals. In this chapter, we discuss specific topics of marine chemical ecology such as marine physico-chemical patterns of marine biodiversity, bioactivity and biomedical applications of marine natural products, chemical mediation of interaction among marine organisms, compounds classes, and other topics.

1. INTRODUCTION

1.1. Definition of Chemical Ecology

Chemical ecology aims at analyzing the intra- and inter-specific interactions between organisms and their environment at the molecular level of chemical substances synthesized by them. By definition, the field is

*Institute of Pharmacy and Molecular Biotechnology, University of Heidelberg, Im Neuenheimer feld 364, 69120 Heidelberg, Germany.

[†]Institute of Pharmacy and Biochemistry, Johannes Gutenberg University, Staudinger Weg 5, 55128 Mainz, Germany.

interdisciplinary in scope and includes both chemical and biological research. Chemical investigations of the compounds produced by organisms include studies of structure, biosynthesis, organic synthesis, and mechanisms of action, while biological studies focus on the ecological consequences of these phenomena.

The chemistry of terrestrial plants and insects has been studied for the past century, and tens of thousands of different natural products have been isolated and chemically defined. This chemical knowledge of terrestrial organisms has contributed greatly to the development of the field of chemical ecology over the past few decades. Chemists have realized that molecules that they isolate and characterize often have potent biological activities and have likely evolved for specific biological functions. Subsequently, biologists and ecologists have realized that chemical substances, particularly the secondary metabolites, play an important role in complex behavioral and ecological interactions among organisms. The field has advanced most rapidly as the result of collaboration among chemists and biologists, including the incorporation of results and ideas from chemical research into biological research, and *vice versa*.

Research on terrestrial chemical ecology has provided a great deal of basic information that has advanced the fields of organic chemistry, biochemistry, ecology, behavior, and evolution. In addition, many practical applications have also developed. Knowledge of plant-insect interactions mediated by defensive compounds and other secondary metabolites has been used to create applications or the control of insect pests and microbial diseases in crop plants. Much of the pharmaceutical industry is based on terrestrial natural products or compounds modeled after these natural products.

So far, several thousand marine natural products have been chemically defined; many of which are biologically active compounds possessing novel functional groups and molecular structures.[1–3] Interest in biotechnological applications for marine natural products has increased over the past decade, as the knowledge of the chemistry of marine organisms has developed. Applied studies have identified applications of these compounds as pharmaceuticals. Moreover, preliminary evidence suggests that marine organisms also provide an untapped resource for future biotechnological applications. Several marine natural products have already been approved for the treatment of cancer and pain and many others are currently being evaluated in clinical trials in the United States and in Europe for the treatment of various cancers.[4] In contrast to terrestrial studies, however, much less is known about the natural functions of these metabolites in the marine environment. Only in the past few years have experimental evaluations been conducted to shed light on the role of marine natural products in the lives of the organisms that produce them.

Studies in marine chemical ecology have increased quantitatively and qualitatively in the last few decades and this has fueled a rapid development of the field. The main reasons for that were the identification of groups of marine plants and animals that are especially rich in secondary metabolites as well as the technological advances in the isolation and characterization of chemical compounds. Interest in marine chemical ecology developed as the knowledge of numerous and diverse natural products found in marine organisms increased. Marine natural product chemists began to ask questions about how the compounds they isolated function in nature. Early experiments addressed the antibiotic, antifungal, and ichthyotoxic effects of these metabolites, but not all of them used marine microorganisms or marine fish to test the hypothesis that these compounds were toxic and deterrent to predators and pathogens. The ecological context of these metabolites and the marine organisms that produce them is now being considered and experiments are more carefully designed and replicated for statistical analysis. For instance, naturally co-occurring predatory fishes and pathogenic microorganisms are now being used to test the deterrent effects of compounds from marine organisms. Progress in experimental design and analysis of these types of bioassays has been made in the past years, especially as marine ecologists have become more interested in chemical ecology.[5]

Manipulative field studies are now common in both terrestrial and marine ecology. In many experiments, competitors or predators are either added to or excluded from a habitat to understand their influence on community structure. Similarly, secondary metabolites can be added to artificial diets and tested as feeding deterrents against natural herbivores or predators in field studies.[6]

The increasing interest in collaborative studies between marine chemists and ecologists has contributed to the development of a new generation of marine chemical ecologists. These individuals incorporate advances from both marine biology and natural products chemistry into their research. As a result of this collaboration, marine biologists and chemists were able to explore the marine environment and could observe ecological phenomena. Thus, more sophisticated research targets came under study with the realization that many observed intra- and inter-specific phenomena might have a molecular basis. Since these molecular basis are involved in all life processes, chemical ecology can justifiably include a large variety of topics (i.e. biochemistry, digestive physiology, biogeochemistry, etc.) making it impossible to include all of them in one chapter.

1.2. Chemical Mediation of Ecological Interactions

Chemical signals identify biologically important targets for those who have the proper receptors.[7] Chemical communication occurs when receptors detect signals released to the environment as a consequence of metabolic activity or tissue damage. All living organisms release metabolites and such release is a potential opportunity for chemoreceptive animals interested in finding food or hosts. For instance, body fluids released from damaged tissues and decay products from dead organisms can be particularly potent signals. Also breath, sweat, urine, feces, and other metabolites released to the environment together with their bacterial and other symbiotic embellishments can be detected and serve as stimuli for predators and parasites.

A variety of critical ecological interactions are chemically mediated and substantial information is available on the many types of biological responses to environmental chemical stimuli. Chemical mediation, for example, strongly influences predation,[8] courtship and mating,[9] aggregation and school formation,[10,11] and habitat selection.[12]

In the marine environment, receptors must be able to distinguish between chemical signals and background of many other compounds dissolved in the water, in order to mediate chemical communication. Recognition is not only achieved through receptor specificity and diversity but also through identification of the intensity and time course of a signal. This allows the receptor to discriminate a true signal against background. In fact, a receptor organ is able to determine by its spectral and temporal tuning what is signal and what is noise. Therefore, an environmental signal can result in different information depending on its distribution. It is important to notice, however, that the acuteness of receptors, and consequently, the perception of different organisms differ between species as well as between organs of a single species.[13]

The nervous system is the filter responsible for the conversion of environmental stimuli into sensory responses, and then into a behavioral task.[14] Recent research on sensory biology has explored relations between the natural physical chemical environments and the properties of animal sensorial systems.[15] Such studies, together with behavioral and population studies, can find out relations between stimulus, behavior and ecological consequences of decisions made by individual organisms. Therefore, a full understanding of chemical ecology has to include the characterization of chemical signals together with their environmental dispersal and degradation patterns that are an intrinsic part of chemosensory transduction and signal processing and lead to the appropriate behavioral responses.

1.3. Chemical Ecology and Evolution

The basis of biological sciences is evolution, which emphasizes the relationship between the most different organisms. Therefore, questions relating to speciation, biogeography, and biodiversity are particularly compelling. The understanding of how chemical signals from the environment, chemical defenses, and other chemical agents intervene in life processes can shed light upon the forces driving the ecology and evolution of organisms.

Chemistry is generally considered as a science concerned in establishing the laws of nature irrespective of particular entities or events. Natural products chemistry, however, is more concerned with what are considered "individuals."

There are many biological species; each of them being unique and original, not only by their exterior appearance, but also by features of their molecular anatomy, polynucleotides, proteins, lipids, polysaccharides, etc. A huge structural molecular variability of subspecies, populations and individuals is known. This biological variability has been known for a long time, but only with recent advances in modern molecular biology has its significance and complexity became clearer.

The concept of the Darwinian Theory of evolution seemed to be weakened, when one tried to interpret statements of molecular biology in it. As a consequence, a theory of non-Darwinian neutral evolution arose, which stated that under natural conditions no factors exist, which can distinguish and select organisms on their molecular structure. Supporting this new theory, one could mention the phenomenon of intraspecific molecular polymorphism. This refers to similar molecules with a common origin and function, but which for some reason differ in details of their structure. How does one explain, for example, the differences in the structures of the blood cells among the various populations of the same species?

According to the Darwinian Theory three phenomena i.e. heredity, inherent variability and natural selection must occur for evolution to happen. In the course of their lives, organisms interact with each other and in so doing individual species and populations permanently come in conflict with each other. These conflicts are resolved through natural selection of forms, where organisms possessing a greater vital capacity under adequate conditions survive during evolution.

Analyses of intermolecular interactions in an evolutionary ecological perspective stretch the idea that each conflict is an example of a general molecular ecological regularity, which provides a basis for the processes of biomolecular evolution. The chemo-ecological factors affecting single organisms and populations are of utmost significance in the evolution of living forms.

To exemplify how ecological interaction is related to evolution, infectious diseases, which may be the main agents of natural selection, are briefly discussed here.

All species inevitably sustain losses due to microbe aggression; there is no single species that can protect itself from it. Microbes (i.e. viruses, bacteria, fungi, protozoa) are spread all over the earth and many of them are parasites of other biological species. In contrast to predators and macroscopic parasites, microbes cannot break through the physical barriers (i.e. integuments or envelopes) of an entire organism or separate protected cells. Still, many microbes overcome these physical barriers by means of cellular and molecular interactions with their host, and are thus able to reproduce and develop in the cells of organisms, thereby affecting chemical structures and functions. In other words, the interactions are mediated exclusively at the molecular level by specialized receptors and other molecular structures. Each microbial species produces distinctive chemical compounds called factors of pathogenicity, e.g. proteins, toxins, lipids and polynucleotides, which can target specific molecular structures in the victim. Such targets simply exist in the organisms of victims and do not have the purpose of interacting with the microbes at all. For example, the target for the tetanus toxin is identical with the receptor for thyrotropin.[16] Choleraic toxin is able to interact only with certain ganglioside macromolecules, namely those that contain subunits such as ceramide, lactose, galactosamine, galactose and the radical of sialic acid joined with the molecule of lactose. Other types of gangliosides (those without the above mentioned subunits) are not able to interact with the choleraic toxin.[17] As a consequence, mutations leading to stereochemical modifications in the victim drastically reduce or even abolish successful intermolecular interactions between microbe and victim, thereby reducing the success of microbial invasion. Interactions of molecular-ecological factors can be prevented by antagonistic interactions and influence the variability of organisms. Genetics and particularities of the constitution of molecular targets are thus a determinant key. Such particularities of the molecular structure of victims can be inherited, thereby abolishing the chemico-ecological interaction of microbial invasion and conferring immunity to their possessors whereas other organisms are still susceptible to microbial attack and die without producing offspring.[7] In the course of such interactions, natural selection of organisms by their characteristics takes place, to which processes of molecular evolution contribute.[18]

It is important to keep in mind that natural chemical signals released by marine organisms, for example, do not evolve any more than does the calcium carbonate that forms the shells of molluscs. What evolves are individual populations and lineages, which change with respect to the properties of the organisms and their parts, such as the enzymes that produce the chemical signals. Selection pressure acts on both the biochemical regulation of signals as well as on the physical features of signal receptors.

There seems to have been a chemical arms race between some organisms. An illustration of this is the evolution of chemical defense in opisthobranch gastropods. A question was posed by researchers, whether the reduction of the shell in these animals preceded the evolution of chemical defense (a post-adaptive scenario) or whether the loss of the shell was made possible by the presence of chemical defense (pre-adaptive scenario).[19] The latter hypothesis was preferred based on the argument that chemical defense is already present in groups in which the shell is relatively well developed. The biological plausibility of the sequence in question is a way of sensing that refers to the Darwinian Evolution Theory. Evolution proceeds in steps and in each step the functioning of the organism as a whole is conserved, but particular functions often succeed one another over time.

2. THE MARINE ENVIRONMENT

2.1. Marine Physico-Chemical Patterns and Biodiversity

Human beings live on land and are consequently most habituated to the terrestrial environment. However, oceans cover more than 70 percent of the earth's surface. In the regions close to the continent, known as continental shelf, oceans reach depths of roughly hundred meters, whereas depths of most oceanic regions are close to 4,000 m. This habitat is populated by many organisms.

At first glance, we may not notice patterns in the marine environment. Often, these patterns are only observed through sample collection and through successive interpretation. In many aspects, marine and terrestrial patterns are similar: on a large scale along latitudinal and altitudinal (depth) gradients, from the coast to ocean/continent centers and from young to old areas (e.g. from the mid Atlantic ridge (new) to the eastern or western Atlantic sea-floor (oldest)). In the shallowest sea regions, organisms' mature and their forms may change significantly, which is called zonation.[20] For example, in warm tropical waters, the type and dominance of corals changes sharply with depth. In Polar Regions the marine abundance and richness changes equally quickly in response to decreasing physical disturbance caused by floating icebergs that scour the seabed. Zonation is most evident at the shore and in this region this phenomenon has been well studied.[21]

A strong zonation pattern also occurs across the ocean's full depth range. In the shallowest regions ranging from a few meters to 200 m depth, termed the Euphotic Zone, incidence of light during the day is enough to supply photosynthesis by primary producers. As a consequence, the upper 200 m of the ocean water column and 1 to 50 m of coastal water is a very different environment from that in the deeper oceanic regions. Of course, even within the Euphotic Zone a strong gradient of light intensity and wavelength can be observed. Beyond a depth of 1,000 m (which covers most of the world's ocean volume), the ocean is effectively lightless. Incident light on the surface of the marine environment warms the surface water, which consequently has a lower density than the cold water below it. The temperature of deep water is relatively uniform around the globe (a few degrees Celsius), different from the temperature of the surface water that varies with latitude and season. In the Polar Regions, temperature of the surface layer remains constantly under 4°C during the whole year. However, this water is well mixed due to the convection of dense cold water together with elevated wave height and greater wind speed at polar latitudes.[22] The global ocean is stratified into an upper mixed layer, a rapidly cooling zone, also termed thermocline, and a deep cold zone. The nature of thermoclines changes from place to place, but in general the tropical thermoclines are around 10°C warmer than those in temperate regions and they are permanent rather than seasonal. This stratification plays an important biological role once the different layers present different concentrations of nutrients, minerals, and dissolved respiratory gases. In a local scale, other gradients, e.g. salinity or oxygen may also stratify water layers. In general, salinity of sea water is around 35 psu. However, in tropical waters increased evaporation on surface causes salinity to be more. In contrast, water salinity is lesser in areas close to continents and in the Arctic due to the influence of greater fresh water flow. Surface currents lead to mixing of water layers, which is a potent driving force that counters stratification in shallower water layers. Marine biologists have been interested in biogeography for more than a century. However, advances in marine biogeography were not that easy, where divergences in different aspects of oceanic biogeographic knowledge have occurred even in the most accessible systems. Nevertheless, since the advancement of our knowledge of underwater instrumentation and marine sampling platforms, our understanding of ocean biogeography has increased remarkably.[20]

Biodiversity refers to the life on the planet and encompasses extinct and living organisms.[23] In the marine environment, similar debates rage over the latitudinal pattern of increasing richness of species towards the equator. However, it is noteworthy that to date only a few marine data sets cover entire hemispheres or even just a number of oceans.

Biodiversity measurements of certain marine habitats or particular groups of marine organisms reveal that biodiversity is neither physically nor temporally static. As in earth regions, some oceanic regions are richer in taxa than others. For instance, the incidence of brachipods in Japanese coastal waters is higher than in other seas. Likewise, the Antarctic continental shelves are rich in sea spiders (pycnogonids). A higher richness of fish, corals, molluscs and other marine organisms is observed in the Indo-West Pacific region, particularly around Indonesia, in contrast to marine richness observed elsewhere.[20] Despite recent advances in biogeography techniques, providing estimates of the total number of species in the ocean is still problematic due to the following reasons: (1) Most of the oceanic environment is rather poorly studied. (2) Some taxa, e.g. nematodes, may be very rich in species, but there are few taxonomists to describe the species. (3) Some previously well-studied taxa appear to be a complex of many cryptic species (reproductively isolated groups of organism but with very similar morphology), whose identification requires extensive genetic studies.

Functional types can be assigned to organisms that constitute marine ecosystems. Pioneer encrusting suspension feeders (e.g. some polychaetes and bryozoans), competitively dominant encrusting suspension feeders (e.g. some sponges and ascidians), benthic zooplankton feeders (e.g. some anemones), deposit feeders (e.g. echiuran worms), scavengers (e.g. amphipods), mobile carnivores, and others exemplify some of them. The number of functional types within a system can be applied in the concept of biodiversity. This would mean that a higher functional diversity of an ecosystem would suggest a higher number of functional types and/or a higher number of members of each functional type within a system.

2.2. Physico-Chemical Aspects in Marine Chemical Ecology

The ocean is one part of the earth system, and it mediates processes in the atmosphere by the transfer of mass and energy through the sea surface.

Organisms living in the ocean use their chemical senses in all aspects of their lives: from reproductive behavior to feeding, habitat selection, and predator avoidance. The ocean's hydrodynamic and physical properties determine the possibilities and limits of chemical communication in water. In contrast to the propagation in a wave or wave-like form of sound, light, and other electromagnetic signals, molecular diffusion and bulk flow are responsible for the dispersal of chemical signals through the marine environment. However, like air, water is a very dynamic medium, where convection and advection may, in some cases, play a more important role in transport of chemicals than diffusion. Diffusion of chemical signals is important only in smaller ranges (submillimeter), while turbulence is typical for larger water masses (above the centimeter range). At small spatial scales (below 10 μm) diffusion is a biologically useful transport mechanism and due to limitations imposed by fluid viscosity sometimes even the only effective one. At larger scales, however, flow is indispensable to obtain metabolic energy (e.g. food particles), for waste elimination (e.g. urine) and the transport of chemical signals. Distribution by water currents is also important because of the fact that compounds of similar molecular weight diffuse four orders of magnitude more slowly in water than in air.

Chemical signals are widespread in the marine environment in different and overlapping scales: they are part of the metabolites released by a single marine bacterium, the odor plumes emitted by a migrating school of fish or the metabolites arising from a whale carcass laying on the ocean floor. Known signals include dissolved gases, functionalized hydrocarbons, peptides,[24] and proteins.[25] Chemical composition, concentration, flux, hydrodynamic transport and water properties (e.g. light intensity, pressure, temperature, salinity, sedimentation, etc.) play decisive roles in chemically mediated ecological interactions between individual and populations of marine organisms. Chemicals are transported via molecular and turbulent diffusion from areas of higher concentration to areas with lower ones. Whereas these processes operate at small (molecular) and large (turbulence) scales, at intermediate spatial scales, undiluted chemical patches can be transported in organized fluid tracks. In such tracks, rotating portions of water (vortices) are caused in the lee of a flow, which create "streets" of odor.[26,27] These serve as chemical trials and can be useful to organisms searching for food, mates and other resources.

Many marine organisms have to sample turbulent water with an irregular odor distribution. For instance, two types of stimulus access ("sniffing") have been distinguished in marine organisms: "cyclosmates," which sample a specific sniff volume in sniffs or flicks, and "isomates," which sample a steady, ciliary-driven water flow. This latter type of stimulus access is found in slow-moving animals such as catfish, eels, dogfish, or mud snails.[28] In turbulent waters, patch boundaries are sharper the more recently the odor was released. When crossing an odor patch, an animal learns about the distance to the odor source from the rate of concentration change.

Animals are able to identify novel compounds and unique mixtures of chemical compounds thereby distinguishing the quality of different stimuli.[29] This sensibility is of paramount importance to organisms because stimulus and behavioral response are intrinsically connected and affect the abundance and spatial/temporal distributions of organisms.

3. SEMINOCHEMICALS

Natural product chemistry is a basic science, involved in the discovery, characterization and cataloging of new chemical substances found in nature. Chemists, biologists, ecologists and other scientists acknowledge that natural products might be able to explain species composition, distribution, and diversity in some locations.[30]

In nature, successful life history strategies employed by organisms intrinsically involves adaptations to their environment. These may include behavioral, physical, or chemical strategies.[6,31] Organisms may produce toxic defensive compounds or metabolites to attract a sexual mate as chemical strategies, thereby utilizing, what chemists refer to as "natural products" literally alluding to products made by living organisms.

The term "natural product" is commonly used as a synonym of "secondary metabolite." In particular, "secondary" is used in the context of functional significance, so that something that is secondary is less important than something that is primary. Indeed, primary metabolites are the prevalent macromolecules that make up the basic machinery of life and cellular function is not possible without them. Primary metabolites are normally ubiquitous; proteins, enzymes and genes from all organisms are built from combinations of the same building blocks (amino acids and nucleotides). Thus primary metabolites drive all important living processes such as supply of energy, structure and reproductive capacity. Secondary metabolites are believed to be at the basis of ecological specialization by affecting species' distribution patterns, community organization and other ecological patterns.[32] They are produced by a wide variety of marine organisms and may present fundamental differences in comparison to terrestrial secondary metabolites. To date, secondary metabolites have been described from sponges, ascidians, soft corals, bryozoans, polychaetes, seaweeds, marine microbes, and other benthic and pelagic organisms.[33] Often secondary metabolites constitute a very small fraction of the total biomass of an organism.[34] One of the ways to identify a secondary metabolite is by establishing that it is not a primary metabolite. For example, C15 halogenated compounds called acetogenins are produced only by some red algae. These compounds are neither building-blocks nor are they ubiquitous. Thus, they are classified as secondary metabolites. However, algae steroid derivates known as phytosterols differ from animal sterols by bearing alkylation at a different carbon.[35] Although they play roles in membrane structure, which is their primary characteristic, they are often species-specific and are, therefore, classified as secondary metabolites. Substances presenting primary roles in some organisms may act as secondary metabolites as well. An example of such a natural product is the phlorotannins, which can have primary roles in brown algal cell wall biosynthesis (primary characteristic) and at the same time confer an odd taste to potential predators (secondary characteristic).[36,37] Thus, in the modern use of the term, mentioning secondary metabolites or natural products, compounds are meant, which are not intrinsically involved in life maintenance and development of an organism, but rather those with limited biological distribution and production, often species-specific, and most often produced to mediate ecological interactions.[38] It is important to note that it is not always clear, which biological role these compounds play. This, however, does not mean that these secondary metabolites are necessarily superfluous. Then, even the failure to assign a particular role to them is not sufficient to argue against their ecological relevance. Therefore, careful investigations of the ecological roles for many natural products remain to be accomplished.

3.1. Natural Product Names

Like in other fields of science involved in characterizing and cataloging, nomenclature of natural products also follows the International Union of Pure and Applied Chemistry (IUPAC) rules. However, natural product nomenclature as stated by IUPAC may be so complex that official natural product names are mostly unpractical for common use.[39–41] Like living organisms, many compounds have common names besides the official one. Such compound names are given at the time a new compound is first described in the literature and are generally more concise, and especially, more easily associated mentally with a chemical structure. As an example, most people know that vitamin C is required for a wide range of essential metabolic processes in animals and plants, but few people fully comprehend the significance of (R)-3,4-dihydroxy-5-((S)-1,2-dihydroxyethyl)furan-2(5H)-one for the metabolic processes. Thus, a common naming system simplifies the life of natural product chemists. Nonetheless, common names assigned to some compounds can be difficult to keep in mind or even to pronounce. When a novel natural product is described, the name assigned to it usually refers either to its species of origin or to a geographic characteristic associated with its acquisition. Dactylomelin-P, for example, was the common name assigned to a protein first isolated from the Brazilian sea hare *Aplysia dactylomela*,[42] while Kahalalide describes a group of compounds isolated from a green alga (and its sacoglossan predator) from the Kahala district of O'ahu, Hawaii.[43] Some people may argue that a natural product isolated from an organism in North American waters may one day be collected from a related organism living in South American waters. In the same way, genus and species names may become obsolete or their members reallocated as biological taxa and undergo constant revision, so that neither of these guidelines offer the most durable or accurate system. According to the IUPAC nomenclature guidelines, the origin of the natural products is described as the "parent" structure. To the parent name a suffix is added, which points out the functional group of the molecule. For instance, the suffix "-ide" of kahalalide denotes a lactone function present in this molecule. The suffix may suggest additional information about the molecule such as biological activity (e.g. "-toxin") or the structural class to which the natural product belongs.

This nomenclature system based on a parent term with an appended suffix accounts for most natural product names. Also, a variety of terms can be employed, when a suffix has to denote multiple functional groups. For example, the compound name caulerpenyne suggests its alkene and alkyne functions and its pronunciation emphasizes the three units forming the name: Caulerp-en-yne. In the same manner, isocyanopupukeanane encompasses its geographical origin (Pupukea, Hawaii) as well as its alkane class.[44]

3.2. Compound Classes

The synthesis of the most different secondary metabolites can be assigned to a limited number of core biosynthetic pathways. Although common biosynthetic pathways provide the framework for secondary metabolites, their bioactivity is mostly provided by enzymes that are in most cases unique to natural products.[45] The addition of functionalities, stereochemical modifications, reduction, and oxidation are often responsible for secondary metabolites' uniqueness and bioactivity.

At first glance, the structures of secondary metabolites seem to be surprisingly diverse. However, most of these compounds are classified into chemical families and present particular structural characteristics arising from their biosynthetic pathway. According to Hanson (2003),[46] the classes of secondary metabolites are: polyketides and fatty acids, terpenoids and steroids, phenylpropanoids, alkaloids, and specialized carbohydrates.

Polyketides are derived from the "building block" acetyl co-enzyme A. They are synthesized by polyketide synthase (PKS) enzymes, which combine acetate (ethanoate) units in a linear manner.[47] Terpenoids (also called isoprenoids) and steroids are derived from C5 isoprene units, which are linked together in a head-to-tail manner. Chemicals containing a phenylpropanoid (C6-C,) unit consist of another group of natural products and are derived from the amino acid phenylalanine.[48] Although amino acids are considered as primary metabolites, some uncommon ones are of restricted occurrence. Penicillins are for instance formed from small peptides. The alkaloids are a structurally diverse group of natural products containing basic nitrogen atoms, which may be derived from amino acids such as ornithine, lysine, tyrosine or tryptophan. Carbohydrates are also normally considered as primary metabolites, nonetheless there are other sugars that are of a much more limited occurrence. Some of these uncommon sugars are bound to natural products as part of a glycoside.

4. BIOACTIVITY AND BIOMEDICAL APPLICATION OF MARINE NATURAL PRODUCTS

Natural products are intrinsically bioactive. Bioactivity is a physiological response to the binding of a molecule to a receptor, which generates downstream cascading consequences. Natural products that spread information between individuals are called semiochemicals by chemical ecologists. Semiochemicals is synonymous with "signal substances," which are divided into two different groups: pheromones and allelochemicals. Pheromones are used for intraspecific communication, while allelochemicals are used for communication between organisms belonging to different species.

Semiochemicals are products of enzymatic processes and have the ability to interact with receptors. The receptors of ecological interactions are still not well understood and as for many other natural products, the roles for many semiochemicals remain to be defined.[12] Most experimental evidence for semiochemcial receptors is derived from biomedical applications and not from the ecological relevance of these compounds. Kahalalide F, a potent cytotoxic compound present in the sacoglossan mollusc *Elysia rufescens* and also in the green alga it feeds upon,[43] has been tested in phase II clinical trials to treat melanoma, hepatocellular carcinoma, and non-small-cell lung cancer.[49] The mode of action of this compound is still not well understood, though cytoplasmatic swelling and DNA clumping properties have been assigned to it.[50]

Up to now, over 15,000 new compounds have been isolated from sponges, ascidians, soft corals, seaweeds, marine microbes, and many other benthic and pelagic organisms, with more being discovered every day.[51]

Natural products from marine organisms need to be highly potent in view of the fact that they are immediately diluted after they are released into water. For this reason and because of the enormous biological diversity, the ocean is recognized as a potential source of a huge number of novel compounds with biological activity that may have medicinal applications.[52] In fact, there are many examples of bioactive compounds, which have been isolated from marine organisms and have entered clinical trials for drug development with promising results. Besides Kahalalide F, bryostatin-1, a protein kinase C inhibitor isolated from the bryozoan *Bugula neritima*, is currently in clinical trials for cancer.[53] Methopterosin, isolated from the soft coral *Pseudopterogorgia elisabethae* is in phase I of clinical trials for the treatment of inflammation/wound.[54] Another compound isolated from a marine organism is the GTS-21 from the nemertini *Amphiporus lactifloreus.* This compound is in phase I of clinical trials for the treatment of Morbus Alzheimer/schizophrenia.[55] A complete list of approved drugs and drug candidates isolated from marine organisms would be too extensive, and novel isolated compounds can be found in some comprehensive reviews.[52,56]

Like chemical ecology of natural products, bioprospecting of novel marine compounds is still in its infancy. However, the rapid advances in technology in underwater exploration and the fact that marine biodiversity is greater than land diversity gives us reason to believe that the approval of drugs that have originated from the marine environment is the beginning of their application in medicaments for the improved treatment of human diseases.

5. CHEMICAL MEDIATION OF INTERACTION AMONG MARINE ORGANISMS

In this section, progress in the field of marine chemical ecology in the last years is discussed with examples on the defensive functions of primary and secondary metabolites from marine organisms against predators, competitors, fouling organisms and, microorganisms as well as their roles in mate recognition, reproduction and larval settlement (Fig. 1).

5.1. Bacteria

The mucus of the Caribbean coral *Acropora palmate* selectively inhibits the growth of some microbes on the surface of the coral at the same time as it selects for some antibiotic-producing strains. This selection was shown to be regulated by marine bacteria associated with the coral mucus, from which 20 percent presented antibiotic activity against other bacteria, including some coral pathogenic strains. Conversely, mucus collected during a summer bleaching event did not present antibiotic activity anymore suggesting

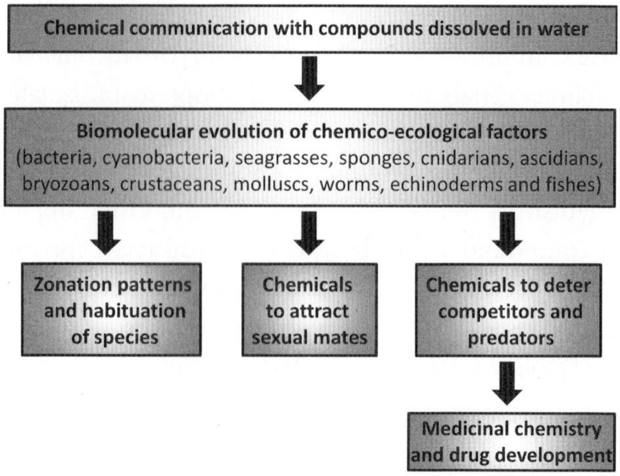

Figure 1 Chemical communication with compounds dissolved in water.

that environmental factors shift the bacterial communities in the mucus.[57] These findings provide an insight into the global problem of coral diseases caused by environmental stress.

Quorum sensing is a chemical communication among bacteria and involves gene expression regulation according to local cell density population.[58] Although less is known about chemical interactions of marine bacteria, this aspect of bacterial communication has been well studied. N-Acylhomoserine lactones (AHLs) are well known chemical signals in Gram-negative bacteria. Recent studies show that AHLs produced by marine bacteria in biofilms attract zoospores of the green seaweed *Ulva*, thereby causing their accumulation at the source of the AHLs.[59] Experiments showed a reduction in zoospore's swimming speed as well as a calcium influx into the spore after detection of AHLs, which could influence flagellar motility. This suggests that AHLs released from marine bacteria in biofilm may act as a settlement cue for zoospores of *Ulva*. Skindersoe *et al.* showed that marine microorganisms can also produce compounds that inhibit quorum sensing in marine bacteria. In a general quorum sensing screen using 284 extracts of marine tropical organisms from the Great Barrier Reef in Australia, 23 percent presented biological activity.[60] Cross-kingdom signaling between marine micro- and macro-organisms has also been reported in the literature. In laboratory assays, the crab

Menippe mercenaria showed a 2.4-fold higher preference for fresh fish against microbe-colonized fish. Consistent with this, fresh fish attracted 2.6-fold more animals into baited traps towards Atlantic menhaden carcasses that were colonized by microbes from marsh sediments.[61] Further feeding assays showed that organic extracts from the microbe-colonized fish presented compounds which deterred feeding by *Menippe mercenaria.* Since food-fall is an important food source in many ecosystems, competition between micro- and macro-organisms could commonly occur among marine organisms.

5.2. Cyanobacteria

Knowledge about compounds synthesized by cyanobacteria and their roles in the ecology of these organisms is growing as more bioactive compounds are being discovered. Natural products isolated from *Lyngbya* spp. deter generalist consumers whereas specialist consumers like some molluscs promptly consume these organisms and even extract and accumulate their secondary metabolites.[62] Through its low palatability to generalist consumers, *Lyngbya* spp. together with other cyanobacteria has a competitive advantage over other benthic species in coastal environments, which indeed allows for blooms formation at opportune environmental conditions.[63]

The most abundant macroalgae that occurs on tropical reefs communities in Panama, *Acanthophora spicifera,* does not present chemical defenses and is highly palatable to herbivores. However, after a mortality event that occurred in the Eastern Pacific due to the El Niño in 1997–1998, *A. spicifera* from these reefs was protected from grazers by epiphytic cyanobacteria. The cyanobacteria allowed growth and the dominance of red algae in the community structure by deterring herbivores.[64]

5.3. Macroalgae

Feeding deterrents belong to the best studied class of natural products in macroalgal (seaweeds) chemical ecology. The presence of feeding deterrent compounds in seaweed is usually closely correlated with their lower susceptibility of being grazed by generalist herbivores. A theory with regard to plant allocation of chemical defenses has been proposed, which says plants allocate these compounds to valuable parts of their body or to those more susceptible to herbivores. This theory is called Optimal Defense Theory (ODT).[65] This theory has been tested in seaweeds. Reproductive parts,[66] vulnerable portions of the thallus,[67] and more exposed parts of plants contained high concentrations of chemical defenses against herbivores.[68]

Some algae may produce and release secondary metabolites in response to external aggression. Inducible defense studies have been mostly conducted with brown algae, because this kind of response to external aggression seems to be common among them.[69] The brown alga *Dictyota dichotoma*, for example, releases volatile compounds after wounding, which are feeding deterrents to some amphypods.[70] Also, waterborne cues alone from grazed brown algae can induce the production of chemical defenses in neighboring conspecifics. This was shown in aquarium assays, where the brown alga *Fucus vesiculosus* significantly reduced its palatability to amphipod grazers in response to.[71] Yet, a chemical basis for inducible defenses has not been demonstrated for most studies. However, when considering inducible defenses a specificity of interactions seems to exist where only certain species of grazers are able to induce this behavior in seaweeds.

Seaweeds producing antimicrobial and antifouling defenses have also been reported in recent literature with some of these substances presenting activity even towards human pathogenic bacteria.[69] When considering the ecological context of the seaweeds, extracts from species living in the Atlantic[72] and Pacific Oceans[73] exerted activity towards at least one marine microorganism they were tested for. Two Atlantic green algae (*Halimeda copiosa* and *Penicillus capitatus*) and two Pacific specimens (green alga *Bryopsis pennata* and red alga *Portieria hornemanni*) showed activity against all microorganisms tested suggesting that chemical compounds with antimicrobial activity are widespread among algae.

In addition, chemical inhibition of fouling on seaweeds by algae and invertebrates has been investigated in numerous studies. An example of chemically mediated antifouling activity was observed with the red alga *Delisea pulchra*, which produces halogenated furanones. Surface extracts of *D. pulchra* prevented settlement by algae even in concentrations lower than those occurring naturally on the alga surface.[74] In another set of experiments, antifouling activity assays compared whole algal extracts *vs.* surface extracts from six different species in inhibiting the settlement and development of ecologically relevant fouling species. All whole extracts tested presented significant activity towards the epiphytic algae, while only two surface extracts (from *Delisea pulchra* and *Caulerpa filiformis*) were able to significantly hinder settlement. Noteworthy, these two species also had significantly lower biofouling cover in the field compared to the other seaweeds.[75]

5.4. Sea grasses

The fist experimental evidence supporting the presence of chemical deterrents towards herbivores in sea grasses was published by Vergés *et al.* (2007).[76] This study reported that organic extracts from *Posidonia oceanica*, the dominant sea grass in the Mediterranean Sea, inhibited feeding of four different sympatric herbivores in agar-based tests. Although feeding deterrence rates differ among consumers, this study showed that sea grasses adopt a chemical defense mechanism that is also observed in many terrestrial plants and marine algae. This emphasizes the importance of chemically mediated interaction between sea grasses and herbivores.

Vergés *et al.* (2007)[77] also explored the palatability of different sea grass tissue to herbivores. Experiments targeting the herbivorous fish *Sarpa salpa* and the sea urchin *Paracentrotus lividus* revealed that the incidence of herbivory on vegetative and reproductive parts of the temperate seagrass *Posidonia oceanica* differed. In the field, a higher percentage of inflorescences were damaged compared to the blades. A further confirmation of this tendency was seen in laboratory feeding assays, where extracts of inflorescences were more deterrent to sea urchins than to extracts of blades. This study illustrated that sea grasses may allocate chemical defenses in different tissues thereby warranting the safety of reproductive important structures.

5.5. Sponges

Among all marine invertebrates, sponges are well known regarding the ecological significance of natural products. Sponges have yielded the highest diversity of compounds. Sponge extracts have been shown to present a broad range of activities, such as inhibition of settlement and metamorphosis of marine larvae, behavioral modifications of adult invertebrates, and feeding deterrence against several fishes.[78]

The production of chemical defenses by sponge subjected to predation may influence their distribution between habitats. In a study conducted with two temperate sponge assemblages, which have different levels of predation by sea urchins, Wright *et al.* (1997)[79] have shown that those producing chemical defenses towards sea urchins populated areas with high predator density. Production of chemical defenses is especially important in habitats characterized by high predation levels. In tropical reefs, predation of sponges towards fish is quite intense.[80] Pawlik *et al.* (1995)[81] conducted surveys of the chemical defenses of large numbers of Caribbean sponges from reef, mangrove, and seagrass habitats toward different consumers in laboratory assays and revealed that 69 percent of sponge extracts were deterrent toward the fish *Thalassoma bifasciatum*, a generalist predator. Furthermore, a relationship between color and deterrency was not observed, suggesting that sponges are not aposematic to visual predators such as reef fishes. In addition, sponges with palatable extracts did not differ from those with deterrent extracts in nutritional value or structural materials. This proves that differences on palatability were due to deterrent compounds produced by the sponge.[82]

Allelopathy also plays ecological roles in sponge colonizing reefs. Sponge compounds may stress the symbiotic coral zooxanthellae. This mechanism is an important ecological process that occurs on reefs, where competition for space is high.[83]

Defense mechanisms can be activated when sponges are wounded. Products of conversion after wounding have been shown to be more active towards microorganisms and fish predators than towards other organisms.[84,85]

5.6. Cnidarians

Cnidarians, especially octocorals, have been a constant source of novel natural products in chemical ecology. These compounds may have roles in the invasion of new habitats, protection from fouling organisms and as feeding deterrents.

Transplantation experiments conducted with the Indopacific octocoral *Stereonepthya* aff. *Curvata* found on the southern coast of Brazil showed that this species was able to stress the native gorgonian *Plyllogorgia dilate.* Colonies of *P. dilate* that were adjacent to *S.* aff. *Curvata* showed signs of necrosis, while no negative effects were observed on the control group. This study shows that allelopathy also plays a role in the competition among cnidarians, allowing invasion into new habitats.[86]

Octocorals also use secondary metabolites to avoid fouling organisms. A study with the South China Sea gorgonian *Junceella juncea* reported 10 novel isolated diterpenoids able to inhibit barnacle larval settlement, but non-toxic to the barnacle *Balanus amphitrite.*[87]

It is known that many cnidarians maintain a symbiotic relationship with zooxanthellae. Laboratory experiments have shown that these dinoflagellates can change swimming behavior in the presence of their host. In the experiment, the presence of the octocoral *Heteroxenia fuscescens* caused zooxanthellae to swim in straighter paths than in the absence of their host. Symbiosis with zooxanthellae is critical for coral survival, and this study showed that zooxanthellae change their swimming behavior to find a host by detecting waterborne cues.[88]

5.7. Ascidians

Both secondary metabolites and inorganic acids have been proposed to protect adult ascidians and its larvae from predation .[89,90] One study investigated the feeding deterrence properties of organic extracts from 17 species of tropical and warm temperate ascidians towards the generalist consumer *Thalassoma bifasciatum.*[91] Sixteen out of 17 extracts revealed deterrent activity. In the same study, feeding deterring pH levels below 3 were found in the tunics of some ascidians. Nine species sequestered inorganic acids in their tunics at levels, which avoided *Thalassoma bifasciatum* feeding. Many species had the combination of deterrent organic extracts with acidic tunics. Besides feeding deterrent compounds, ascidians may also synthesize potent anti-fouling secondary metabolites.[92] However, ecological roles for ascidians natural products have been poorly investigated in comparison to other benthic invertebrates.

5.8. Bryozoans

Sharp *et al.* (2007)[93] have highlighted that bryozoans have been neglected in the natural products literature. Out of approximately 8,000 species, only 32 have been reported to synthesize secondary metabolites. However, approximately 200 compounds have been identified in these 32 species, indicating how promising this group is for natural product research and chemical ecology.

The bryozoan *Alcyonidium nodosum* represents a good example of an organism conferring chemical defenses to its host by symbiotic association. This bryozoan was shown to protect the South African whelk *Burnupena papyracea* against feeding by the lobster *Jasus lalandii* when encrustated to its shell. Food pellets with freeze-dried *A. nodosum* as well as their organic extracts added to food pellets deterred feeding by the lobster as compared to control.[94]

The bryozoan *Bugula neritina* is the best studied species among this phylum due to the synthesis the anti-cancer compounds, bryostatins. Bryostatins are produced by the bacterial symbiont *Candidatus Endobugula sertula*, which is present throughout all life stages of *B. neritina*. Besides their pharmacological application, bryostatins have been assigned to protect *B. neritina* larvae towards predation by fish.[95] This was also shown in a more recent work, where changes in bryostatin concentrations during *B. neritina* life cycle were correlated with different palatability of these organisms to the pinfish *Lagodon rhomboids*.[96] *B. neritina* larvae and post-settlement juveniles were unpalatable to *L. rhomboids* in laboratory experiments. Adults that had ovicells were less eaten than adults without eggs. Biochemical analysis of *B. neritina* in these different life stages revealed significantly higher concentrations of bryostatin in larvae and up to 1-day post-settlement juveniles than in the 5- and 7-day post-settlement juveniles and in the ovicell-free adult tissue. In addition, carbonate and chitin (putative structural defense) were measured over the same bryozoan life stages course. The structural material increased in these organisms, as they developed from larvae into the adult forms, suggesting an exchange between chemical and structural defense during the development of these bryozoans.

Another ecological aspect studied in bryozoans is their change in response to settlement cues. In laboratory experiments, where newly released larvae of *W. subtorquata* and *B. neritina* were given the option between a petri dish treated with algae extracts from *Delisea pulchra* and *Dilophus marginatus* and a petri dish treated with solvent only (control), larvae settled at higher rates on the control dish. However, selectivity decreased, when larvae were held for six hours. Older larvae settled at higher frequency in algae treated plates relative to the newly released larvae. This indicates that bryozoans become less selective to their substrate as larvae get older.[97]

5.9. Crustaceans

The ability to detect food in the surroundings is one of the best studied ecological aspects in crustaceans. The crab *Callinectes sapidus* is an animal used in laboratory tests to investigate how chemical signals affect crab behavior. In flume experiments, different boundary-layer turbulence was created by using four different types of benthic substrates (sand, small gravel, pea gravel and large gravel). Three-dimensional mixing of the plume is influenced by the bed substrate, with generally more mixing in the rougher substrates. The addition of food odor cues to the flume changed crab behavior, which was able to detect and to track towards prey odors with different success rates depending on the concentration of the odor cue and bed substrate used. This experiment exemplifies how animals' ability to detect and track potential preys that release water-soluble cues can be influenced by the physical environment.[98] Odor cues are also used by amphipods and copepods to direct their behavior. After detecting an amino acid solution in water, the amphipod *Scopelocheirus onagawae* is able to slow its swimming speed and change its direction into the odor plume. This response suggests an adaptation of these crustaceans to finding food in its environment.[99]

Behavioral responses against velocity and density gradients and biological parameters (i.e. against phytoplankton and the supernatant from phytoplankton cultures) were tested in the copepods *Temora longicornis* and *Acartia tonsa* in flume experiments.[100] Both species increased swimming speed and the time spent in the layer with the cue as a response to increased velocity gradients and chemical exudates.

Copepods and shrimps may also use pheromones to find their mates. Males of *Oithona avisae* circle around the pheromone patch after detecting a trail released by *O. avisae* females.[101] The shrimp *Lysmata wurdemanni* can detect and approach females that are ready to couple. For this purpose, both the antennae and the antennules (their sensory organs) cooperate for successful copulation.[102]

5.10. Molluscs

Many molluscs are consumed by other species from diverse taxa, such as fish, crustaceans, echinoderms and cnidarians. For their part of the evolutionary arms race, they have developed a series of strategies to escape predation. Some molluscs are protected by shells, while others such as opisthobranch gastropods and cephalopods are not and, thus, rely on nonmechanical defense strategies.

The gastropods are among the best studied of all molluscs. Special efforts have been invested in unraveling their chemical defense strategies in ink, mucus, skin, and digestive glands. Sea hares are the most studied marine organisms in terms of chemical defense due to the lack of an apparent protection structure such as a shell and their apparent vulnerability to predators.

The ink from the sea hare *Alpysia californica* contains compounds, which serve as alarm cues. The detection of ink is sufficient to induce a flight response in conspecifics.[103] *A. californica* also presents a flight response, when exposed to ink and opaline from other sea hare species (*A. dactylomela and A. juliana*). Flight responses were also observed in response to ink from the cephalopods *Loliguncula brevis* and *Octopus vulgaris*, suggesting conserved chemical alarm cues in mollusc ink. Sea hare mucus may also contain chemical cues, thereby increasing the persistence of the chemicals and reducing their diffusion into water.[104] Skin and mucus of snails, for example, contain feeding deterrent components, which avoid predators.[105] The opisthobranchs *Sagaminopteron nigropunctatum* and *S. psychedelicum* contain diet-derived chemical defenses that are located in vulnerable body parts.[106] Many of the deterrent compounds found in molluscs are diet-derived and not synthesized *de novo*. Many of them are terpenoids, especially sesquiterpenoids and diterpenoids.[107–109]

The gastropod nudibranch *Tritonia diomedea* is able to detect water-soluble cues in flow and actively moves towards preys and conspecifics.[110]

Water flow consistently influences the ability of *T. diomedea* to detect prey and conspecifics, whereby its perception is mediated by the rhinophores (chemosensory organs).

The chemical ecology of molluscs' larvae has also been investigated. Larval metamorphosis of the gastropod *Alderia* sp. changes as they get older. Botello and Krug (2006)[111] showed that larvae metamorphosis rates were positively related to increased concentrations of the metamorphic inducer excreted by the alga *Vaucheria longicaulis.* A higher sensibility was found in older larvae (3-day old) as compared to younger larvae (1-day old). Also, as larvae aged metamorphosis rates increased after shorter times of exposure period to the algae. Further, larvae that were not fed had higher metamorphosis rates than fed ones. These results are in agreement with the 'desperate larva' hypothesis, which postulates that lecithotrophic larvae become less discriminating in their settlement requirements over time, due to depletion of energy reserves.

Filter feeders can change their behavior and even induce physical defenses after sensing water-soluble cues emitted by potential predators. Behavioral changes in response to water-soluble cues from predators were observed for the bivalves *Macoma balthica* and *Cerastoderma edule.* Seawater originated from tanks with the crab *Carcinus maenas* induced burrowing responses on both bivalves. After removal of predator water, *M. balthica* returned to a similar burial depth than the controls. Parallel experiments showed that burial depth has ecological consequences and bivalves that bury deeper in the substrate are in fact less vulnerable to predation by the crab. This behavioral response represents an effective escape strategy for these bivalves towards *C. maenas.*

The mussels *Perna viridis* and *Brachidontes variabilis* were shown to increase the number and diameter of their byssal threads in response to water-soluble cues from the crabs *Thalamita danae* within 48 hours. This suggests a defense response towards predators in a way that mussels become harder to detach and be eaten by the crab.[112] Another example of induction of physical defenses after exposure to water-soluble cues from potential predators was observed in two oysters from Chesapeake Bay: *Crassostrea virginica* and the non-native *C. ariakensis* were exposed to water cues of crustaceans that fed on them. After 54 days, an increase of both shell density and compression force was observed in both oysters.[113]

Mytilus edulis was able to induce specific morphological characters depending on which kind of predator it was exposed to. Water-soluble cues from crushing predators such as the crab *C. maenas* and from the starfish *Asterias vulgaris* induced the development of thicker shells and heavier adductor muscles, respectively. Increased shell thickness meant increased handling time by *C. maenas*, and increased muscle mass made the mussel more difficult for the starfish to consume .[114]

5.11. Worms

It is known that marine worms play critical roles in trophic interactions and affect biogeochemical cycles. Despite the ecological importance of this group, little is known about their chemical ecology. In the most recent literature, greater attention has been paid to the chemical ecology of marine worms with focus on their palatability to and defenses towards consumers.

Chemical defense mechanisms have been discussed in a comprehensive study on marine worms.[115] In this study, out of 81 worm species, 37 percent were deterrent to at least one predator out of three fishes and one crab.

However, worm toxins may also play offensive ecological roles. For instance, a still undescribed flatworm (Planocerid sp. 1) fed on a variety of mobile molluscs. Since flatworms usually feed on sessile benthic prey toxic compounds were investigated in Planoceridsp. 1, and a deadly neurotoxin (tetrodotoxin) and its analog 11-nortetrodotoxin-6(S)-ol (TTXs) were identified.[116]

Conspecific communication among worms can be mediated by chemical signaling. Water extracts from the polychaete *Nereis virens* induced a reduction in feeding and foraging activity levels. Extracts from other worm species did not have behavioral effects on *N. virens*, suggesting a water soluble alarm cue specific to conspecifics.[117]

An interesting case of chemical compounds playing ecological roles was described in the simultaneous hermaphroditic polychaete *Ophryotrocha diadema*. This worm suppressed the female function in the population in the presence of reproductive competitors or multiple partners.[118] Cocoons produced by *O. diadema* females were used as a proxy for the number of females in the population. In larger populations, lower female allocation was observed, which was triggered by conspecific water-soluble compounds. Water-soluble compounds from the allopatric worm *O. hartmanni* did not have the same effect.

5.12. Echinoderms

Sea stars also take advantage of feeding deterrent compounds to avoid predation. The Antarctic Peninsula sea stars *Granaster nutrix* and *Neosmilaster georgianus* were reported to deter feeding by the generalist sea star *Odontaster validus*, a common predator of other Antarctic sea stars. Freeze-dried krill food was preferred by the predator instead of live intact individuals and body-wall tissues from both sea stars. Indeed, when methanol extracts were incorporated to the freeze-died krill, they were rejected more by *O. validus* than by control pellets.[119]

Sea urchin larvae also use chemical cues in the selection of their settlement habitat. Natural biofilms microbes, which colonize the surface of coralline algae induced metamorphosis of the generalist herbivore *Heliocidaris erythrogramma.*[120] *H. erythrogramma* reduced its metamorphosis rates, when algae were wiped or soaked in antibiotic seawater, suggesting that microbes colonizing these algae mediated settlement of this sea urchin.

Marine invertebrates use chemical cues as an indicator of settlement habitat; however, little is known about how larvae sense the presence of potential predators in the environment. Water-soluble cues from the ctenophore *Bolinopsis infundibulum* were used to test, whether larvae of the sea urchin *Strongylocentrotus droebachiensis* could detect potential predators.[121] In laboratory experiments, the vertical distribution of the sea urchin larvae showed that *S. droebachiensis* could detect and avoid the presence of the ctenophore trapped in the water at the top of the experimental columns. In the absence of *B. infundibulum*, sea urchin larvae presented a higher proportion on the top of the container. Such behavioral avoidance of predators may increase the survival rates of larvae.

Natural products from echinoderms can also mediate conspecifics communication. Ovary homogenates of the brittle stars *Ophiocoma dentata* and *O. scolopendrina* contain spawn-inducing pheromones. This effect was always induced by ovary homogenates and never by seawater or testis homogenates. However, the compound responsible for this mediation still remains to be identified.

5.13. Fish

Most fish possess highly developed sense organs and specialized chemoreceptors that are responsible for extraordinary senses of taste and smell. Larval fish can use hearing and vision to detect a suitable habitat, while others can use chemical cues for this purpose.[122]

Others, such as the cardinal fish *Cheilodipterus quinquelineatus* showed preference to water conditioned from home reef conspecifics *versus* water from another place and is also attracted to artificial reefs, which had previously contained conspecifics, suggesting that cardinal fish uses chemical cues from conspecifics for its homing instinct.[123]

Fish may be able to detect sex pheromones released by potential partners. Traps baited with spermiating male sea lampreys attracted ovulating females, while empty traps and traps with non-sperming males did not.[124]

Fish behavior might be influenced by the detection of food. In the Gulf of Mexico, high concentrations of dimethylsulfoniopropionate (DMSP) were correlated with an increased abundance of carangid fish.[125] DMSP is produced by marine algae and its presence is associated with areas of high primary and secondary productivity in the ocean. This suggests that specific habitat-associated chemical cues in the ocean might control pelagic fish local abundance.

Defense behaviors can also be mediated through chemical cue detection. The fish *Asterropteryx semipunctatus* was observed in laboratory and field experiments to sense cues from damaged conspecifics and flee from them, but no alarm response was observed in response to extracts from damaged fish from a different taxonomic family.[126] This study shows a clear evidence for the use of chemical alarm cues in a marine fish in their natural environment.

6. CONCLUSION

With the growing understanding of molecular physiology of marine organisms together with the ecological functions of natural products released by them, chemical ecology has developed in the past decades into a field, where chemical and biological aspects are simultaneously investigated using ecologically realistic conditions. The roles of marine natural products in predator-prey interaction, courtship and mating, aggregation and school formation, and habitat selection have been well studied in ecology.

The knowledge of marine chemical ecology has been built up by the isolation and characterization of many compounds that mediate chemical interaction among organisms. Some of these compounds may be difficult to isolate and to characterize either due to their presence in trace amounts in organisms or due to their low stability. However, the characterization of ecologically relevant compounds is essential for the understanding of ecological interactions and how these compounds affect physiological and neurobiological pathways. Further, structurally characterized compounds allow additional research on their biosynthesis, localization and distribution in the organisms and in the water column, in case they are released.

The advances in molecular techniques and the interdisciplinary exchange of knowledge between natural product chemists, marine biologists, physiologists, microbiologists and biochemists are key approaches for further advances and growth of chemical ecology of marine organisms.

Meanwhile, marine chemical ecology has become a field that is capable of describing many basic processes that explain patterns of behavioral, population, and community ecology.

REFERENCES

1. Blunt JW, Copp BR, Hu WP, *et al.* (2008) Marine natural products. *Nat Prod Rep* **25**: 35–94.
2. Scheuer PJ. (1978–1983) *Marine Natural Products: Chemical and Biological Perspectives*. Vols. 1–5. New York: Academic Press.
3. Scheuer PJ. (1990) Some marine ecological phenomena: chemical basis and biomedical potential. *Science* **248**: 173–177.
4. Molinski TF, Dalisay DS, Lievens SL, Saludes JP. (2009) Drug development from marine natural products. *Nat Rev Drug Discov* **8**: 69–85.
5. Peterson CH, Renaud PE. (1989) Analysis of feeding preference experiments. *Oecologia* **80**: 82–86.
6. Hay ME, Fenical W. (1988) Chemically-mediated seaweed herbivore interactions. *Annu Rev Ecol Syst* **19**: 111–145.
7. Lancet D, Pace U. (1987) The molecular basis of odor recognition. *Trends in Biochem Sci* **12**: 63–66.

8. Leonard GH, Bertness MD, Yund PO. (1990) Crab predation, waterborne cues, and inducible defenses in the blue mussel Mytilus edulis. *Ecology* **80**: 1–14.
9. Hardege JD, Bentley MG, Beckmann M, Mueller C. (1996) Sex pheromones in marine polychaetes: volatile organic substances (VOS) isolated from Arenicola marina. *Mar Ecol Prog Ser* **139**: 157–166.
10. Hamner WM, Hamner PP, Strand SW, Gilmore RW. (1983) Behavior of antarctic krill, Euphausia superba: chemoreception, feeding, schooling and molting. *Science* **220**: 433–435.
11. Ratchford SG, Eggleston DB. (1998) Size and scale-dependent chemical attraction contribute to an ontogenetic shift in sociality. *Anim Behav* **56**: 1027–1034.
12. Pawlik JR. (1992) Chemical ecology of the settlement of benthic marine invertebrates. *Oceanogr Mar Biol Annu Rev* **30**: 273–335.
13. Atema J. (1995) Chemical signals in the marine environment: dispersal, detection, and temporal signal analysis. *Proc Natl Acad Sci USA* **92**: 62–66.
14. Carr WES, Derby CD. (1986) Chemically stimulated feeding behavior in marine animals: implications of chemical mixtures and involvement of mixture interactions. *J Chem Ecol* **12**: 989–1011.
15. Finelli CM, Pentcheff ND, Zimmer-Faust RK, Wethey DS. (1999) Odor transport in turbulent flows: constraints on animal navigation. *Limnol Oceanogr* **44**: 1056–1071.
16. Lee G, Consiglio E, Habig W, *et al.* (1978) Structure: function studies of receptors for thyrotropin and tetanus toxin: lipid modulation of effector binding to the glycoprotein receptor component. *Biochem Biophys Res Communs* **83**: 470–477.
17. Foster JW, Kinney DM. (1984) ADP-ribosilating microbial toxins. *CRC critical reviews in microbiol* **11**: 273–298.
18. Rumyantsev SN. (1997) Chemical ecology and biomolecular evolution. *Acta Biotheoretica* **45**: 65–80.
19. Faulkner DJ, Ghiselin MT. (1983) Chemical defense and the evolutionary ecology of dorid nudibranchs and some other opisthobranch gastropods. *Mar Ecol Prog Ser* **13**: 295–301.
20. Kaiser MJ, Attrill MJ, Jennings S, *et al.* (2005) *Marine Ecology: Processes, Systems, and Impacts*, 1st edn. New York: Oxford University Press Inc, pp. 1–30.
21. Barnes DKA, Brockington S. (2003) Zoobenthic biodiversity, biomass and abundance at Adelaide Island, Antarctica. *Mar Ecol Prog Ser* **249**: 145–155.
22. Bentamy A, Grima N, Quilfen Y, *et al.* (1996) An atlas of surface wind from ERS-1 scatterometry measurements. IFREMER, DRO/OS, BP 70, 29280 Plouzané, France. p. 229. IFREMER Publication.

23. Gaston KJ, Spicer JI, editors. (2004) Biodiversity: an introduction. 2nd ed. Oxford: Blackwell Science.
24. Rittschof D, Cohen JH. (2004) Crustacean peptide and peptide like pheromones and kairomones. *Peptides* **25:** 1503–1516.
25. Hallmann A, Godl K, Wenzl S, Sumper M. (1998) The highly efficient sex-inducing pheromone system of Volvox. *Trends Microbiol* **6:** 185–189.
26. Doall MH, Colin SP, Strickler JR, Yen J. (1998) Locating a mate in 3D: the case of Temora longicornis. *Philos Trans R Soc Lond B* **353:** 681–689.
27. Yen J. (2000) Life in transition: balancing inertial and viscous forces by planktonic copepods. *Biol Bull* **198:** 213–224.
28. Atema J. (1986) Review of sexual selection and chemical communication in the lobster, Homarus americanus. *Canad J Fisheries and Aquatic Sci* **43:** 2283–2290.
29. Carr WES. (1988) The molecular nature of chemical stimuli in the aquatic environment. In: *The Sensory Biology of Aquatic Animals.* New York: Springer-Verlag, pp. 3–27.
30. Harborne JB. (1989) Recent advances in chemical ecology. *Nat Prod Rep* **6:** 85–109.
31. Stachowicz JJ. (2001) Mutualisms, positive interactions, and the structure of ecological communities. *BioScience* **51:** 235–246.
32. McClintock JB, Baker BJ. (2001) Marine chemical ecology. Series in marine science. Volume 25. London England: CRC Press.
33. Paul VJ, Puglisi MP, Ritson-Williams R. (2006) Marine chemical ecology. *Nat Prod Rep* **23:** 153–180.
34. Cannell RJP. (1998) How to approach the isolation of a natural product. In: Cannell RPJ (ed.), *Natural Products Isolation, Methods in Biotechnology.* Vol. 4. Totowa, New Jersey: Human Press, pp. 1–51.
35. Parish EJ, Nes WD, editors. (1997) Biochemistry and function of steroids. Boca Raton, Florida: CRC Press Inc. p. 288.
36. Ragan MA, Glombitzka KW. (1986) Phorotannins, brown algal polyphenols. *Prog phycol Res* **4:** 129–241.
37. Amsler CD, Fairhead VA. (2006) Defensive and sensory chemical ecology of brown algae. *Adv Bot Res* **43:** 1–91.
38. Williams DH, Stone MJ, Hauck PR, Rahman SK. (1989) Why are secondary metabolites (natural products) biosynthesized? *J Nat Prod* **52:** 1189–1208.
39. International Union of Pure and Applied Chemistry (IUPAC). (1976) Nomenclature of organic chemistry. Section F: natural products and related compounds (Recommendations 1976). IUPAC Information Bulletin, Appendices on provisional nomenclature, symbols, terminology, and conventions. No. 53.

40. Giles PM. (1999) Revised section F: natural products and related compounds — IUPAC Recommendations. *Pure Appl Chem* **71**: 587–643.
41. Favre HA, Giles PM Jr, Hellwich KH, *et al.* (2004) Revised section F: natural products and related compounds (IUPAC Recommendations 1999). Corrections and modifications. *Pure Appl Chem* **76**: 1283–1292.
42. Melo VM, Duarte AB, Carvalho AF, Siebra EA, Vasconcelos IM. (2000) Purification of a novel antibacterial and haemagglutinating protein from the purple gland of the sea hare. Aplysia dactylomela rang, 1828, *Toxicon* **38**: 1415–1427.
43. Hamann MT, Scheuer PJ. (1993) Kahalalide F, a bioactive depsipeptide from the sacoglossan mollusk Elysia rufescens and the green alga Bryopsis sp. *J Am Chem Soc* **115**: 5825–5826.
44. Burreson BJ, Scheuer PJ, Finer J, Clardy J. (1975) 9-Isocyanopupukeanane, a marine invertebrate allomone with a new sesquiterpene skeleton. *J Am Chem Soc* **97**: 4763–4764.
45. Hertweck C, Luzhetskyy A, Rebets Y, Bechthold A. (2007) Type II polyketide synthases: gaining a deeper insight into enzymatic teamwork. *Nat Prod Rep* **24**: 162–190.
46. Hanson JR. (2003) Natural products: the secondary metabolites. Royal Society of Chemistry, Thomas Graham House, Cambridge, UK, p. 153.
47. Khosla C, Gokhale RS, Jacobsen JR, Cane DE. (1999) Tolerance and specificity of polyketide synthases. *Annu Rev Biochem* **68**: 219–253.
48. Hahlbrock K, Scheel D. (1989) Physiology and molecular biology of phenylpropanoid metabolism. *Annual Rev Plant Physiol and Plant Mol Biol* **40**: 347–369.
49. Hamann MT. (2004) Technology evaluation: Kahalalide F, PharmaMar. *Curr Opin Mol Ther* **6**: 657–665.
50. Janmaat ML, Rodriguez JA, Jimeno J, *et al.* (2005) Kahalalide F induces necrosis-like cell death that involves depletion of ErbB3 and inhibition of Akt signaling. *Mol Pharmacol* **68**: 502–510.
51. Bhakuni DS, Rawat DS. (2005) Bioactive marine natural products. New York: Springer-Verlag, New York and Anamaya Publishers, New Delhi, India. — XV. p. 382.
52. Haefner B. (2003) Drugs from the deep: marine natural products as drug candidates. *Drug Discov Today* **8**: 536–544.
53. Dickson MA, Schwartz GK. (2009) Development of cell-cycle inhibitors for cancer therapy. *Curr Oncol* **16**: 36–43.
54. Mayer AM, Jacobson PB, Fenical W, *et al.* (1998) Pharmacological characterization of the pseudopterosins: novel anti-inflammatory natural products

isolated from the Caribbean soft coral, Pseudopterogorgia elisabethae. *Life Sci* **62**: 401–407.

55. Kem WR. (2000) The brain alpha7 nicotinic receptor may be an important therapeutic target for the treatment of Alzheimer's disease: studies with DMXBA (GTS-21). *Behav Brain Res* **113**: 169–181.
56. Singh R, Sharma M, Joshi P, Rawat DS. (2008) Clinical status of anti-cancer agents derived from marine sources. *Anticancer Agents Med Chem* **8**: 603–617.
57. Ritchie KB. (2006) Regulation of microbial populations by coral surface mucus and mucus-associated bacteria. *Mar Ecol Prog Ser* **322**: 1–14.
58. Waters CM, Bassler BL. (2005) Quorum sensing: cell-to-cell communication in bacteria. *Annu Rev Cell Dev Biol* **21**: 319–346.
59. Joint I, Tait K, Wheeler G. (2007) Cross-kingdom signalling: exploitation of bacterial quorum sensing molecules by the green seaweed *Ulva*. *Philos Trans R Soc London* **362**: 1223–1233.
60. Skindersoe ME, Ettinger-Epstein P, Rasmussen TB, *et al.* (2008) Quorum Sensing Antagonism from Marine Organisms. *Mar Biotechnol* **10**: 56–63.
61. Burkepile DE, Parker JD, Woodson CB, *et al.* (2006) Chemically-mediated competition between microbes and animals: microbes as consumers in food webs. *Ecology* **87**: 2821–2831.
62. Paul VJ, Arthur KE, Ritson-Williams R, *et al.* (2007) Chemical defenses: from compounds to communities. *Biol Bull* **213**: 226–251.
63. Luesch H, Pangilinan R, Yoshida WY, *et al.* (2001) Pitipeptolides A and B, new cyclodepsipeptides from the marine cyanobacterium Lyngbya majuscule. *J Nat Prod* **64**: 304–307.
64. Fong P, Smith TB, Wartian MJ. (2006) Protection by epiphytic cyanobacteria maintain shifts to macroalgal-dominated communities after the 1997–98 ENSO disturbance on coral reefs with intact herbivore populations. *Ecology* **87**: 1162–1168.
65. Rhoades DF. (1985) Offensive-defensive interactions between herbivores and plants their relevance in herbivore population dynamics and ecological theory. *Am Nat* **125**: 205–238.
66. Carlson DJ, Lubchenco J, Sparrow MA, Trowbridge CD. (1989) Fine-scale variability of lanosol and its disulfate ester in the temperate red alga Neorhodomela larix. *J Chem Ecol* **15**: 1321–1333.
67. Meyer KD, Paul VJ. (1995) Variation in secondary metabolite andaragonite concentrations in the tropical green seaweed Neomeris annulata: effects on herbivory by fishes. *Mar Biol* **122**: 537–545.

68. Meyer KD, Paul VJ. (1992) Intraplant variation in secondary metabolite concentration in three species of Caulerpa (Chlorophyta: Caulerpales) and its effects on herbivorous fishes. *Mar Ecol Prog Ser* **82**: 249–257.
69. Paul NA, de Nys R, Steinberg PD. (2006) Chemical defence against bacteria in the red alga *Asparagopsis armata*: linking structure with function. *Mar Ecol Prog Ser* **306**: 87–101.
70. Wiesemeier T, Hay M, Pohnert G. (2007) The potential role of wound-activated volatile release in the chemical defence of the brown alga *Dictyota dichotoma*: blend recognition by marine herbivores. *Aquatic Sciences* **69**: 403–412.
71. Yun HY, Cruz J, Treitschke M, *et al.* (2007) Testing for the induction of anti-herbivory defences in four Portuguese macroalgae by direct and water-borne cues of grazing amphipods. *Helg Mar Res* **61**: 203–209.
72. Engel SE, Puglisi MP, Jensen PR, Fenical W. (2006) Antimicrobial activities of extracts from tropical Atlantic marine plants against pathogens and saprophytes. *Mar Biol* **149**: 991–1002.
73. Puglisi MP, Engel S, Jensen PR, Fenical W. (2007) Antimicrobial activities of extracts from Indo-Pacific marine plants against marine pathogens and saprophytes. *Mar Biol* **150**: 531–540.
74. Dworjanyn SA, de Nys R, Steinberg PD. (2006) Chemically mediated antifouling in the red alga *Delisea pulchra. Mar Ecol Prog Ser* **318**: 153–163.
75. Nylund GM, Gribben PE, de Nys R, *et al.* (2007) Surface chemistry versus whole-cell extracts: antifouling tests with seaweed metabolites. *Mar Ecol Prog Ser* **329**: 73–84.
76. Vergés A, Becerro MA, Alcoverro T, Romero J. (2007) Experimental evidence of chemical deterrence against multiple herbivores in the seagrass *Posidonia oceanica. Mar Ecol Prog Ser* **343**: 107–114.
77. Vergés A, Becerro MA, Alcoverro T, Romero J. (2007) Variation in multiple traits of vegetative and reproductive seagrass tissues influences plant–herbivore interactions. *Oecologia* **151**: 675–686.
78. Thompson JE, Walker RP, Faulkner DJ. (1985) Screening and bioassays for biologically-active substances from forty marine sponge species from San Diego, California, USA. *Mar Biol* **88**: 11–21.
79. Wright JT, Benkendorff K, Davis AR. (1997) Habitat associated differences in temperate sponge assemblages: the importance of chemical defence. *J Experim Marine Biol Ecol* **213**: 199–213.
80. Wulff JL. (1994) Sponges in time and space. In: van Soest RWM, van Kempen TMG, Braekman JC, (eds.), Rotterdam: AA Balkema, p. 265.

81. Pawlik JR, Chanas B, Toonen RJ, Fenical W. (1995) Defenses of Caribbean sponges against predatory reef fish. I. Chemical deterrency. *Mar Ecol Prog Ser* **127**: 183–194.
82. Chanas B, Pawlik JR. (1995) Defenses of Caribbean sponges against predatory reef fish. II. Spicules, tissue toughness, and nutritional quality. *Mar Ecol Prog Ser* **127**: 195–211.
83. Pawlik JR, Steindler L, Henkel TP, *et al.* (2007) Chemical warfare on coral reefs: sponge metabolites differentially affect coral symbiosis in situ. *Limnol Oceanogr* **52**: 907–911.
84. Teeyapant R, Proksch P. (1993) Evidence for an induced chemical defense of the marine sponge *Verongia aerophoba. Naturwissenschaften* **80**: 360–369.
85. Ebel R, Brenzinger M, Kunze A, *et al.* (1997) Wound activation of protoxins in marine sponge aplysina aerophoba. *J Chem Ecol* **23**: 1451–1462.
86. Lages BG, Fleury BG, Ferreira CEL, Pereira RC. (2006) Chemical defense of an exotic coral as invasion strategy. *J Exp Mar Biol Ecol* **328**: 127–135.
87. Qi SH, Zhang S, Qian PY, *et al.* (2006) Ten new antifouling briarane diterpenoids from the South China Sea gorgonian *Junceella juncea. Tetrahedron* **62**: 9123–9130.
88. Pasternak Z, Blasius B, Abelson A, Achituv Y. (2006) Host-finding behaviour and navigation capabilities of symbiotic zooxanthellae. *Coral Reefs* **25**: 201–207.
89. Stoecker D. (1980) Relationships between chemical defense and ecology in Benthic Ascidians. *Mar Ecol Prog Ser* **3**: 257–265.
90. Young CM, Bingham BL. (1987) Chemical defense and aposematic coloration in larvae of the ascidian Ecteinascidia turbinate. *Mar Biol* **96**: 539–544.
91. Pisut DP, Pawlik JR. (2002) Anti-predatory chemical defenses of ascidians: secondary metabolites or inorganic acids? *J Exp Marine Biol Ecol* **270**: 203–214.
92. McClintock JB, Amsler MO, Amsler CD, *et al.* (2004) Biochemical composition, energy content and chemical antifeedant and antifoulant defenses of the colonial Antarctic ascidian Distaplia cylindrical. *Mar Biol* **145**: 885–894.
93. Sharp JH, Winson MK, Porter JS. (2007) Bryozoan metabolites: an ecological perspective. *Nat Prod Rep* **24**: 659–673.
94. Gray CA, McQuaid CD, Davies-Coleman MT. (2005) A symbiotic shell-encrusting bryozoan provides subtidal whelks with chemical defence against rock lobsters. *Afr J Mar Sci* **27**: 549–556.

95. Lopanik N, Lindquist N, Targett N. (2004) Potent cytotoxins produced by a microbial symbiont protect host larvae from predation. *Oecologia* **139**: 131–139.
96. Lopanik NB, Targett NM, Lindquist N. (2007) Ontogeny of a symbiont-produced chemical defense in Bugula neritina (Bryozoa). *Mar Ecol Prog Ser* **327**: 183–191.
97. Gribben PE, Marshall DJ, Steinberg PD. (2006) Less inhibited with age? Larval age modifies responses to natural settlement inhibitors. *Biofouling* **22**: 101–106.
98. Jackson JL, Webster DR, Rahman S, Weissburg MJ. (2007) Bed roughness effects on boundary-layer turbulence and consequences for odor-tracking behavior of blue crabs (Callinectes sapidus). *Limnol Oceanogr* **52**: 1883–1897.
99. Ide K, Takahashi K, Omori M. (2007) Direct observation of swimming behaviour in a shallow-water scavenging amphipod *Scopelocheirus onagawae* in relation to chemoreceptive foraging. *J Exp Mar Biol Ecol* **340**: 70–79.
100. Woodson CB, Webster DR, Weissburg MJ, Yen J. (2007) Cue hierarchy and foraging of calanoid copepods: ecological implications of oceanographic structure. *Mar Ecol Prog Ser* **330**: 163–177.
101. Kiørboe T. (2007) Mate finding, mating, and population dynamics in a planktonic copepod *Oithona davisae*: There are too few males. *Limnol Oceanogr* **52**: 1511–1522.
102. Zhang D, Lin J. (2006) Mate recognition in a simultaneous hermaphroditic shrimp, *Lysmata wurdemanni* (Caridea: Hippolytidae). *Anim Behav* **71**: 1191–1196.
103. Kicklighter C, Germann MW, Kamio M, Derby C. (2007) Molecular identification of alarm cues in the defensive secretions of the sea hare *Aplysia californica. Anim Behav* **74**: 1481–1492.
104. Kicklighter CE, Shabani S, Johnson PM, Derby CD. (2005) Sea hares use novel antipredatory chemical defenses. *Curr Biol* **15**: 549–554.
105. Denny MW. (1989) Invertebrate mucous secretions: functional alternatives to vertebrate paradigms. *Symp Soc Exp Biol* **43**: 337–366.
106. Becerro MA, Starmer JA, Paul VJ. (2006) Chemical defenses of cryptic and aposematic gastropterid molluscs feeding on their host sponge *Dysidea granulose. J Chem Ecol.* **32**: 1491–1500.
107. Avila C. (1995) Natural products of opisthobranch molluscs: a biological review. *Oceanogr Mar Biol Annu Rev* **33**: 487–559.
108. Cimino G, Fontana A, Gavagnin M. (1999) Marine opisthobranch molluscs: chemistry and ecology in sacoglossans and dorids. *Curr Organic Chem* **3**: 327–372.

109. Kamiya H, Sakai R, Jimbo M. (2006) Bioactive molecules from sea hares. In: Cimino G, Gavagnin M, (eds.), *Molluscs: From Chemo-ecological Study to Biotechnological Application, Progress in Molecular and Subcellular Biology. Marine Molecular Biotechnology.* Berlin: Springer, pp. 215–239.
110. Wyeth RC, Willows AOD. (2006) Odours detected by rhinophores mediate orientation to flow in the nudibranch mollusc, *Tritonia diomedea. J Exp Biol* **209**: 1441–1453.
111. Botello G, Krug PJ. (2006) 'Desperate larvae' revisited: age, energy and experience affect sensitivity to settlement cues in larvae of the gastropod *Alderia sp. Mar Ecol Prog Ser* **312**: 149–159.
112. Cheung SG, Luk KC, Shin PKS. (2006) Predator-Labeling effect on byssus production in Marine mussels *Perna viridis* (L.) and *Brachidontes variabilis* (Krauss). *J Chem Ecol* **32**: 1501–1512.
113. Newell RIE, Kennedy VS, Shaw KS. (2007) Comparative vulnerability to predators, and induced defense responses, of eastern oysters *Crassostrea virginica* and non-native *Crassostrea ariakensis* oysters in Chesapeake Bay. *Mar Biol* **152**: 449–460.
114. Freeman AS. (2007) Specificity of induced defenses in *Mytilus edulis* and asymmetrical predator deterrence. *Mar Ecol Prog Ser* **334**: 145–153.
115. Kicklighter CE, Hay ME. (2006) Integrating prey defensive traits: contrasts of marine worms from temperate and tropical habitats. *Ecol Monogr* **76**: 195–215.
116. Ritson-Williams R, Yoshi-Yamashita M, Paul VJ. (2006) Ecological functions of tetrodotoxin in a deadly polyclad flatworm. *Proc Natl Acad Sci USA* **103**: 3176–3179.
117. Watson GJ, Hamilton KM, Tuffnail WE. (2005) Chemical alarm signalling in the polychaete Nereis (Neanthes) virens (Sars) (Annelida: Polychaeta). *Anim Behav* **70**: 1125–1132.
118. Schleicherová D, Lorenzi MC, Sella G. (2006) How outcrossing hermaphrodites sense the presence of conspecifics and suppress female allocation. *Behav Ecol* **17**: 1–5.
119. McClintock JB, Amsler MO, Amsler CD, Baker BJ. (2006) The biochemical composition, energy content, and chemical antifeedant defenses of the common Antarctic Peninsular sea stars Granaster nutrix and Neosmilaster georgianus. *Polar Biol* **29**: 615–623.
120. Huggett MJ, Williamson JE, de Nys R, *et al.* (2006) Larval settlement of the common Australian sea urchin Heliocidaris erythrogramma in response to bacteria from the surface of coralline algae. *Oecologia* **149**: 604–619.

121. Metaxas A, Burdett-Coutts V. (2006) Response of invertebrate larvae to the presence of the ctenophore *Bolinopsis infundibulum*, a potential predator. *J Exp Mar Biol Ecol* **334**: 187–195.
122. Arvedlund M, Takemura A. (2006) The importance of chemical environmental cues for juvenile Lethrinus nebulosus Forsskål (Lethrinidae, Teleostei) when settling into their first benthic habitat. *J Exp Mar Biol Ecol* **338**: 112–122.
123. Døving KB, Stabell OB, Östlund-Nilsson S, Fisher R. (2006) Site fidelity and homing in tropical coral reef cardinalfish: are they using olfactory cues? *Chem Senses* **31**: 265–272.
124. Johnson NS, Siefkes MJ, Li W. (2005) Capture of ovulating female sea lampreys in traps baited with spermiating male sea lampreys. *N Am J Fish Manage* **25**: 67–72.
125. DeBose JL, Nevitt GA. (2007) Investigating the association between pelagic fish and dimethylsulfoniopropionate in a natural coral reef system. *Mar Freshwater Res* **58**: 720–724.
126. McCormick MI, Larson JK. (2007) Field verification of the use of chemical alarm cues in a coral reef fish. *Coral Reefs* **26**: 571–576.

CHAPTER 5

Natural Products from Terrestrial Microbial Organisms with Cytotoxic Cell Cycle Inhibitors

Theresa Dreis, Caroline Gartner, Julia Krebs and Mathias Schneider

ABSTRACT

Many microbial products have the potential for therapeutic application, e.g. as anticancer drugs. This chapter presents natural products from terrestrial microbial organisms that attack various targets in tumor cells, e.g. cell cycle inhibitors (inhibitors of phosphoinositide-3-kinase, checkpoint kinase inhibitors, inhibitors of dual specificity phosphatases, inhibitors of cyclin-dependent kinases), inhibitors of histone deacetylases, topoisomerases, or growth factors as well as inhibitors of the proteasome, of heat shock protein 90, nuclear trafficking, or ATPase.

New cancer drugs are most likely be used as part of combination therapy regimens. Attacking different targets in tumor cells may kill tumor cells resistant to one target. The potential of novel target-directed compounds to eradicate otherwise drug-resistant cells is an attractive perspective, which deserves further investigation.

1. INTRODUCTION

Microorganisms comprise all single-celled organisms, including prokaryotic cyanophytes and eukaryotic fungi. Many microbial products can be used for therapeutic purposes, for instance as anticancer drugs. In fact, microorganisms make these products for self-protection. However, many of

Institute of Pharmacy and Biochemistry, Johannes Gutenberg University, Staudinger Weg 5, 55128 Mainz, Germany

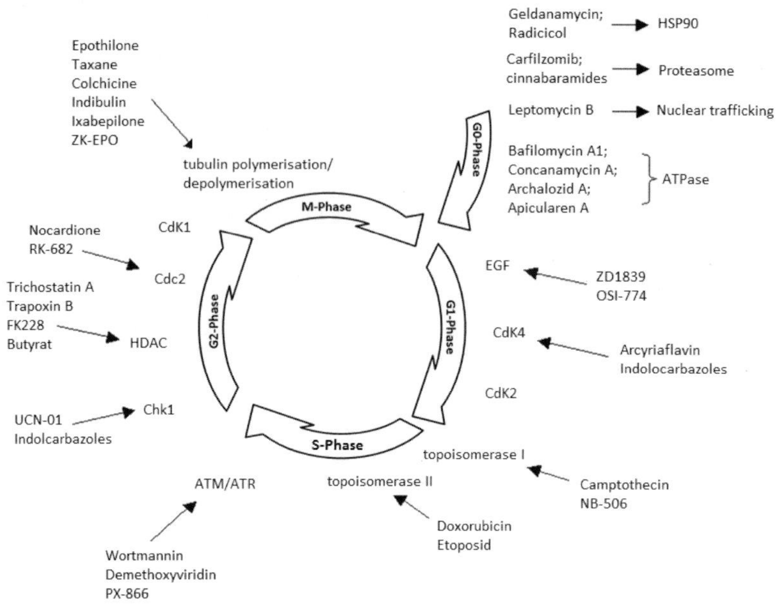

Figure 1 Overview of natural products and their molecular targets in the cell cycle.

these compounds are effective in the human body. Various natural products attacking different targets in tumor processes are presented in the Chapter 6, and an overview is depicted in Fig. 1. There are many signal transduction pathways, which are differently regulated in tumors as compared to normal tissues. Such pathways provide promising targets for cancer therapy if they can be influenced by microbial inhibitors.

2. INHIBITORS OF PHOSPHOINOSITIDE-3-KINASES (PI3K)

Phosphoinositide-3-kinases (PI3Ks) are a class of enzymes phosphorylating lipids. In 1988, Cantley[1] found that PI3Ks phosphorylate the 3-hydroxy group in the myoinositol ring of phosphatidylinositol and produce phosphatidylinositol-3-phosphate. These lipids are members of eukaryotic cell membranes, where they play important roles in lipid and protein trafficking. The PI3Ks are divided into classes (I, II and III). The best known family member is the catalytic domain called p110α, which belongs to class Ia[2] (Fig. 2).

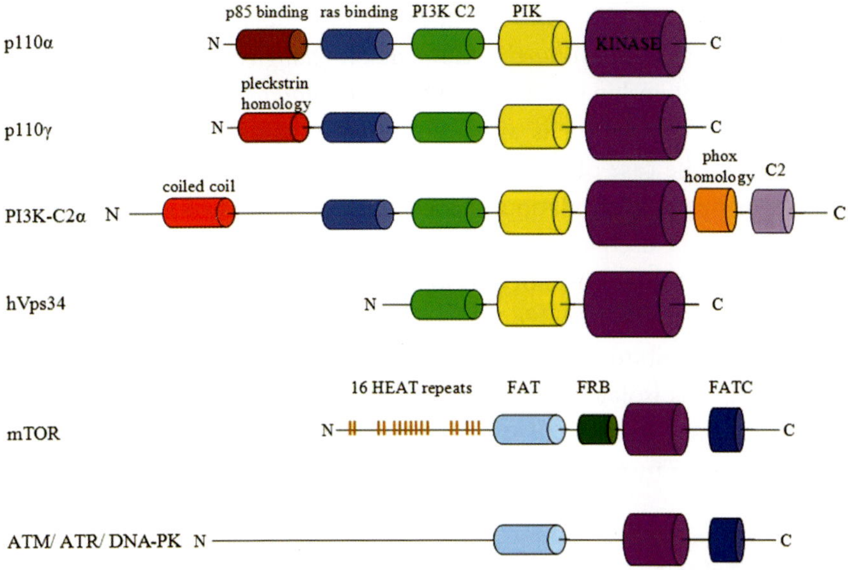

Figure 2 Classes of cell cycle inhibitors.

Class I: p110α, p110γ; Class II: PI3K-C2α; Class III: hVps34; Class IV: mTOR, ATM, ATR, DNA-PK

HEAT: a protein-protein interaction structure of two tandem anti-parallel a-helices found in Huntingtin, elongation factor 3 (EF3), the PR65/A subunit of protein phosphatase 2A and the lipid kinase Tor

FAT: a domain structure shared by FRAP, ATM and TRRAP

FRB: FKBP12/rapamycin binding domain

FATC: FAT C-terminus

PI3Ks are negatively-regulated by PTEN (*phosphatase and tensin homologue*), which is deleted, mutated or repressed in cancer cells. Hence, the persistent activation of PI3K downstream pathways stimulates proliferation by phosphorylation of Akt (*protein kinase B*). Akt down-regulates the expression of death genes and activates NF-κB (*nuclear factor kappa B*) and other proteins important for cell survival. Akt2 is overexpressed in ovarian, breast and pancreatic cancers and Akt3 is increasingly found in breast and prostate tumors.

A PI3K-related family, also called PI3K superfamily or sometimes class IV,[3] are the serine/threonine protein kinases, which consist of a similar sequence to PI3Ks (Fig. 2).[4] Members of this superfamily include ATM (*ataxia telangiectasia mutated*), ATR (*ataxia telangiectasia related*), DNA-dependent protein kinases and mTOR (*mammalian target of rapamycin*). Protein phosphorylation is an important mechanism in regulating the cell cycle. In the case of DNA damage, ATM and ATR activate p53 either directly or by phosphorylation of checkpoint kinases (Chk). Various PI3K inhibitors can be used to substitute for PTEN as a regulator and inhibit Akt activation and Chk phosphorylation. PI3K inhibitors interact with PI3K-related proteins as well (Table 1).

The first inhibitor derived from microbial origin was wortmannin (**1**) (Fig. 3).[5] It was isolated from *Penicillum funiculosum* and belongs to the class of viridins, which consist of a steroid-like skeleton with a lactone in the A ring and a furan connected to the A and B ring.

Table 1 Inhibitor-reactive Lysine in Different P13Ks

Protein	Binding Amino Acids	Sequence Length	Molar Mass/kDa
p110α	Lys-802	1068 aa*	110
p110γ	Lys-833	1102 aa	110
hVps34	Lys-636	887 aa	100
mTOR	Lys-2187	2549 aa	280
DNA-PK	Lys-3752	4128 aa	450
ATM	Lys-3016	3056 aa	350
ATR	Lys-2589	2644 aa	300

*Amino acids = aa

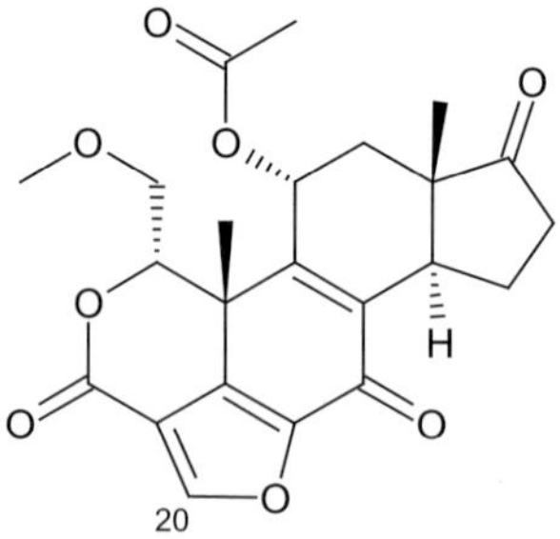

(**1**)

Figure 3 Structure of wortmannin (1).

Wortmannin is a potent inhibitor of p110α, but also inhibits PI3Ks from other classes as well as similar proteins such as DNA-dependent protein kinase, ATM, ATR, mTOR and phosphoinositide-4-kinases. It has not yet been tested in clinical studies because of its weak selectivity. In addition, it exhibits liver toxicity and is insoluble in water. The low therapeutic index is also a result of instability. Wortmannin's short half-life is due to early ring cleavage at C-20. Covalent, irreversible bonds are then formed between the ring-opened C-20 of wortmannin and Lys^{802} in the ATP binding site of p110α (Fig. 4).[6]

Because of the toxicity and instability of wortmannin, many derivatives have been synthesized. One possibility is to replace the lactone by the initial steroid ring A. The acetoxy group at C-11 and methyl group at C-13 are

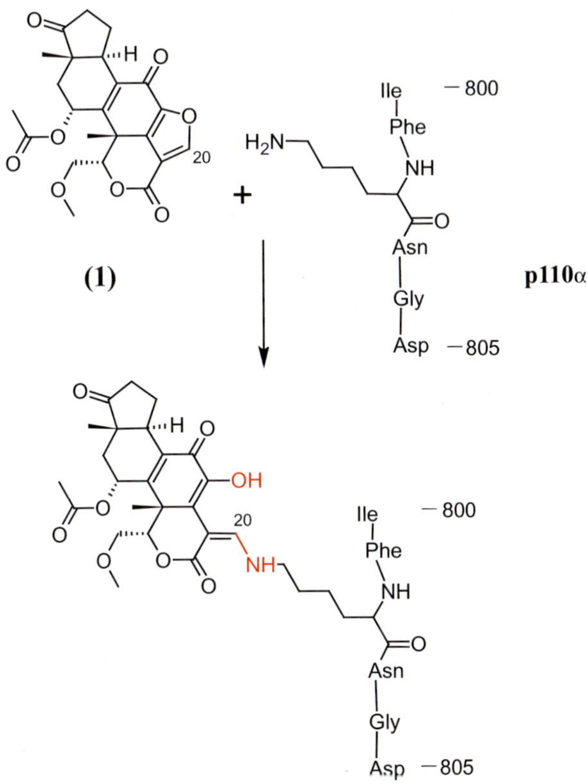

Figure 4 Inhibitory effect of wortmannin at Lys^{802} of p110α.

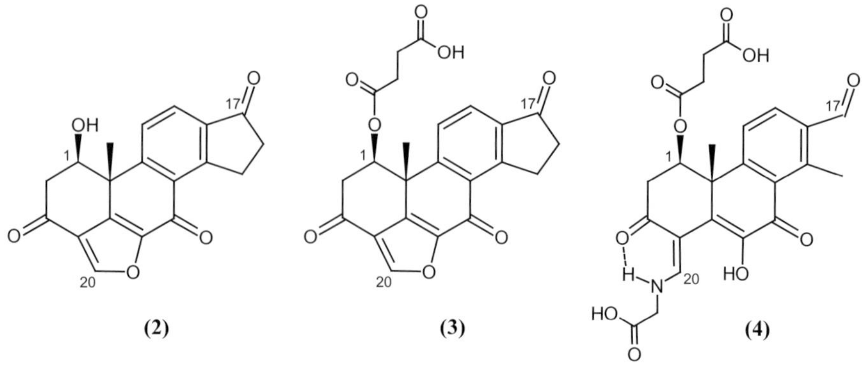

Figure 5 Structure of demethoxyviridin (2) and the substituted analogues (3) and (4).

deleted. Hence, the conjugated system of the furan and B ring is expanded with an aromatic C ring. At C-1, the methoxy-methyl group is replaced by a free hydroxyl group. The resulting compound is demethoxyviridin (2) (Fig. 5) and can be found in the fungus *Nodulisporium hinnuleum* as a natural source.[7]

Demethoxyviridin is as unstable as wortmannin, because of the same furan ring structure. However, the half maximum inhibitory concentration (IC_{50}) is 12-fold lower than that of wortmannin ($IC_{50} = 12$ nM). Therefore, alterations were performed at C-1 and C-20 to improve the half-life. At C-1 an esterification with succinic acid lead to enhanced stability in buffer solution. However, in liquid culture, medium stability decreased to about half a minute, a time similar to the half-life of wortmannin. In addition, the IC_{50} of esterified demethoxyviridin (3) is 1.7 times higher than that of its parent compound. The C-1 substitution thus neither showed increased activity against PI3Ks nor improved stability.

Additional alterations at C-20 of demethoxyviridin should lead to open-ring inhibitors of PI3Ks. A derivative formed by adding glycine by its N-terminus to the furan was investigated (4). Like wortmannin, analogues react with the moiety of Lys^{802} of p110α (Fig. 6). This compound with glycine results in improved stability in the presence of other free amino acids in cell plasma; the half-life increases to 64 minutes. Similar to the effect observed with esterified demethoxyviridin, the IC_{50} of the C-20-altered molecule is 3.6 times higher than that of wortmannin. The substance is less potent, but can still interact with lysine in the enzyme.

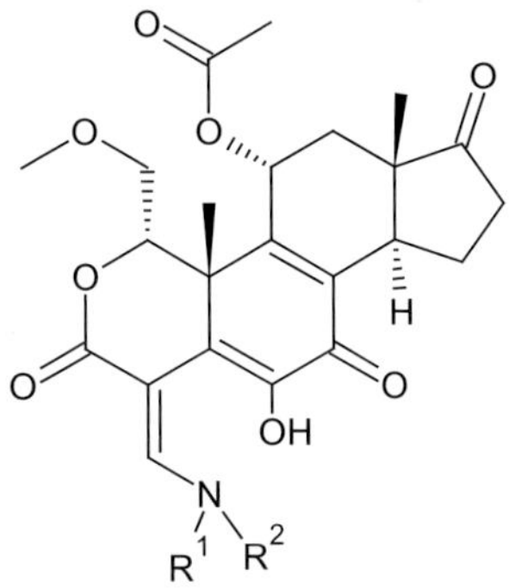

(**5**) $R^1 = -C(CH_3)_3$ $R^2 = -(CH_2)_2-OH$

(**6**) $R^1 = -(CH_2)_3-N(CH_3)_2$ $R^2 = -CH_3$

Figure 6 Structure of two wortmannin analogues with high therapeutic indices.

To conclude, the problem posed by wortmannin's instability[8] can be solved by using open-ring analogues. These modified structures do not lose their potency as PI3K inhibitors.[9] Inhibition is mediated by an exchange of the coupled amine for lysine in the ATP-binding site of the kinase. The toxicity of all ring-cleaved analogues is decreased in comparison to the parent substance, wortmannin, but there exists a wide range of different modifications that show the same inhibitory structure-activity relationship (SAR). By attachment of primary or secondary amines to C-20 the half-life in cells can be increased and reactions with free amines in plasma can be slowed. Esterase activity in cells deacetylates C-11. An analogue (5) with a tertiary butylethanolamine moiety was not a viable substrate for esterase. The derivative with a free hydroxyl group at C-11 was 20 to 100-fold less potent than its acetoxy analogue. This finding reveals that the furan ring is not the only site on the molecule important for inhibition. The C-11 substitution must interact with the pocket of ATP-binding site as well. Furthermore, primary amines are weak inhibitors, while secondary amines exhibit a 20-fold lower IC_{50}.

This higher potency can be explained by the fact that secondary amines can be easily substituted by primary ones under physiological conditions, like amino acids of the inhibited proteins. The primary amines are more reactive than the secondary ones.

For nucleophilic attack of lysine, it is irrelevant whether the amine moieties are polar or aprotic. Furthermore, SAR demonstrates that cyclic amines are weak inhibitors in comparison to aliphatic ones.

Another important point is that diamines are significantly more potent than their monoamine-substituted analogs. In summary, substance (**6**) is a more potent inhibitor than wortmannin, demethoxyviridin or other ring-opened analogues. Substance (6) has an aliphatic nitrogen moiety and contains a diamine and an acetylated form of C-11. It inhibited PI3K pathways and tumor growth in PTEN-negative glioma cells. The inhibitory effect was 426-fold more selective for PI3K than mTOR.[6]

Another ring-opened wortmannin analogue (PX-866) contains a diallylamine substitutent on C-20 (7) (Fig. 7).

PX-866 displayed comparable stability in solution to other ring-cleaved substances (Table 2). It inhibited Akt phosphorylation and decreased cell motility. In mice, growth of ovarian and lung tumors was inhibited without cytotoxicity. Liver toxicity was 65 percent less than the toxicity of wortmannin and renal toxicity disappeared. PX-866 completely blocked the growth of glioblastoma cell lines at a concentration of 10 nM and already 1nM of the compound showed a decrease in cell growth while higher concentrations of wortmannin are necessary to reach the same effects. Still, wortmannin was a better inhibitor of PIP3 (phosphatidylinositol (3, 4, 5)-trisphosphate)

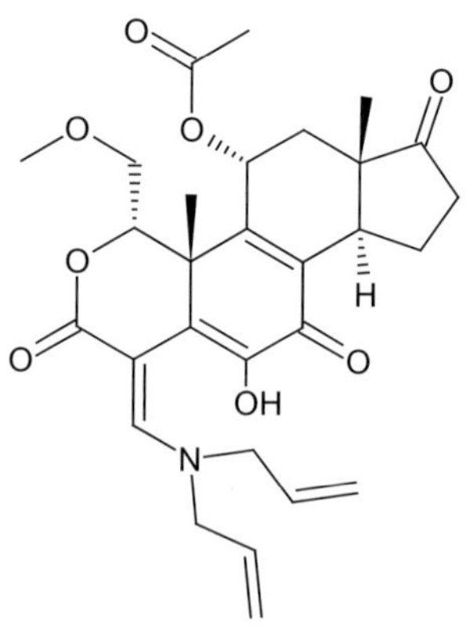

(7)

Figure 7 Structure of PX-866.

Table 2 Inhibition of P13Ks by Wortmannin and Related Compounds

Substance	IC_{50}	Plasma Stability
Wortmannin	12 nM	No
Demethoxyviridin	1 nM	No
esterified demethoxyviridin (substance **3**)	20 nM	No
esterified and open-ring demethoxyviridin (substance **4**)	44 nM	Yes
17-hydroxywortmannin	2.7 nM	No
substance **5**	1110 nM	Yes
substance **6**	6.4 nM	Yes
PX-866	0.1 nM	Yes

synthesis in PTEN-deficient cells than PX-866 was. For cell cycle regulation, however, lipid phosphorylation is less important than protein phosphorylation. Akt phosphorylation was most inhibited by Wortmannin in the first two hours of treatment. In comparison, PX-866 is effective over a longer period due to its higher stability. This viridin analogue is cytostatic rather than cytotoxic. It has been tested in combination with the established agent, cisplatin, and supra-additive effects have been found against cancer cells.

3. CHECKPOINT INHIBITORS (ChkIs)

There are many safeguards to avoid and repair DNA damage in the cell cycle. Cell cycle checkpoints allow cells to recognize DNA damage or other genotoxic stress and to initiate DNA repair or apoptosis. Checkpoint kinases (Chk) are serine/threonine kinases, which are activated by ATM or ATR. They are key regulators for cell cycle progression, and act by phosphorylating Cdc25 and controlling cyclin dependent kinases (CdK) to prevent progression of the cell into mitosis (G2 checkpoint) or synthesis phase (G1/S checkpoint) if DNA damage is present (Fig. 8).

In the case of DNA damage, normal cells stop cell cycle progression, which leads to p53 dependent G1 arrest (Fig. 9). However, p53 is frequently inactive in tumor cells and the only way to arrest cell division in the presence of DNA damage is by activating Chks, which results in mitotic catastrophe or apoptosis.

DNA damaging agents such as the topoisomerase I inhibitor camptothecin or the topoisomerase II inhibitor doxorubicin induce DNA damage

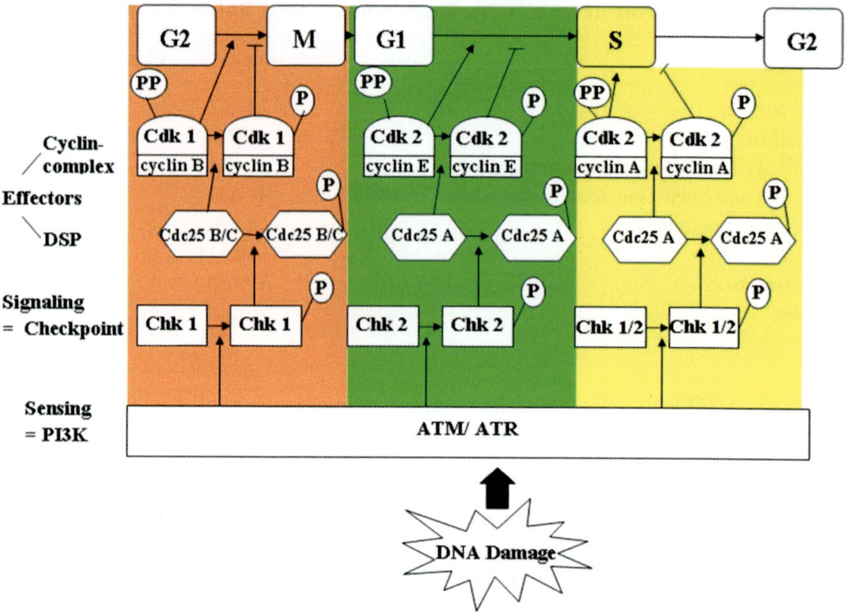

Figure 8 Signal cascade of cell cycle arrest after DNA damage.

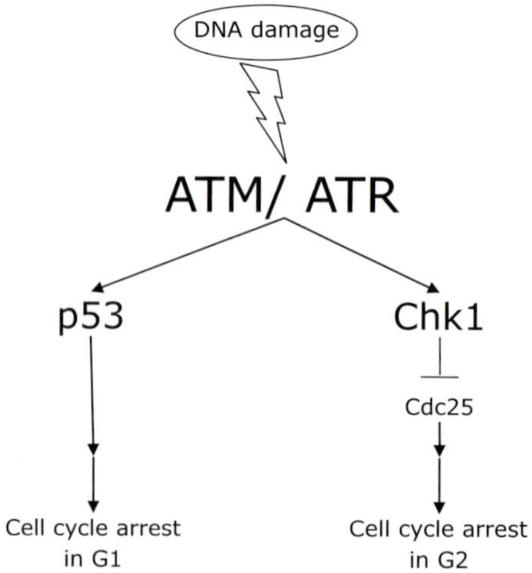

Figure 9 Cell cycle control after DNA damage.

during mitosis, but tumor cells are often drug-resistant. If checkpoint kinases are inhibited, drug resistance can be abolished and efficacy of these compounds can be increased.

Chk1 and Chk2 are structurally different but have similar functions. Chk1 is the important mediator for cell cycle arrest in the G2 phase. Inhibitors targeting both Chk1 and Chk2 decrease the adjuvant treatment effect of Chk1 inhibition by concomitant Chk2 inhibition. Hence, inhibitors against the major mediators of cell cycle control must be very selective.[10]

Another difficulty is that cyclin-dependent kinases (CdKs) have similar structures to Chks. Hence, unselective Chk inhibitors can also target CdK2. CdK inhibition can oppose/counter the desired effects of a Chk inhibitor.

The kinase domain of Chk1 consists of two subdomains connected by an elongated linker region (residues 84–89). The N-terminal domain consists of 88 amino acids and the C-terminus is made up of 210 residues. While the N-terminus predominantly consists of β-sheets, the secondary structure of the C-terminal subdomain comprises mostly helices. The enzyme is partly flexible in its inactive form but has a stable conformation upon binding of substrates or inhibitors. The nucleotide-binding pocket lies in the glycine rich loop (residues 16–23) in the N-terminal domain, while the activation loop (amino acids 148–175) is located in the C-terminus and is stable in the apoenzyme.[11]

Roughly 80 percent of the secondary structure of CdK2 is similar to that of Chk1. The differences between these protein kinases are essential for their individual selective activities. CdKs lack an activation loop and alanine is substituted for Ser.[147] In CdK2, the Asn is located in position 132 instead of position 135 as in Chk1, which changes the conformation of the nucleotide-binding pocket.

These differences can be used to generate selective inhibitors. The largest microbial class of Chk and CdK inhibitors is the indolocarbazoles, which are exemplified by staurosporine (**8**) (Fig. 10), a compound derived from *Streptomyces longisporoflavus.* Many indolocarbazoles inhibit several cell cycle enzymes by ATP-competitive kinase inhibition. Therefore, it is of great interest to identify indolocarbazoles with increased selectivity.

Another important kinase inhibitor is 7-hydroxystaurosporine (UCN-01) (**9**) (Fig. 10), which is less potent against CdKs than staurosporine and hence more selective. UCN-01 has an IC_{50} of 7 nM. It interacts with the

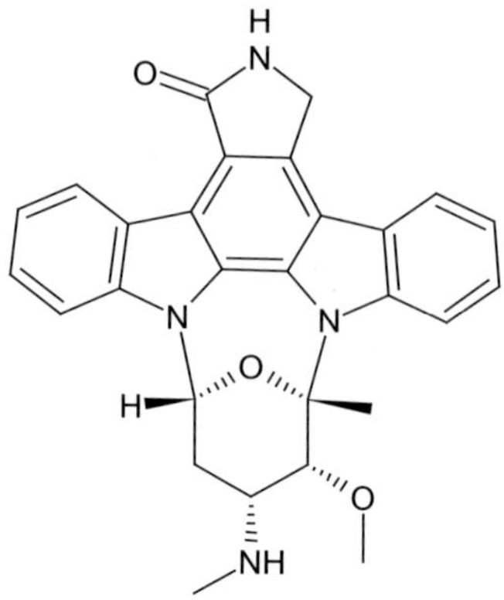

(8)

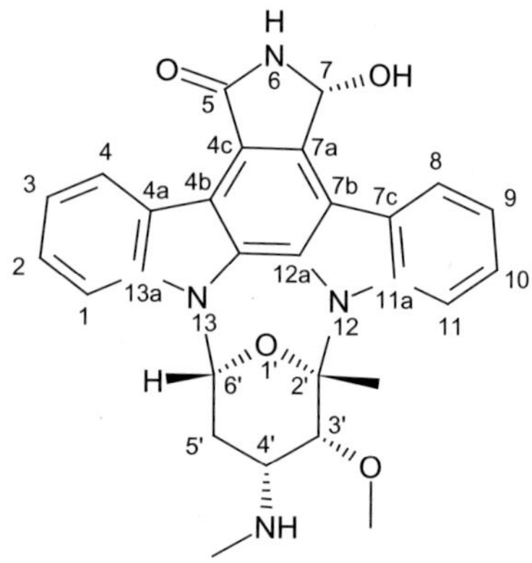

(9)

Figure 10 Structures of staurosporine (**8**) and UCN-01 (**9**).

enzyme via van der Waals forces and hydrogen bonds. In complex, the ATP binding site is exposed and the enzyme assumes a more stable conformation. The indolocarbazole's skeleton forms van der Waals linkages to the enzyme. One of the two phenyl groups interacts with the N-terminal backbone and the other site binds to N- and C-terminal hydrophobic side chains. The glycosyl moiety of the inhibitor is, as expected, in the ribose-binding site. In the linker region, hydrogen bonds connect the amide nitrogen (N-6) to the oxygen of the Glu^{85} backbone (Fig. 11). The carbonyl at C-5 forms hydrogen bridges with the amide nitrogen of Cys^{87}. These two linkages can be found in CdK interactions as well. These features as well as the methylamine group on the 4′ of tetrahydropyran, which is positioned on the side chain carboxyl group of Glu^{91} and main chain of Glu,^{134} act similar to staurosporine in the enzyme complex. The only difference, and the most striking characteristic for selectivity, is the hydroxyl group on C-7 of UCN-01. It interacts with the side chain of Ser^{147} in Chk1's nucleotide-binding pocket.[11] As already stated this amino acid is replaced by a hydrophobic moiety in CdK2 and thus eliminates the opportunity to build a further hydrogen bond.[12]

This inhibition was tested in combination with DNA-damaging agents to examine the adjuvant effect. p53 deficient tumor cells were treated with γ rays or cisplatin, and afterwards UCN-01 was added. Tumor cells with wild-type p53 activity were less sensitive to UCN-01 additional treatment. After γ

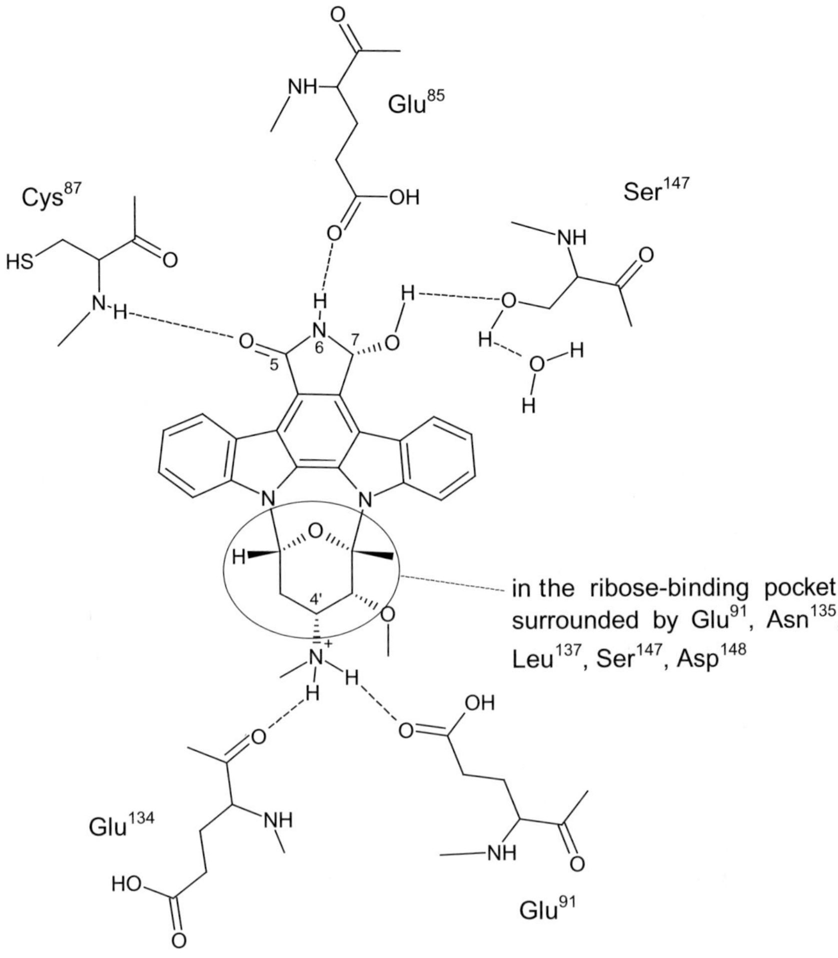

Figure 11 Interaction of UCN-01 with the ATP-binding pocket of Chk1.

irradiation or cisplatin treatment, most cells were arrested in the G2 phase of the cell cycle, which triggers resistance in tumor cells. UCN-01 overcame the G2 arrest and led to mitotic catastrophe. For complete escape from G2 arrest, 300 nM UCN-01 was necessary after irradiation; 50 nM UCN-01 led to mitosis after cisplatin treatment. Radiotherapy normally suppresses CdK1 activity, but UCN-01 treatment increased kinase activity back to normal levels by reduction of hyperphosphorylation. The combination therapy enhanced cytotoxicity in comparison with irradiation alone. *In vivo* experiments in mice with lymphoma or colon cancer also demonstrated 1.7-fold higher cytotoxicity for the combination therapy.[13]

Cisplatin (2 mM) cytotoxicity increased tenfold in combination with 50 nM UCN-01 as compared to cisplatin alone.[14]

Since the development of UCN-01, various other indolocarbazoles have been synthesized. Most were not derived from staurosporine as a starting compound, but rather from arcyriaflavin, which lacks staurosporine's glycosyl group and contains maleimide instead of lactam. Staurosporine and other bisindole alkaloids were obtained from *Actinomycetes* such as *Streptomyces* species. However, the fruiting bodies of Myxomycetes such as *Arcyria* species also produce indolocarbazoles. Arcyriaflavin has been isolated from *Arcyria denuadata* and *Arcyria obvelata.*[15] A possible method of synthesis would be to start with Arcyriaflavin A (**10**) skeleton, replace one indole moiety by an azaindole, and then alter the substituents on positions 9, 10 and on the maleimide nitrogen (Fig. 12).

The sugar moiety of UCN-01 is thought to bind to the ribose-binding pocket of Chk1, but aglycones also inhibit Chk1. Other substituents may be more important for binding. Substituents on the maleimide like N-methyl or N, N-dimethylaminoethyl are worth mentioning (e.g. compound (**11**)), because these analogues reveal considerable cytotoxicity, but lack inhibitory effects. A free maleimide is necessary for binding to the enzyme. Steric configuration hindered the binding of 9-Benzyloxy-azaindolocarbazole (**12**) to

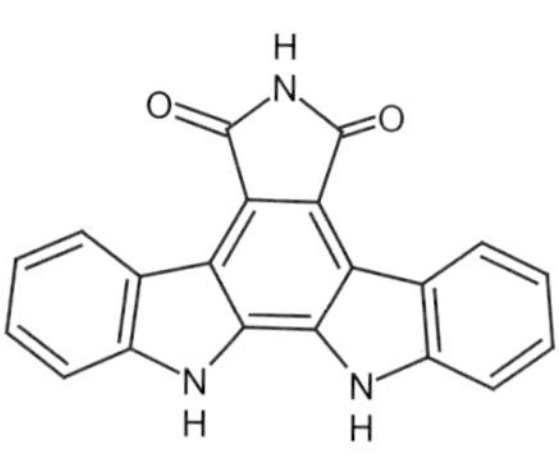

(**10**)

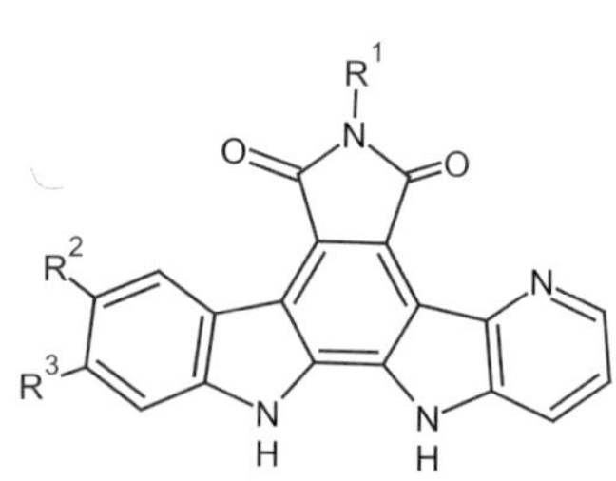

(**11**) R^1= $(CH_2)_2N(CH_3)_2$ R^2=H, R^3=$OCH_2C_6H_5$

(**12**) R^1= H R^2=$OCH_2C_6H_5$, R^3=H

(**13**) R^1= H R^2=H, R^3=$OCH_2C_6H_5$

(**14**) R^1= H R^2=H, R^3=OH

Figure 12 Structure of arcyriaflavin A (**10**) and 5-azaindole-analogues (**11–14**).

the ATP-binding pocket of Chk1, and 10-benzyloxy substituted azaindole-arcyriaflavin (**13**) was more potent than the unsubstituted azaindole. The 10-hydroxyanalogue (**14**) showed the best inhibitory activity and was half as cytotoxic (IC_{50} =14 nM) as UCN-01 (7 nM).[16] (Fig. 12).

Other indolocarbazoles such as isogranulatimide and its analogues can be found in marine invertebrates, e.g. *Didemnum* species.

4. INHIBITORS OF DUAL SPECIFICITY PHOSPHATASES (I-DSPs)

Phosphorylation and dephosphorylation are the main processes by which cell cycle proteins are activated or inactivated. The cell division cycle 25 enzymes (Cdc25) belong to class III protein tyrosine phosphatases (PTPs) and are dual specificity phosphatases, which can dephosphorylate proteins at both serine and threonine residues. Cdc25 proteins are activated by checkpoint kinases and dephosphorylate phosphotyrosine and phosphothreonine. Cdcs have oncogenic properties, e.g. Cdc25B is highly expressed in cancer cells. Inhibition of Cdc25B has the opposite effect to inhibition of checkpoint kinases. It arrests transition from G2-phase to mitosis.[17] Historically, Cdc25A was thought to dephosphorylate CdK1 complexed with cyclin E and Cdk2-cyclin A. The current understanding is that it induces G1/S-transition, while Cdc25B and Cdc25C activate CdK1-cyclin B complexes and induce G2/M phase-transition.[18]

Competitive inhibitors bind to phosphotyrosine or phosphothreonine by hydrogen bonding with cysteine and arginine in the catalytic loop.[19] Quinones or other compounds with numerous carbonyl groups or oxygen atoms fit into this loop because of their similar polarity to phosphate. Examples include tricyclic polyketide ortho-quinones such as nocardione A (**15**) (Fig. 13).

Nocardione A contains a naphto[1,2-b]furan-4,5-dione skeleton and is derived from a *Nocardia* species that is a gram-positive, rod-shaped and pathogenic bacteria.

Inhibitory activities against tyrosine phosphatases have been detected with extracts of both the mycelial cake and broth filtrate. Nocardione A inhibited Cdc25B with an IC_{50} of 7 μM, but was unselective against PTP1B at similar concentrations.[17]

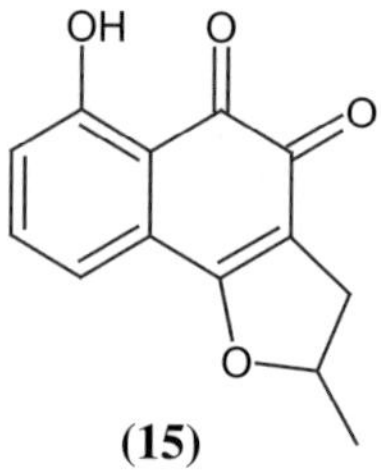

(15)

Figure 13 Structure of nocardione A (15).

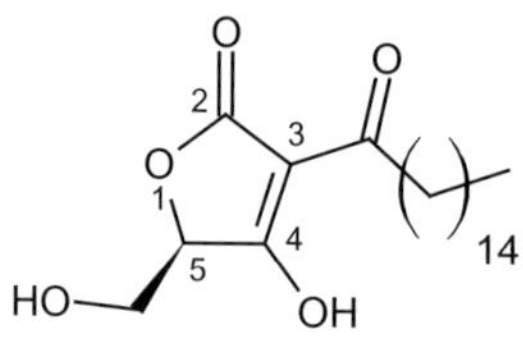

(16)

Figure 14 Structure of RK-682 (16).

A highly selective inhibitor is RK-682 (**16**) (Fig. 14). It is a tetronic acid with a saturated fatty acid at C-3 and a 5-hydroxymethyl moiety.

RK-682 (3-Hexadecanoyl-5-hydroxymethyltetronic acid) is derived from a *Streptomyces* species. It inhibits the activation of CD45 and VHR.[20] CD45 is a PTP while VHR is a DSP that inactivates ERK (extracellular-signal-regulated-kinase). ERK is a classical MAP-kinase and plays an important role in cell proliferation.[21]

A derivative with an unsaturated side chain at C-3 is able to inhibit Cdc25A and Cdc25B with an IC_{50} of 34 μM, while a derivative with a cinnamoyl moiety lacks inhibitory activity against PTPs and DSPs. Hence, C-3 modifications have considerable influence over the compound's inhibitory activity.[20]

The free hydroxyl at C-4 is also important for inhibition, while the stereochemistry of the C-5 hydroxymethyl moiety is irrelevant. In structure activity relationship analyses, carboxylate or phosphonate were first thought to mimic the substrate, but tetronic acid also fits into the loop.

Because all PTPs contain this loop, tetronic acid is unspecific. Since the neighboring domains differ among PTPs, alterations of C-3 and C-5

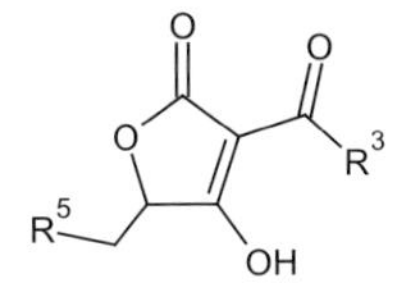

(16) $R^3 = (CH_2)_{14}CH_3$ $R^5 = OH$

(17) $R^3 = O(CH_2)_{13}CH_3$ $R^5 = OH$

(18) $R^3 = O(CH_2)_7CH_3$ $R^5 = OH$

(19) $R^3 = (CH_2)_{14}CH_3$ $R^5 = OOC\text{-}(CH_2)_8CH_3$

(20) $R^3 = (CH_2)_{14}CH_3$ $R^5 = OOC\text{-}C_6H_4\text{-(m)-}COC_6H_5$

(21) $R^3 = (CH_2)_{14}CH_3$ $R^5 = OOC\text{-}CN_2\text{-}COOCH_2CH_3$

Figure 15 Structures of Cdc25B inhibitors with tetronic acid cores.

moieties of tetronic acid increase its selectivity as an inhibitor. Length and saturation of hydrophobic side chains at C-3 are not important for inhibition of Cdc25B. Fatty acid moieties at C-3 (**17**) produce inhibitors which are less potent than acyl moieties (**16**) and a fatty acid chain with more than 7 carbons at C-5 results in a better inhibitor than shorter moieties (**18**) (Fig. 15).

If the hydroxymethyl group at C-5 is changed into an acetyl or long chain fatty acid (**19**), the IC_{50} value decreases from micromolar to nanomolar concentrations. The best inhibitors for Cdc25B possess benzoyl groups (**20**) and have IC_{50} values around $0.5\,\mu M$. A tetronic acid with a diazomalonyl group on C-5 (**21**) showed considerable inhibition ($IC_{50} = 0.4\,\mu M$) in comparison to other compounds with a tetronic acid core and was 30 times more selective for Cdc25B than for VHR,[19] other DSPs, or phosphatases such as PTP-S2[22] (Fig. 15).

5. INHIBITORS OF CYCLIN-DEPENDENT KINASES (CdKs)

Cyclin-dependent kinases are serine/threonine kinases regulating the cell cycle. Similar to checkpoint kinases, they contain an N-terminal region with β sheets and a C-terminal region with α helices. The ATP-binding pocket is located in the linker region between the two lobes. CdKs complex with cyclins for activation; the kinase domain represents the catalytic subunit in the complex, while the cyclin is the regulatory subunit. At present, 9 CdKs and 11 cyclins have been identified. The different complexes fulfill different functions; CdKs 1, 2, 4 and 6 are cell cycle-related CdKs, while the others are involved in transcriptional functions.[23] CdK2 complexed with cyclin E regulates G1/S transition, but CdK2 acts during S-phase if complexed with cyclin A.[24] The various complexes are regulated by phosphorylation and dephosphorylation. To active a CdK2/cyclin complex, Thr^{160} is phosphorylated by H-CdK7/MAT1, and the dual specificity phosphatase Cdc25 dephosphorylates Thr^{14} and Thr^{15} of CdK2.

CdK2 is an intensively studied cyclin-dependent kinase, but it is not suitable as a therapeutic target because it is not essential to cellular function. If CdK2 is not expressed, cyclin E can target another CdK that can replace CdK2 in function.[25] In addition, few CdK2 inhibitors are selective enough to be effective due to CdK2's similarity to other kinases, e.g. Checkpoint Inhibitors (ChkIs).

A better target may be CdK4, which plays a role in mid-G1 phase regulation. In normal cells, a CdK4/cyclin D complex phosphorylates retinoblastoma protein (Rb) causing the transcription factor E2F/DP to stay active. Rb is a tumor suppressor that binds to E2F/DP and stops entry into S phase.[26]

In cancer cells, CdKs or/and cyclins are often overexpressed or mutated. In the CdK4 pathway, the Rb gene can be mutated or deleted. There are also endogenous inhibitors such as INK4 (inhibitor of CdK4) or CIP-KIP (CdK inhibitor protein-kinase inhibitor protein) which are frequently mutated, deleted or silenced in tumor cells.

As already mentioned, indolocarbazoles are potent kinase inhibitors that function by competitive blockage of ATP-binding sites. The interaction in the binding pocket via hydrogen bonds is similar to the one shown in Fig. 11.

To test the inhibitory effects of the indocarbazoles against D1-CdK4, phosphorylation of Rb was measured[27] Selective inhibitors were identified by derivatization from indolocarbazoles, but without the sugar moiety found in UCN-01 or staurosporine. Both maleimide and carbonyl groups are important for enzyme interaction. Reduction to an alcohol or methylene group at one carbonyl decreased potency.

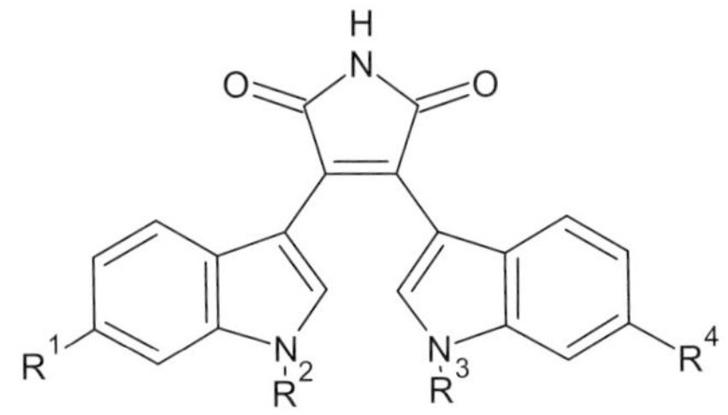

(22)

Figure 16 Structure of bis-indolylmaleimides.

Arcyriaflavin A (**10**) was frequently used as parent compound. It is a naturally occurring substance that is potent against CdK4, inhibits cell cycle progression in G1 phase and is selective against E-CdK2.

To sum up, indolocarbazoles are better inhibitors than bis-indolylmaleimides (**22**) (Fig. 16) and exert stronger antiproliferative activity. Moreover, bis-indolylmaleimides are less selective against E-CDK2 and inhibit G2/M transition rather than G1 arrest. Consequently, it may be possible that bis-indolylmaleimides inhibit targets other than D1-CdK4. Therefore, the Arcyriaflavin skeleton was used as a starting point to generate better inhibitors. Alkylation at one indole nitrogen leads to increased inhibition of CdK4. Polar groups at the alkyl moiety give rise to even more potent inhibitors. The most potent alkylated indolocarbazoles can inhibit other kinases as well, so they are unselective. Arcyriaflavin A and compounds (**23**)–(**25**) (Fig. 17) were CdK4 inhibitors with IC_{50} values between 80 and 180 nM.[26]

Furthermore, the indole ring can be replaced by another aryl or heteroaryl ring. The easiest alteration is to use naphthalene, which can be connected in three different ways (Fig. 18).

Only the first compound, in which a naphthyl-moiety is linked to maleimide, revealed good inhibitory effect and selectivity; the other two substances were less potent. Compounds with a phenyl ring, or tetrahydronaphthalene or heteroaryl rings, such as imidazole, pyridine or thiophene, were less potent as well.[28] Inhibitory activities were also observed with up-angular naphthyl[2,1-a]carbazoles such as isoquinolinyl[6,5-a]pyrrolo[3,4-c]carbazole (IC50 = 69 nM) and carbazoles with a nitrogen

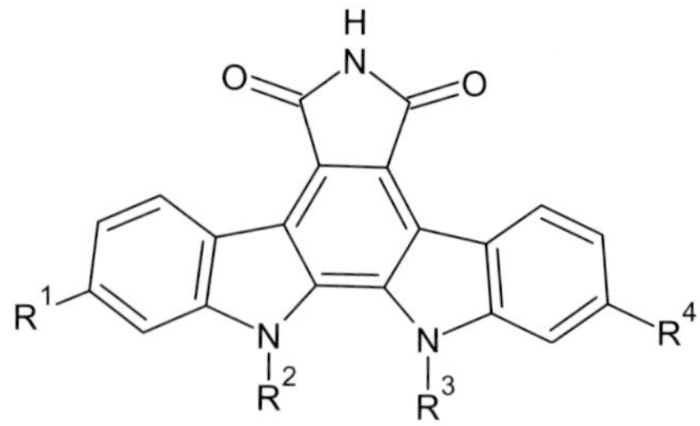

(10) R^1= H R^2= H R^3= H R^4= H Arcyriaflavin A

(23) R^1= H R^2= CH_3 R^3= H R^4= H

(24) R^1= H R^2= H R^3= $(CH_2)_3N(CH_3)_2$ R^4= H

(25) R^1= OCH_3 R^2= H R^3= H R^4= OCH_3

Figure 17 Structures of indolo-[2,3-a]pyrrolo[3,4-c]carbazoles.

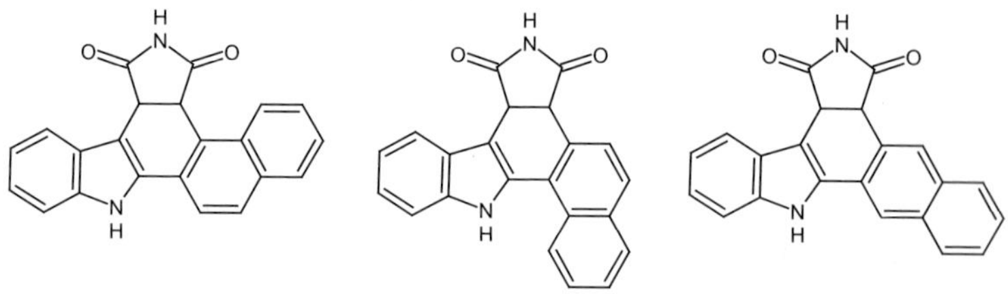

against D1-CdK4: IC_{50} = 45nM IC_{50} = 430nM IC_{50} = >10μM

Figure 18 Three ways to replace indole by naphthalene.

in 4-position of the isoquinoline (**26**) (Fig. 19). Quinolinyl[a]pyrrolo [3,4-c]carbazoles and other isoquinolinylcarbazoles are weaker inhibitors with micromolar IC_{50} values.[29]

In another investigation, the indole ring was connected to the maleimide not by the pyrrole but by a benzene ring and substituted on both indole nitrogens and on the phenyl (Fig. 20).

Substituents larger than ethylene at N-5 decrease the compound's ability to inhibit D1-CdK4. C-9 and C-10 substitutions also lower the potency except in the case of fluorine, but this derivate lacks an antiproliferative effect. The most tolerant position for substituents with minor effects on CdK4 inhibition is C-11. Hydroxyalkylation at C-12 results

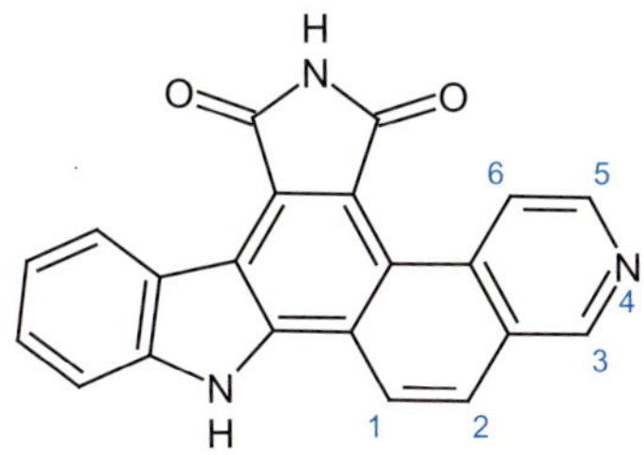

(26)

Figure 19 Structures of isoquinoliny l[6,5-a]pyrrolo[3,4-c]carbazole.

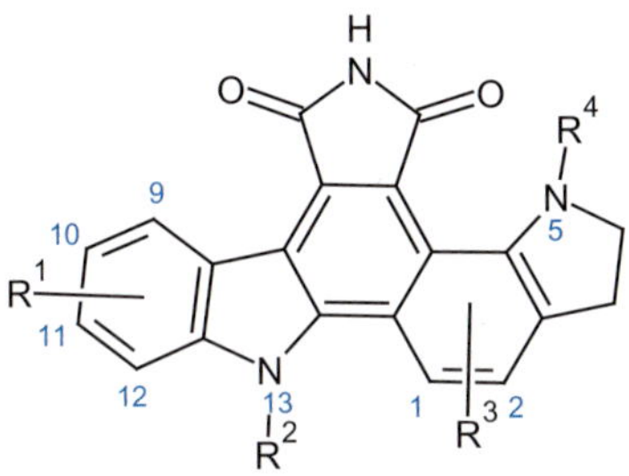

Figure 20 Structure of indolo-[6,7-a]pyrrolo[3,4-c]carbazoles.

in potent inhibitors, which are selective towards E-Cdk2 and PKA. However, hydroxalkylation at N-13 leads to E-Cdk2 inhibitors. This demonstrates that the bond with C-12 substituents in the ATP-binding pocket is very specific yet does not interact with the same substituents if they are at N-13. In addition, if there is an alkyl moiety at C-1 and also a substituent at N-13, the inhibition is also more selective for E-Cdk2 versus D1-Cdk4. All these mentioned compounds, however, lack aqueous solubility and hence show little antiproliferative effect in cells. Therefore, more polar C-12 aminoalkylated compounds possessing more than 20-fold better selectivity against CdK2 were designed. These compounds are also more selective towards 36 other kinases, although they do not show specificity for two lymphocytic kinases Lyn and Blk. Aminoalkylated compounds such as (27) (Fig. 21) are antiproliferative and the aqueous solubility is further increased by converting them into their methanesulfate salts, which are quite stable in solution.[27]

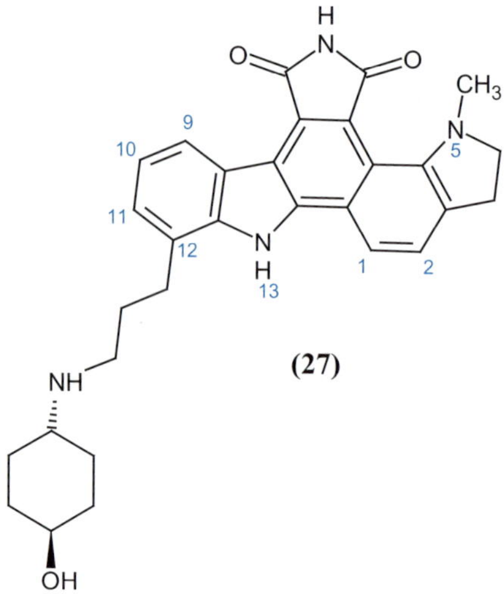

(27)

Figure 21 Structure of 12-aminoalkylated indolo-[6,7-a]pyrrolo[3,4-c]carbazoles.

6. INHIBITORS OF HISTONE DEACETYLASES (HDACs)

Histone deacetylase (HDAC) is a multi-protein complex that suppresses gene expression. In cooperation with its counterpart histone acetyltransferase (HAT), it regulates the acetylation of histone proteins and other substrates, influencing chromatin structure and regulating gene availability for transcription.[30]

HDAC catalyses the cleavage of the acetyl group from lysine residues on histones, leaving the residues with a positive charge. The negatively charged DNA strands are then attracted to the positively charged histones. This close interaction between the histones and the DNA can prevent gene transcription by limiting transcription factors' spatial access to the DNA. In this way, HDACs can regulate the availability of genes for transcritption.[31]

The balance of histone acetylation is maintained by two groups of enzymes, HDAC and HAT. The process of lysine acetylation and deacetylation is post-translational and reversible.[30] The acetylation of histone lysine residues is important both in normal and cancer cells because it causes active and inactive gene segments to develop. The 11 classical human HDACs are classified by sequence and domain relationships into four classes according to their homology to yeast proteins. Histone deacetylases participate in various protein and co-repressor complexes.[30]

Class I histone deacetylases (HDACs 1, 2, 3 and 8) are homologous to the yeast protein RPD3. They are expressed in many tissues, including heart, liver and skeletal muscle.[32,33]

Class II HDACs (4, 5, 6, 7, 9 and 10) are homologous to the yeast protein HDA1 (histone deacetylase 1). They are expressed in a tissue-specific manner.[34] HDAC4 and HDAC5 are expressed mainly in brain, heart and skeletal muscle, while the expression of HDAC6 is high in heart, liver, kidney and pancreas.

Class III, includes SIRT1-7, which differs structurally from other HDACs. It is named after its yeast-homolog SIR2 (silencing information regulator 2), and requires NAD^+ as a cofactor. Classical HDAC inhibitors cannot inhibit SIR2.[35]

Class IV consists of HDAC11, which regulates interleukin-10 expression in antigen-presenting cells.

Increased HDAC activity is observed in cancer cells, which leads to changes in local chromatin structure, alterations in gene transcription and impaired differentiation. Therefore, HDAC inhibitors (HDI) are novel candidates for cancer treatment. They lead to an increase in histone acetylation and thus convert the DNA in a more open and transcriptionally active conformation. This in turn increases gene activation, ultimately resulting in cell differentiation and/or apoptosis.

By crystallographic methods, it was found that the active site of histone deacetylases contains a zinc ion.[36] All HDAC structures have a long, hydrophobic binding pocket channel with the zinc ion at one end. The zinc is coordinated by two aspartate residues and a histidine residue.[36] It is directly involved in the catalytic mechanism of deacetylation and chelation of the zinc ion plays an essential role in the binding of inhibitors.[37] Apart from coordinating the zinc ion, HDIs can also spatially block the binding pocket channel so that the natural substrate cannot be bound. HDIs thus inhibit HDACs reversibly via a competitive mechanism.[38,39]

The already known HDAC inhibitors have been categorized into five different groups based on their chemical structure: hydroxamic acids, epoxides (cyclic tetrapeptides with AOE (2-amino-8-oxo-9,10-epoxydecanoic

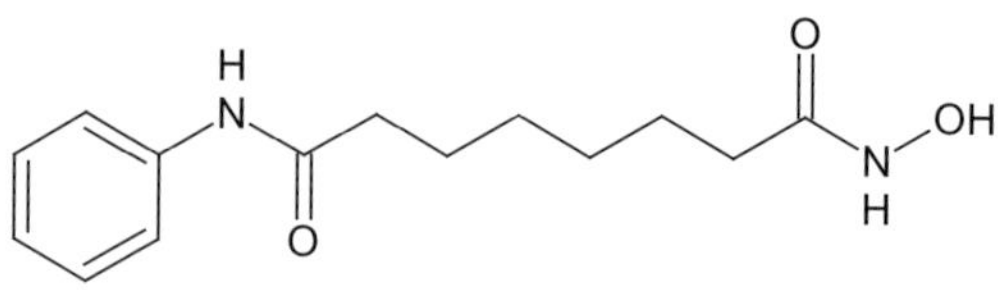

Figure 22 Structure of trichostatin A.

acid) group), cyclic tetrapeptides without AOE radical, short-chain fatty acids and benzamides.

Trichostatin A (TSA) is a member of the hydroxamic acid (Fig. 22), the largest group of HDIs. TSA is a product of *Streptomyces hygroscopicus* and was initially developed as a fungicide in 1990. It acts at nanomolar concentrations and is considered to be one of the most potent HDIs.[40,41]

TSA induces cell cycle arrest both in G1 and in G2 phases via the following mechanisms: HDAC inhibitors induce expression of the CDKN1A gene, which codes for the cyclin-dependent kinase inhibitor p21 (cdk = cyclin dependent kinase).[42] p21 causes cell cycle arrest in the G1 phase. SP1 binds directly to HDAC1 and co-mediates suppression of gene expression. Sp1 enhances its own expression by participating in a NFκB/HDAC complex.[43] The fact that TSA induces p21 indicates that the binding of TSA directly activates the CDKN1A promoter.[44,45] SP1 is also necessary for transactivation of hTERT, a gene that encodes catalytic subunits of telomerase. The enzyme telomerase protects the ends of chromosomes (telomeres) from degradation. The inhibition of transcription of hTERT promoter by HDACs completely abolishes expression of telomerase. TSA induces telomerase activity in normal cells but not in cancer cells. The HDAC inhibitor activates the hTERT promoter in normal cells, in which Sp1 plays a key role.[46] SP1 interacts not only with HDACs, but also with other transcription activators. For example, recent studies have shown that SP1 interacts with p300, a histone acetyltransferase.[47] SP1-dependent gene regulation is thus controlled by a competition between HDACs and other activating factors.

Trichostatin A has a number of limitations as an HDAC inhibitor. It cannot inhibit class III HDACs, since they have a different mechanism of action. For clinical use, TSA is not suitable because of its toxicity and low specificity. It is used rather as a reference in the pre-clinical research.

Amide analogues of TSA, such as MD85, oxamflatin or scriptaid are also inducers of differentiation and/or apoptosis in low micromolar concentrations.

Suberoylanilide hydroxamic acid (SAHA) is another member of the hydroxamic acids (Fig. 23). SAHA inhibits HDACs at micromolar concentrations and causes G1 and G2 cell cycle arrest.[48,45] The replacement of the hydroxamic acid structure of SAHA by a carboxylic acid or an amidoxime structure results in an inactive compound. N-methylation or the introduction of a methyl group leads to a loss of activity. The optimal length of the spacer was determined to be six methylene groups.

The natural product trapoxin B is an isolate from *Helicoma ambiens* (Fig. 24). It is an epoxy and irreversibly inhibits histone deacetylases in a nanomolar range leading to cell cycle arrest in the G1 and G2 phases.[40] The effect of this cyclic tetrapeptide is attributed to the epoxy-keto structure at the end of the hydrophobic chain, which covalently binds to HDAC. HDAC6 is resistant to trapoxin B.[49]

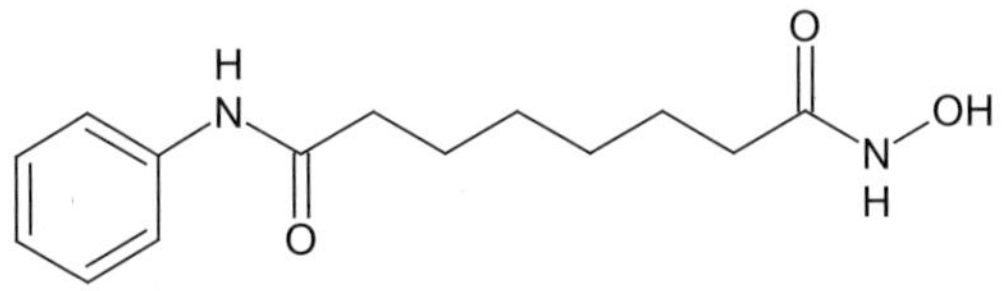

Figure 23 Structure of suberoylanilide hydroxamic acid.

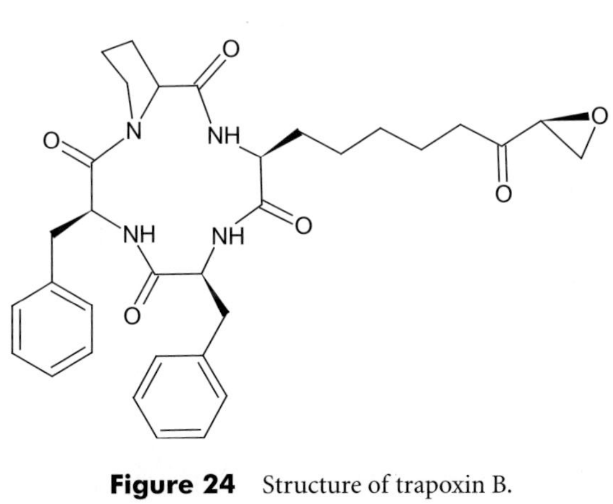

Figure 24 Structure of trapoxin B.

Trapoxin B has several analogues that inhibit HDAC in a different way.[50] HC-toxin, a natural product from *Cochliobolus carbonum*, inhibits reversibly despite its epoxy-keto structure. The epoxy-keto structure was initially considered to be essential for inhibition because compounds without it were ineffective. Apicidin from *Fusarium pallidoroseum* and Apicidin A, which have a simple keto-enol function, are also potent inhibitors. In these compounds, the tryptophan residue is important for the inhibitory effect.

FK228, better known as romidepsin or depsipeptide, belongs to the group of cyclic tetrapeptides without an AOE radical (Fig. 25). It is a natural product of *Chromobacterium violaceum*. FK228 nitiates cell cycle arrest in the G1 and G2-phases.[51] Garlic ingredients and their metabolites such as diallyl sulfide and allylmercaptan are also HDAC inhibitors, which suggest that the zinc ion in the active site of an HDAC is complex after the opening of the disulfide bond by thiol groups.

Butyrate belongs to the group of short-chain fatty acids and is a reversible HDI (Fig. 26). It induces cell cycle arrest in the G1 phase, a hyper-acetylation of histone H4 [52] In addition, butyrate induces apoptosis.[53,54]

The therapeutic value of butyrate is limited by its low activity (in millimolar concentrations) and the rapid *in vivo* metabolism of the compound.

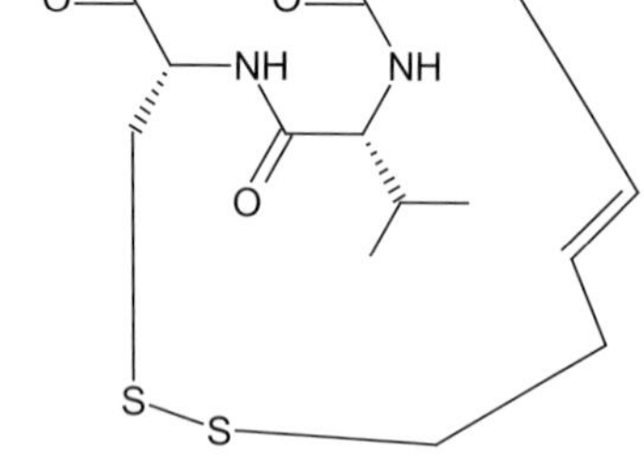

Figure 25 Structure of FK228.

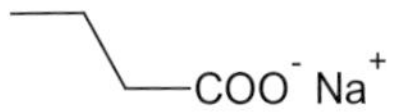

Figure 26 Structure of butyrate.

Administration of prodrugs, such as tributyrin or pivaloyl oxymethyl butyrate, may circumvent this problem.

7. TOPOISOMERASE INHIBITORS

Topoisomerases were discovered in 1979.[55] They are DNA-cutting enzymes that change the topology, e.g. the degree of twisting in the DNA strand. During replication, recombination, and chromatin aggregation, topoisomerases reduce the tension in the twisting helix by nicking the phosphodiester backbone and allowing the helix to unwind. Type I and Type II topoisomerases use a tyrosine to form temporary phosphate esters and are structurally and evolutionarily related to each other.[56,57] During DNA-splicing, a covalent DNA-protein bond is formed between the tyrosine-hydroxyl group of the topoisomerase and the phosphate of DNA.

Type I topoisomerases are monomers that cause a DNA single strand break. They can be divided into two groups: the predominant prokaryotic type Iα (Topo III) and the eukaryotic split-type Iβ (Topo I).[58] Type Iα can be subdivided into Topo IIIα and Topo IIIβ. Type Iβ has its subunit Topo I. Type Iα enzymes are only able to relax negatively twisted superhelical DNA. They form an intermediate bond with the 5′ end of the cleaved strand and require Mg^{2+} for activation. Type Iβ topoisomerases relax both positive and negative superhelices and form an intermediate bond with the 3′ end of the cleaved strand. Type Iβ topoisomerases do not require Mg^{2+}.[57]

Type-II isomerases consume ATP and break both DNA strands. They are protein dimers that form covalent bonds with the 5’ end of DNA. There exist two isoforms: topoisomerase IIα and topoisomerase IIβ.[59] Both are present as homodimers and relax both positive and negative superhelices.[57] Topoisomerase II enzymes consist of an N-terminal domain for binding and hydrolysis of ATP, a central domain for the separation and subsequent ligation of the DNA strands and a C-terminal region. The C-terminal domain is not involved in enzymatic activity, but it likely regulates the localization of the enzyme to the DNA. The Topo II dimer forms two large pockets into which the DNA double helices are positioned.[60] An important mechanistic aspect of topoisomerase activity is the formation of covalent protein-DNA complexes.[61]

Inhibitors of topoisomerase initiate apoptosis of cancer cells during the S-phase. They stabilize the DNA-topoisomerase complex that the two DNA strands are not able to reconnect again. Thus, the DNA is no longer functional or able to be replicated. Cancer cells divide much more frequently than do healthy cells. Therefore, topoisomerases inhibitors initiate apoptosis more quickly in cancer cells.[62,63]

A predominantly hydrophobic pocket of topoisomerase-I was identified in the protein-DNA complex as the major site for interaction with inhibitors. This pocket is situated opposite the DNA's minor groove, and inhibitor molecules of various sizes interact with the hydrophobic pocket on the one side and the minor groove on the other side.[64]

Several alkaloids can act as topoisomerase-I inhibitors. Camptothecin (CPT), a quinolinalkaloid derived from the tree *Camptotheca acuminata*, inhibits topoisomerase I in cells in the S-phase (Fig. 27). It consists of a pentacyclic ring system that includes a pyrrole-(3, 4 β)-quinoline-unit and an asymmetric center in the α-hydroxyl-lactone-ring with 20S-configuration. The alkaloid connects via hydrogen bonds to the DNA on one side and to the topoisomerase I on the other side. The connection to the DNA is between the lactam ring of camptothecin and a cytosine of DNA. On the other hand, the amino acid asparagine and arginine residues of topoisomerase can bind to the ring E of camptothecin. Frequently used CPT analogues are the water-soluble topotecan and irinotecan.[65,66]

The indolocarbazoles, which are related to the antibiotic rebeccamycin, represent a second class of topoisomerase I inhibitors.

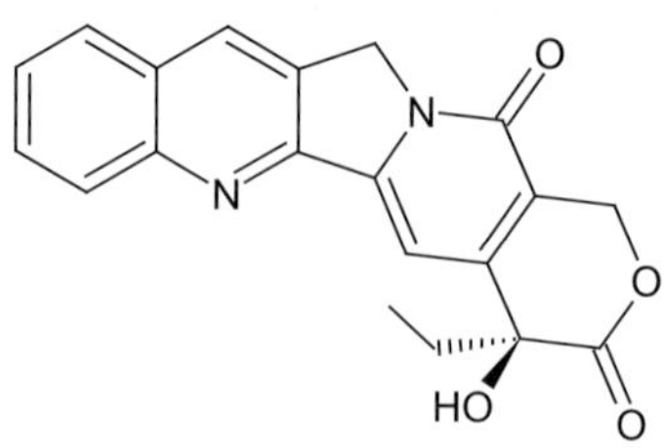

Figure 27 Structure of camptothecin.

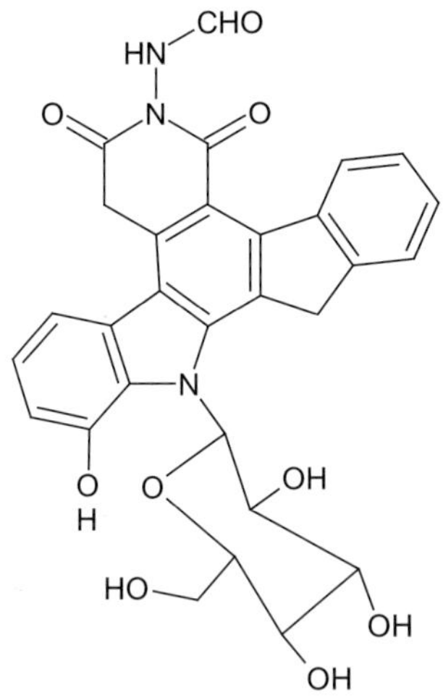

Figure 28 Structure of NB-506.

Camptothecin and indolocarbazoles have similar binding sites, which leads to cross-resistance. NB-506 (6-N-formylamino-12,13-dihydro-1,11-dihydroxy-13-(D-glucopyranosyl)-5H-indolo [2,3-a] pyrrolo [3,4-c] carbazole-5, 7-(6H)-dione) is water-soluble and is the most active example of this group (Fig. 28). This indolcarbazole preferentially intercalates within GC-rich sequences of DNA.[67]

Compounds such as benzimidazole and indenoisochinoline are also topoisomerase I inhibitors.[68] Although they display good membrane permeability, they lack the desired level of specificity. Their cytotoxicity correlates with their DNA-binding affinity.[69] Other topoisomerase-I-inhibitors, such as boswellic acid derivates and thiazole-containing peptides, which both bind to DNA, and directly inhibit the topoisomerase I.[70]

The topoisomerase II inhibitors are divided into two classes. The anthraquinone derivatives daunorubicin, doxorubicin, idarubicin and epirubicin, the anthacendione mitoxantron function through indirect inhibition, and the epipodophyllotoxins etoposide and teniposide function through direct inhibition (Fig. 29).[71]

Doxorubicin intercalates into the DNA in a sequence-dependent manner. Its position between bases of the double helix is stabilized by hydrogen

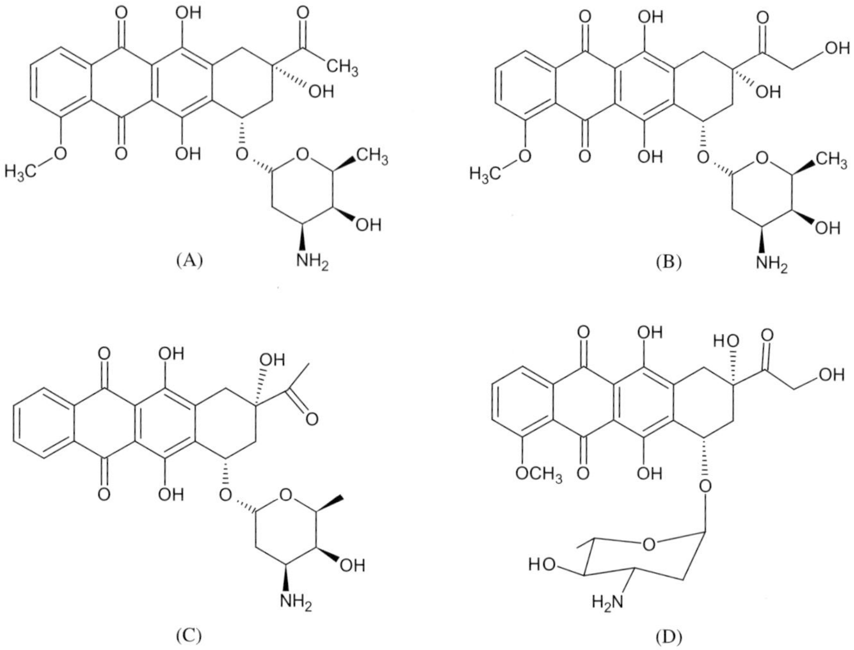

Figure 29 Structures of daunorubicin (A), doxorubicin (B), idarubicin (C) and epirubicin (D).

bonds. The formation is induced when the DNA-base adjacent to doxorubicin and the NH2-group of the angled amino-sugar come into contact with the minor groove of the DNA. If topoisomerase II encounters intercalated doxorubicin, the two together form a cleavage complex. Due to replication or transcription processes, the topoisomerase is released, leaving a double-stranded DNA break.[72]

Anthracyclines inhibit topoisomerase II in an indirect manner. However, epipodophyllotoxins, such as etoposide (Fig. 30), are able to inhibit the enzyme directly. They "capture" the cleavage complex and stabilize it at the stage in which the double strand break is set.[73] The resulting ends of the DNA strands are covalently linked to topoisomerase II. Hence, the passage of the second strand through the fist and subsequent ligation of the strand break cannot take place. As with anthracyclines, topoisomerases bound to epipodophyllotoxins can be removed from the DNA by replication

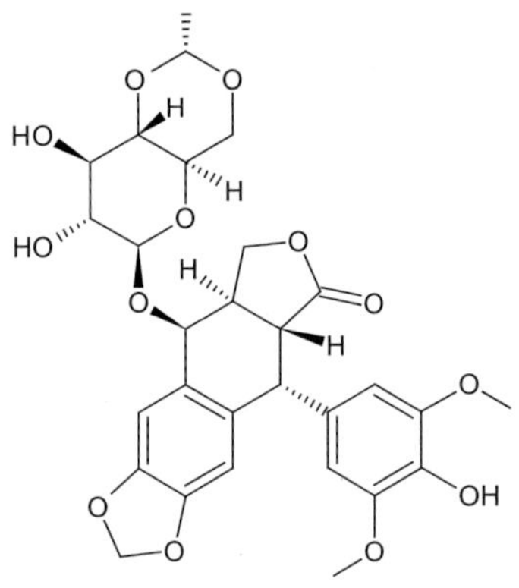

Figure 30 Structure of etoposide.

or transcription machinery. This, however, also leaves double strand breaks.[74,75]

8. GROWTH FACTORS

Growth factors are a diverse group of proteins. They have widely varying tasks and effects throughout the body, from influencing cell proliferation to promoting differentiation of progenitor cells. They can also initiate cell cycle arrest and apoptosis, and interact with both stem cells and somatic cells.

Types of growth factors include epidermal growth factor (EGF), which controls cell growth; vascular endothelial growth factor (VEGF), which controls blood vessel development; platelet derived endothelial growth factor (PDGF), which influences the cell growth and blood vessel development; and fibroblast growth factor (FGF), which controls cell growth.[76]

Activation of glycoprotein growth factor receptors (tyrosine kinases) activates intracellular signalling cascades that induce gene expression in target tissues.

Overexpression of either receptors or growth factors stimulates proliferation, which can lead to tumor development. Autocrine growth factors provide independent self-stimulation of tumor cells.[77]

There are different ways in which growth factor inhibitors can influence cancer cells.

The epidermal growth factor (EGF) plays a well-known role in tumor development. It is also responsible for epidermal proliferation and wound healing. Therapy strategies focus either on inhibiting the activation of the signalling cascade by using monoclonal antibodies that compete with EGF for the active site of the EGF receptor (EGFR), or on interfering with the signal cascade after receptor activation by treatment with small molecules. Classification of the different inhibitors is based on the protein group they inhibit: tyrosine kinase inhibitors (TKI) and proteasome inhibitors.[78] Tyrosine kinases inhibitors stop cell division and cause cell cycle arrest.[79]

Several TKIs competitively inhibit ATP-binding to EGFR. Thus, autophosphorylation of the receptor cannot take place and as a result tumor growth stagnates or regresses. Examples are ZD1839 (gefitinib) (Fig. 31) and OSI-774 (erlotinib),[80] which block cell cycle progression in the G1 phase and induce apoptosis. They belong to the group of 4-anilino-quinazoline.[81,82]

Two other groups of TKIs are 4-[Ar(alk)ylamino] pyridopyrimidine and 4-phenylaminopyrrolo pyrimidines. Other TKIs irreversibly inhibit EGFR kinases. One example is CI-1022, which inhibits all four receptors of the EGFR family by covalently binding to a cysteine located near the ATP-binding site.[83]

The proteasome is a multi-enzyme complex of more than 30 different proteins. It regulates protein turnover in eukaryotic cells by recognizing and destroying ubiquitinated peptides.[84] The proteasome is particularly important in breaking down peptides responsible for the regulation of the cell cycle and apoptosis. The 26S proteasome plays a crucial role, degrading over 80 percent of all cellular enzymes.[85]

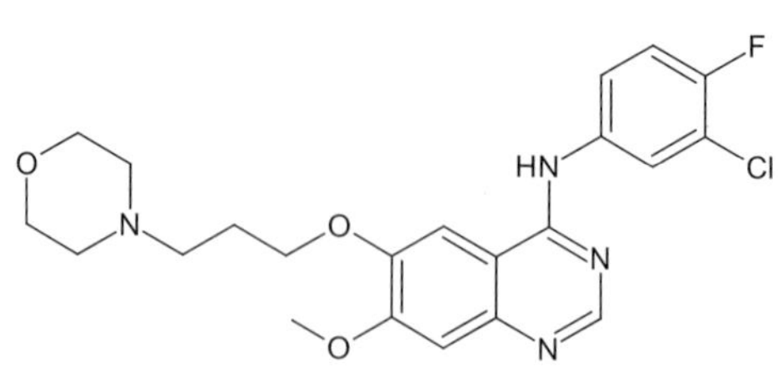

Figure 31 Structure of ZD1839.

The multienzymes are stabilized and accumulated by protease inhibitors. This results in contradictory and uncoordinated signals leading to cell cycle arrest and apoptosis.[86]

9. ANTIMITOTIC AGENTS

During mitosis chromosomes are duplicated and distributed into the newly generated daughter cells. Mitotic checkpoints control the progression of this process. Errors in DNA replication that are not repaired can lead to apoptosis. Mutations in the mitotic checkpoint machinery may eventually cause deregulated cell proliferation and the development of cancer.

Antimitotic drugs are anti-tumor agents that inhibit the organization of microtubules. Microtubules, together with actin microfilaments and intermediate filaments, constitute the cytoskeleton of eukaryotic cells. They control cell shape, cell motility, chromosome movement and intracellular transport of molecules and organelles.[87] Microtubules are cylindrical structures consisting of heterodimer subunits of alternating α- and β-tubulins.[88]

Longitudinal microtubule-associated proteins (MAPs) are located on the surface of α- and β-tubulins and are responsible for polymerization (MAP1, MAP2, tau protein) and stability (MAP1, MAP2) of the microtubules. One end of a microtubule is positively charged, while the other is negatively charged. During polymerization, each tubulin molecule binds two molecules of GTP. The GTP is broken down into GDP and Pi, and the energy released is used to link the tubulin dimer to the positively charged end of the growing microtubule. Microtubules are involved in the construction of the mitotic spindle in the cell. Substances inhibiting microtubules are important for cancer therapy.[89]

There are three main classes of antimitotic agents. (1) Taxanes, e.g. paclitaxel and docetaxel; (2) *Vinca* alkaloids (vincristine, vinblastine, vindesine, and vinorelbine) and (3) colchicins. All these classes inhibit polymerization of tubulin.[90]

Taxanes from *Taxus brevifolia* bind to the β-subunit of tubulin polymers. They prevent the growth of cancer cells by stimulating aggregation, inhibiting depolymerization of microtubule proteins, and stabilizing the mitotic spindle. The overall result is cell cycle arrest in mitosis. Docetaxel

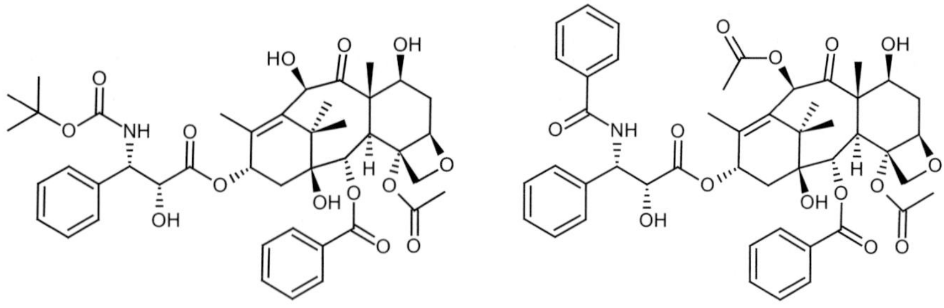

(A) (B)

Figure 32 Structures of docetaxel (A) and paclitaxel (B).

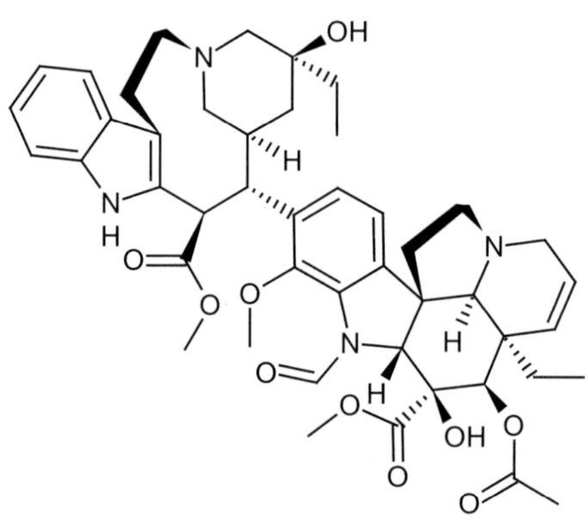

Figure 33 Structure of vincristine.

(Taxotere) is a partially synthetic derivative of paclitaxel and is much more potent (Fig. 32).[90]

Vinca alkaloids such as vincristine (Fig. 33), isolated from the shrub/plant/flower *Catharanthus roseus*, bind to the β-tubulin subunit and block its dimerization with an α-tubulin subunit, thus preventing polymerization of microtubules. This causes cell cycle arrest in metaphase due to the duplicated chromosomes' inability to align along the division plate. The ultimate fate of such cells is apoptosis.[91]

Colchicine alkaloids, from *Colchicum autumnale*, are the third group of antimitotic agents. Colchicine binds to the β-subunit of tubulin homodimers, inhibits mitosis and causes apoptosis. Because of its high toxicity, it is not used for cancer therapy.[92]

Inhibitors of mitotic kinesin motors are a new type of antimitotic agents.[93,94] Kinesin spindle protein (Eg5/KSP),[95] a microtubule-associated motor protein essential for cell cycle progression and the proper partitioning of sister chromatids, is overexpressed in many cancer cells.

Monastrol, an inhibitor of Eg5/KSP, arrests cells in mitosis with monoastral spindles.[96] Chromosomes in monastrol-treated cells frequently have both sister kinetochores attached to microtubules extending to the center of the monoaster. Due to the lack of tension between sister kinetochores a SCP-mediated induce mitotic arrest.

Mitotic arrest–deficient protein 2 (Mad2) localizes to a subset of kinetochores, suggesting the activation of the spindle assembly checkpoint in these cells. Mad2 localizes to some kinetochores that have attached microtubules in monastrol-treated cells, indicating that kinetochore microtubule attachment alone may not satisfy the spindle assembly checkpoint.[93,96,97]

Monastrol represents the prototype of Eg5/KSP inhibitors and is currently used in experimental research.[93] The mitotic arrest due to monastrol is rapidly reversible.

Its relatively low cellular activity is the reason why it is searched for more KSP inhibitors.[98] Eg5/KSP inhibitors that are in clinical development include Ispinesib,[99] a follow-up derivate SB-743 921 with a five time higher activity, MK-0731[100] and ARRY-520[101] also with a five time higher biochemical and cellular potency. In a prolonged treatment with Eg5/KSP inhibitors enter the mitotic "slippage" and lead to the initiation of apoptosis.

For the efficient activation of proapoptotic signaling pathways that lead to the induction of apoptosis activation of the SCP and the mitotic "slippage" is required.

The advantage compared to microtubule-binding agents is the exclusive efficacy in mitosis and proliferation of specificity.

Epothilones are another new class of anti-mitotics. These 16-member ring macrolides with a methylthiazole side chain were isolated from the myxobacterium *Sorangium cellulosum* (Fig. 35). Naturally occurring epothilones are classified as epoxides A, B, E or F. They bind to microtubules and suppress microtubule dynamics comparable to paclitaxel. However, epothilones are much more potent than taxanes and drug resistance is less frequent than with taxanes or vinca alkaloids.[102]

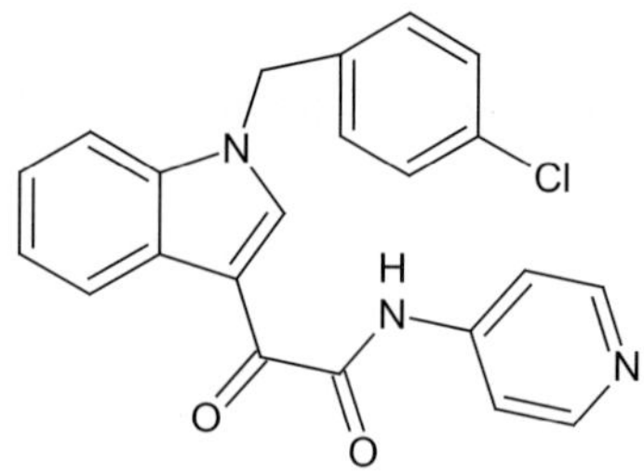

Figure 34 Structure of indibulin.

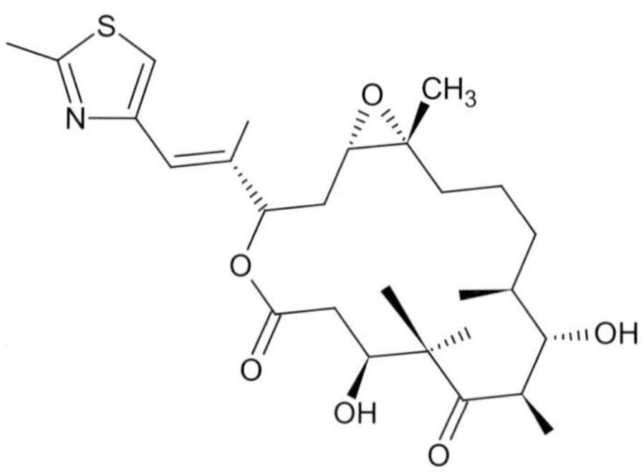

Figure 35 Structure of epothilone B.

Epothilone B and epothilone D are especially commonly used for cancer treatment. Ixabepilone is a second generation analogue of epothilone B. Rational design by modification of a lactone to a lactam side group resulted in greater metabolic stability.[103] ZK-EPO is a fully synthetic, third generation epothilone that crosses the blood brain barrier (Fig. 36). It has greater antitumor potency than other epothilones and retains activity even in multidrug-resistant tumor cells.[102,104]

10. PROTEASOME INHIBITORS

Proteasome is an enzyme complex located in the nucleolus and cytoplasm that is required for the degradation of proteins, including those needed for cell cycle, apoptosis and angiogenesis. For example, cyclins, cyclin-dependent kinase inhibitors, tumor suppressors and transcription factors are critical regulatory proteins that are selectively eliminated by proteasome.

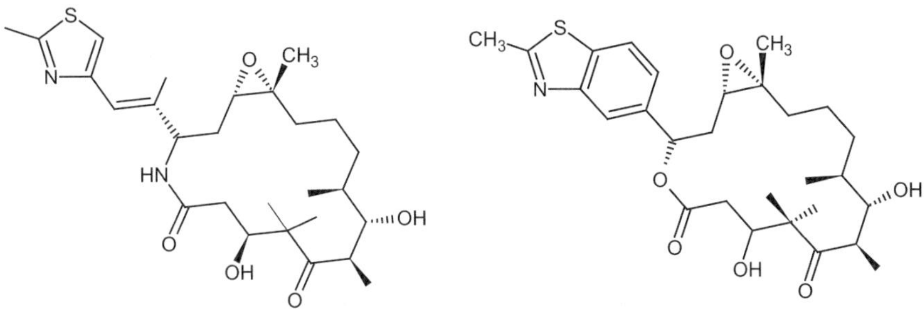

Figure 36 Structure of ixabepilone and ZK-EPO.

Proteasome is essential for maintaining cell homeostasis through degradation of damaged proteins and, hence, is essential for cell survival.[105] The 26S proteasome consists of a 20S core subunit and a 19S regulatory particle.[106] The core subunit is a barrel-like structure composed of two α- and two β-rings, each consisting of seven polypeptide units. The α-rings on either end block direct access to the inner β-rings where the protease is located. Furthermore, the α-units permit or restrict access to the catalytic core and bind the regulatory subunits.[107]

The first step in the ubiquitin proteasome pathway is the attachment of polyubiquitinated moieties to target proteins. Three enzymes are needed in this process: the ubiquitin-activating enzyme E1 binds ubiquitin and transfers it to the ubiquitin-conjugating enzyme E2. In the next step, the ubiquitin ligase E3 catalyzes the transfer of polyubiquitinated tails from E2 to lysine residues on the target protein.[108] The regulatory particle of the proteasome recognizes the ubiquitin markers, unfolds the polyubiquitinated target proteins and allows them access to the catalytic core. This process occurs via an ATPase, which harnesses energy from ATP to create access to the interior of the proteasome.[109] Inside the proteasome, degradation of the protein occurs. A threonine residue mediates the hydrolysis of each proteolytic site. These proteolytic domains are classified as chymotrypsin-like, trypsin-like, or post-glutamyl peptide hydrolase-like.[107]

The ubiquitin proteasome pathway and the resulting cellular protein homeostasis are necessary for the survival of both malignant and nonmalignant cells. Inhibition of this pathway leads to the accumulation of damaged and unneeded proteins, the consequences of which include cell cycle arrest (supported by p53, p21 and p27) and programmed cell death (by activation of pro-apoptotic Bax).[110] Normally IκB (Inhibitor of NF-κB) is degraded under cellular stress. An abnormal rate of degradation, as seen in cancer cells, leads to a release of NF-κB, which induces the transcription of growth and survival factors such as IL-6, IL-8, Bcl-xL and Bcl-2 homolog A1/Bfl-1.[111] Proliferating cells are more susceptible to proteasome inhibition than quiescent cells. Therefore, proteasome inhibitors show potential as anti-cancer drugs. One reason for their elevated action against cancer cells could be that the transcription rate in cancer cells is higher, and therefore more proteins accumulate. This would make cancer cells more dependent on proteasomal degradation than normal cells.[112] In addition, inhibition of the proteasome can make cancer cells more sensitive to conventional chemotherapeutics. For example, the combination of bortezomib and CPT-11 led to synergistic induction of apoptosis in a colon cancer xenograft model; treatment with CPT-11 alone activated NF-κB, whereas the combination therapy led to complete NF-κB inhibition.[113] The success of the inhibitor bortezomib in the treatment of multiple myeloma, mantle cell lymphoma and other selected subtypes of non-Hodgkin's lymphoma resulted in the development of new proteasome inhibitors, some of which were of natural origin.

Epoxomycin was isolated from an *Actinomycetes* strain by Hanada *et. al.*[114] and belongs to the epoxyketone class of proteasomal inhibitors. It showed antitumor properties towards a B16 murine melanoma tumor model. Epoxomycin inhibits the proteasome by binding selectively and irreversibly to the chymotrypsin-like proteolitic active site.[115] Carfilzomib is a derivate of epoxomycin that also irreversibly inhibits proteasome (Fig. 37). Clinical trials showed that carfilzomib targets both forms of proteasome: the constitutive proteasome and the immunoproteasome, both of which are expressed in hematological tumor cells. In multiple myeloma, non-Hodgkin's lymphoma and leukemia, carfilzomib was able to induce antitumor response while showing minimal cytotoxic effects on normal cells. Specific chemical inhibitors of only one form of proteasome showed no antitumor effect, but the combined inhibition of the subunits β5 (constitutive proteasome) and LMP7 (immunoproteasome) had an effect.[116] Furthermore, carfilzomib induced higher levels of caspase-8 and caspase-9 activity than did bortezomib, ultimatey leading to apoptosis.

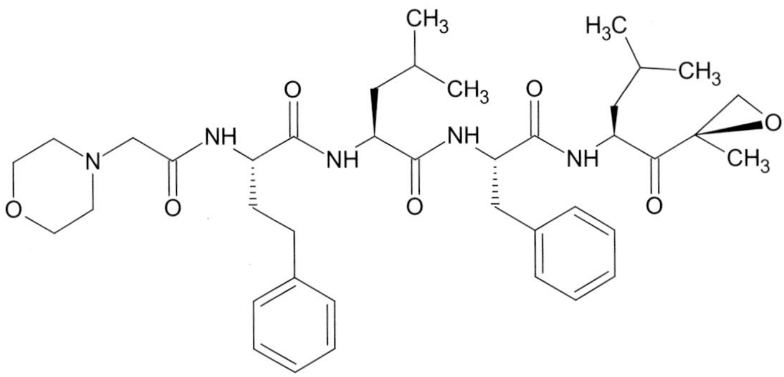

Figure 37 Structure of carfilzomib.

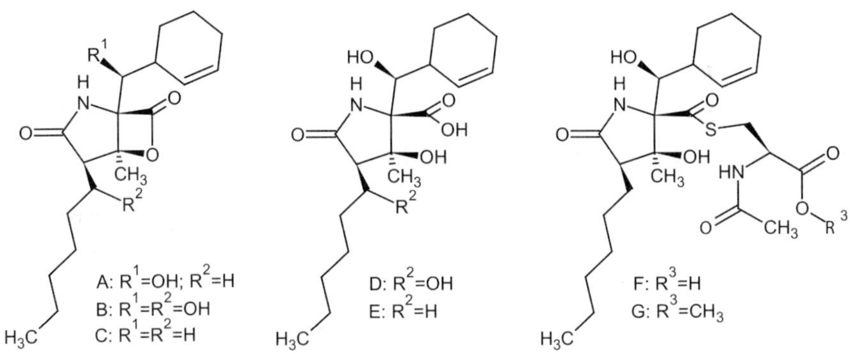

Figure 38 Structure of the cinnabaramides A-G.

Cinnabaramides A-G were also used as proteasome inhibitors (Fig. 38). They were isolated from a strain of *Streptomyces* and showed potent and selective activity towards the human 20S subunit. Chemically, they are related to salinosporamide A, a leading proteasome inhibitor isolated from marine actinomycetes. Cinnabaramides inhibited the proteasome, but did not inhibit other proteases such as trypsin or chymotrypsin.[117]

The natural product syringolin A is isolated from *Pseudonomas syringae* pv, which belongs to the syrbactin family (Fig. 39). Syringolin A is synthesized by a non-ribosomal peptide synthetase/polyketide synthetase. It consists of a tripeptide part that includes a twelve-membered ring with an N-terminal valine joined to a second valine by a highly unusual ureido group.[118] Syringolin A is a promising anticancer drug candidate; however,

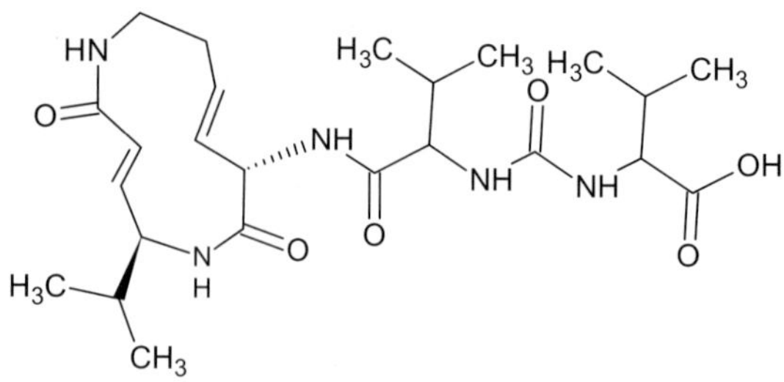

Figure 39 Structure of syringolin A.

it has also shown unwanted side effects. Syringolin A cross-reacts with other nucleophiles, e.g. cysteine proteases. Syringolin A selectively blocks $\beta 5$ and $\beta 2$, while Bortezomib targets $\beta 1$ and $\beta 5$. Bortezomib resistant cells depend on $\beta 2$ that can be efficiently targeted with syringolin A.[119]

11. INHIBITORS OF HEAT SHOCK PROTEIN 90

Heat shock proteins (HSPs) are abundant, conserved proteins. They can be constitutively or inductively expressed. The induced expression is triggered by physiological or environmental insults. These stress stimuli include hyperthermia, oxidative stress, inhibition of tyrosine kinases, ligation of the Fas/Apo-1/CD95 death receptor, radiation or addition of cytotoxic drugs. The HSPs are divided in classes according to their molecular weight. The families are HSP90, HSP70, HSP60 and small HSPs (15–30 kDa) that include HSP27.[120]

Many HSPs function as chaperones, a role critical for cell survival. Their cytoprotective activity includes both protein folding and protein holding. In protein folding, proteins undergo molecular interactions in chambers in HSPs. The HSP60 'chaperonin' family and several other related chaperonin proteins, self-associate to form these chambers. Hence, HSP chaperones ensure that the substrate proteins obtain their correct tertiary structures.[121] When HSPs bind to hydrophobic elements of the substrate proteins it is known as protein holding. This happens during translation to prevent early maturation of peptides. Furthermore, HSPs bind to and stabilize proteins during heat shock, when proteins expose hydrophobic structures. During heat shock, proteins synchronously unfold and are prone to aggregation.

HSPs' function in protein holding is necessary to prevent this aggregation. In addition, some HSPs are constitutively bound to proteins.[122]

HSP90 exists in two isoforms (HSP90a and HSP90b) and works in a multi-chaperone complex in association with several co-chaperones and client proteins. It is a homodimer with three relevant domains. The N-terminal domain has an adenine-binding pocket and carries out the ATPase-activity, the middle linker domain binds to co-chaperones and client proteins, and the C-terminal end is responsible for dimerization.[123]

The client proteins of HSP90 are important in mitosis and the cell cycle. HSP90 stabilizes various regulatory enzymes in the normal cell growth pathway and maintains proteins triggered by growth signals in their activated form. HSP90 facilitates extracellular signals required for cellular development and renewal.[124]

The client protein interaction is driven by ATP hydrolysis and nucleotide exchange at the HSP's amino terminal domain. If the association is disrupted, the client protein becomes degraded. A number of co-chaperones, such as Aha1, p60Hop, p50cdc37 and p23, can regulate the ATPase activity both positively and negatively.[125] In cancer cells, HSP90 is often over-expressed. In breast cancer cells, HSP90 over-expression is associated with elevated expression of estrogen receptors and HER2/erbB2.[126] Furthermore, HSP90 over-expression is associated with mutated epidermal growth factor receptors in lung cancer.[127] Cancer cells are dependent on HSPs to maintain mutated enzymes in an activated form and prevent them from degradation. For example, HSP90 is needed to stabilize telomerase, allowing cancer cells to bypass telomeric crisis (the point at which telomeres become too short for further cell division).[128]

HSP90 client proteins are involved in cell growth, proliferation, adhesion and motility. This makes HSP90 an interesting target for anticancer therapy, and many inhibitors of HSP90 have been described. Some natural products of terrestrial bacteria inhibit HSP90.[129]

Groups of natural HSP90 inhibitors include the benzoquinones geldanamycin, radicicol and herbimycin.[130] Radicicol was the first antibiotic found to reverse v-Src transformation (Fig. 40A). It can be isolated from the fungus *Monosporium bonorden*. Similar to geldanamycin, it binds to the N-terminal ATPase pocket of HSP90 and causes degradation of client proteins even in 17-(allylamino)-17-demethoxygeldanamycin resistant cells

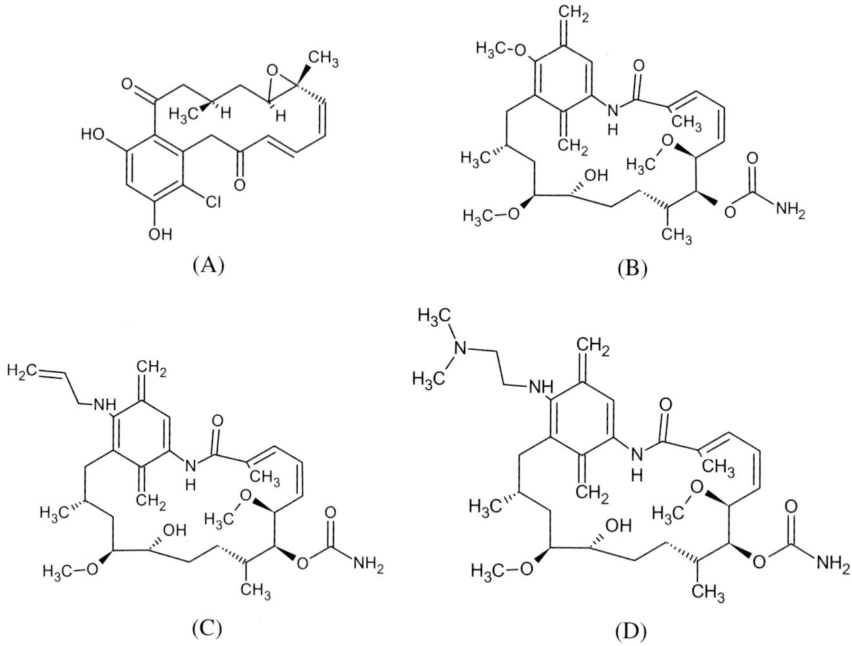

Figure 40 Structure of radicicol (A), geldanamycin (B), 17-AAG (C) and 17-DMAG (D).

(17-AAG, Fig. 40C). Although radicicol showed cytotoxic activity *in vitro*, no results have been obtained *in vivo* due to its instability. Its breakdown leads to the development of oxim-derivates.[131]

Geldanamycin (GA; Fig. 40B) isolated from *Streptomyces hygrocopicus* was the next asanamycin antibiotic discovered with a benzoquinone moiety. As mentioned earlier, GA binds to the N-terminal ATPase pocket of HSP90. It showed anticancer properties *in vitro*, but it is also hepatotoxic *in vivo*. The toxicity is caused by the benzoquinone moiety, which acts as a 'Michael-acceptor' and interacts with nucleophilic protein ends.[130]

When GA binds to the N-terminal domain of HSP90, it disrupts the association with the client protein. This leads to a degradation of client proteins as may occur via the proteosomal pathway. GA-mediated degradation inhibits the network for G_1/S transition. Several growth factors and their receptors (e.g. IGF-I, EGF and PDGF receptors, and HER-2) as well as several kinases (Src, Lyn, Lck, Raf-1 and Cdk-4) are inactivated by GA.

GA induces depletion of the mutated Bcl-Abl kinase in leukemia cells and sensitizes cells to apoptosis.[132]

Because of GA's toxicity and limited water solubility, a number of analogous has been developed. 17-AAG has poor water solubility and a lack of oral bioavailability. The N, N-dimethylethylamino analogue of 17-AAG (17-DMAG; Fig. 40D), was proven to be potent and to have improved water solubility. It also demonstrated a tolerable toxicity in clinical trials. In breast cancer, 17-DMAG sensitized TRAIL-induced apoptosis through down-regulation of RIP. 17-DMAG is also known as alvespirmicin.[133]

Herbimycin A was isolated from *Streptomyces hygroscopicus*, a soil bacterium with herbicidal activity towards mono- and dicotyledonous plants (Fig. 41). Chemical modifications of the ansa-chain moiety and 17- or 19-amino substitutions have been carried out leading to analogues with various levels of cytotoxicity towards cancer cells. In addition to interacting with HSP90, herbimycin A may bind to the sulfhydryl-group of the tyrosine kinase $p60^{v\text{-}Src}$ through Michael addition to the benzochinone nucleus. $p60^{v\text{-}Src}$ is a product of the v-src oncogene.[134] Herbimycin causes cell cycle arrest in G_1 by down-regulating cyclin D, which inhibits the phosphorylation of Rb.[135]

The coumarin antibiotic novobiocin (Fig. 42) binds to the carboxyl-terminus of HSP90 and destabilizes client proteins such as RAF-1, mutant p53, v-SRC and ERBB2. In addition, it down-regulates RAF-1 expression in murine splenocytes *in vivo*.[136]

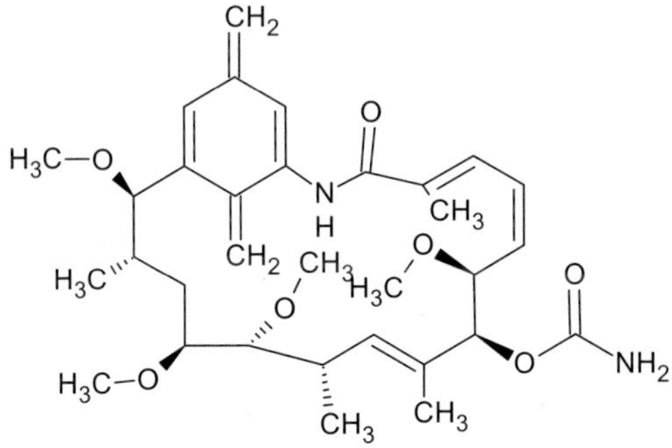

Figure 41 Structure of herbimycin A.

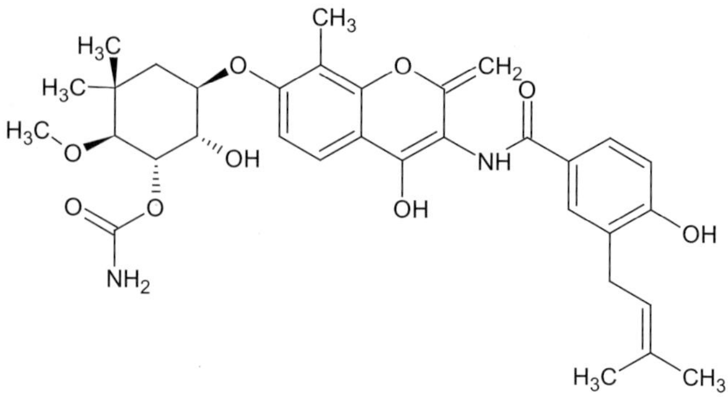

Figure 42 Structure of novobiocin.

Clinical trials of HSP90 inhibitors did not produce clear results. One hypothesis to explain the mixed results is that HSP90 inhibitors may induce the expression of HSP70, a protective protein that could counteract pro-cell death effects.[120]

12. INHIBITORS OF NUCLEAR TRAFFICKING

Some proteins, such as transcription factors, can only carry out their function if they are translocated from the cytoplasm to the nucleus. The transport of a variety of molecules up to 120 kDa occurs through the nuclear pore complex in the nuclear envelope. The nuclear pore complex recognizes two types of signals: nuclear localization signals (NLS) and nuclear export signals (NES). Ran, a Ras-like GTPase, is required to ensure the correct direction of transport. As molecules pass through the nuclear pore complex, the RanGTP concentration in the nucleus is increased, because the Ran GTPase activation protein (RanGAP) is located in the cytoplasm and the Ran Guanine nucleotide exchange factor (RanGEF) in the nucleus. Once a complex of NES-containing cargo is formed, the exportin CRM1 and RanGTP are translocated from the nucleus to cytoplasm. RanGAP then induces the hydrolysis of RanGTP to RanGDP.

The adaptor protein importin α recognizes NLS-containing proteins tagged for import. This adaptor protein is then bound by the receptor

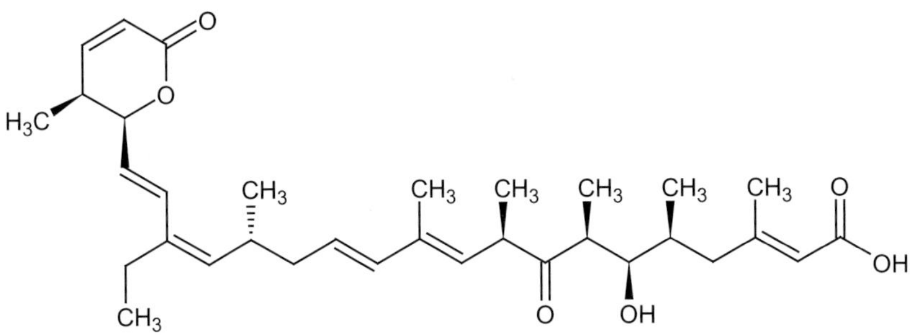

Figure 43 Structure of leptomycin B.

importin β. The entire complex is translocated into the nucleus. The cargo is released when/and RanGTP binds to importin β in the nucleus.

Leptomycin B, isolated from *Streptomyces*, inhibits nuclear export (Fig. 43). Furthermore, it acts as an antibiotic and antifungal agent and causes cell-cycle arrest in the G1 and G2 phases in yeast and mammalian cells. It was identified in a screening for export inhibitors of HIV Rev. The target of leptomycin B is CRM1. Mutant CRM1 conferred leptomycin B resistance in yeast. In these mutants, cysteine-529 was replaced by a serine in CRM1. When leptomycin B binds to this cysteine, CRM1 can no longer interact with NES-containing cargos.[137] In phase 1 of the clinical trial, leptomycin B showed high toxicity. Patients suffered from nausea, vomiting, anorexia, etc.[138] Despite these results, understanding the interaction of leptomycin B with CRM1 and nuclear pore complexes may lead to the development of similar but less toxic compounds for cancer therapy.

In chronic myeloid leukemia (CML), fusion of the cellular breakpoint cluster region gene (Bcr) with the Abelson murine leukemia oncogene (Abl) results in a constitutively active non-receptor tyrosine kinase (Bcr–Abl) that stimulates several signal transduction pathways. CML can be treated with STI571 (Gleevec), an inhibitor of Bcr–Abl.[139] Concurrent treatment of cells with STI571 and leptomycin B results in the nuclear entrapment of Bcr–Abl. Furthermore, leptomycin B serves as an important tool for understanding the molecular mechanisms of nucleo-cytoplasmic transport of proteins and holds potential as a therapeutic drug for diseases caused by mislocalization of regulatory proteins. Some derivatives of leptomycin

B such as kazusamycins, anguinomycins, leptolstatin, and callystatins were also reported to exert potent cytotoxic activity.[140]

13. ATPASE INHIBITORS

Vacuolar ATPases (V-ATPases) are a class of enzymes that are found in many tissues and cells. They are located in vacuole membranes and gain their energy from ATP hydrolysis. This energy is used to pump protons from the outer side of the membrane to the inner side. V-ATPases provide energy for transport processes and regulate the pH of the cytoplasm and organelles. In the membrane of kidney cells, ATPases maintain acid-base-balance. V-ATPase mutations can lead to metabolic acidosis. Furthermore, control of cellular pH is crucial for receptor-mediated endocytosis, because receptor-ligand-complexes require acidification of endosomes for dissociation and recycling.[141] Osteoclast V-ATPases acidify the bone surface and activate secreted hydrolases, and mutations lead to defects in osteoclast function.[142]

ATPases are composed of two domains: the cytosolic V_1 domain and the transmembrane V_0 domain. The V_1 domain consists of three A subunits, three B subunits, two G subunits, and the C, D, E, F and H subunits, while the V_0 domain consists of five different subunits (a, c, c′, c″, and d).[143] The A subunit of V_1 is the catalytic site where ATP is hydrolyzed. ATP hydolysis leads to a rotational movement of the central stalk subunits D, F and C, and results in proton translocation across the membrane.[144]

Many cancer cells secrete lyosomal enzymes required for metastatic invasion into the extracellular matrix. These enzymes are most active at low pH, and V-ATPases are essential in creating the microenvironmental acidification that is necessary for tumor progression and metastasis. Because V-ATPases play such a key role in cancer development, V-ATPase inhibitors are promising candidates in cancer therapy.[145]

The first specific inhibitors of V-type ATPases were the plecomacrolide antibiotics concanamycin A and bafilomycin A1 (Fig. 48). Their chemical structure consists of an 18-membered lactone ring.[146] Bafilomycin A1 was first isolated from *Streptomyces grisseus* on the basis of its antifungal and antibacterial activity.[147] Concanamycin A was first identified as an immunosuppressive compound.[148] At low concentrations, bafilomycin A1

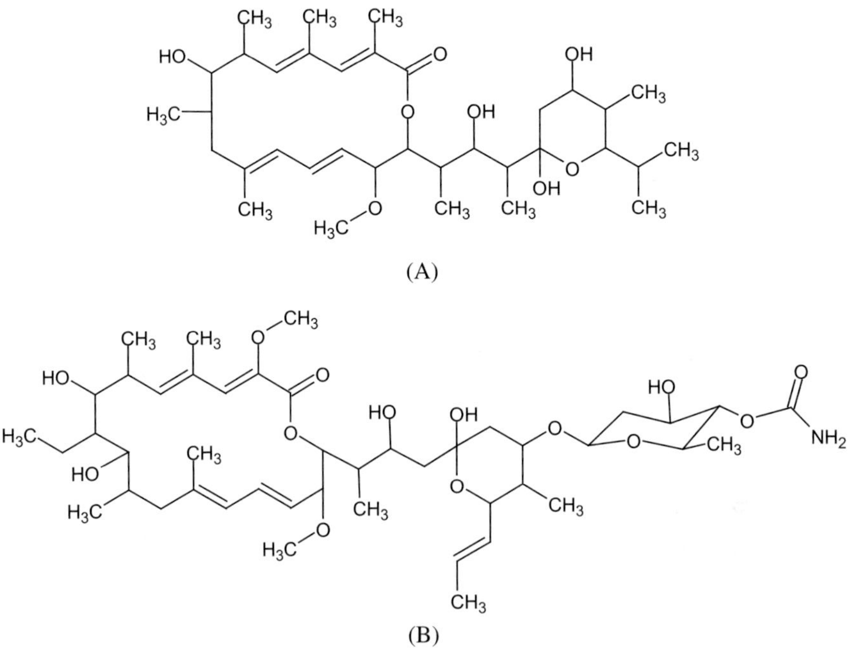

Figure 44 Structure of bafilomycin AI (A) and concanamycin A (B).

and concanamycin A inhibit the growth of *Neurospora crassa*. The inhibitors produced no effect in strains without V-ATPase, indicating that plecomacrolides are specific for these enzymes.[149] Bafilomycin A1 was approximately 80,000-fold more selective for V-ATPase than for P-type ATPase. These properties make the plecomacrolides ideal probes with which to study the function of V-ATPase.[150]

Bafilomycin A1 inhibited proton flux through the V_0 subunit in studies using bovine clathrin-coated vesicle V-ATPase. The addition of excess V_0 subunit restored the proton pump activity of coated vesicles. Hence, it was hypothesized that the V_0 subunit harbors the binding site for bafilomycin A1.[151] Amino acid exchanges ($Phe^{136} \rightarrow Leu^{136}$; $Thr^{32} \rightarrow Ile^{32}$; $Tyr^{143} \rightarrow Asn^{143}$ or His^{143}) in subunit c of the V_0 domain caused bafilomycin resistance.[152] These mutants, however, were not cross-resistant to concanamycin. Recent photoaffinity experiments with concanamycin derivatives in tobacco hornworm preparations clearly demonstrated that both classes of macrolides bind to V_0 subunit c [Huss *et al.*, 2002].[153]

Plecomacrolides inhibit proton transport by preventing the rotation of $ATP6V_0C$ multimer causing intracellular acidosis.[154] Structure-activity relationships showed a toleration of acetylation at the 7- and 21-positions of bafilomycins, marked reduction of activity with tetrahydropyran ring opening, and total loss of activity with macrolide ring opening. Although bafilomycin and concanamycin were found to be too toxic in clinical trials, the structure activity research on bafilomycin has provided new possibilities for the development of less toxic V-ATPase inhibitors.[155]

Archazolid A (Fig. 45) and B are novel substances for V-ATPase inhibition. They are produced by the myxobacteria *Archangium gephyra* and *Cytobacter violaceus.*[151] Archazolides, another class of macrolactones, bind to the c subunit of the membrane integrated V_0-domain.[156] Chemical modifications of archazolid A led to improved IC_{50} values in the nanomolar range. In fact, the derivates archazolid C and D, which are linked to a sugar residue, revealed 100-fold lower IC_{50} values than archazolid A.[151]

Other novel ATPase inhibitor drug candidates are the benzolactone enamides apicularen A (Fig. 46) and B, which were isolated from a variety of strains of the myxobacterial genus *Chondromyces* (i.e., *C. apiculatus, C. lanuginosus, C. pediculatus,* and *C. robustus*). Apicularens inhibit proton transport and V-ATPase-dependent ATP hydrolysis. Apicularen B is the

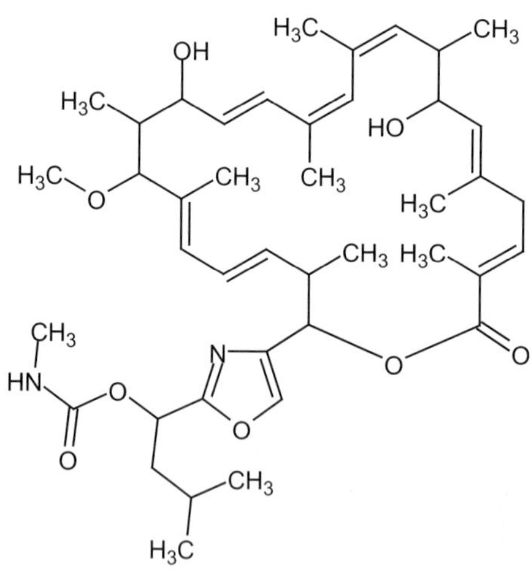

Figure 45 Structure of archazolid A.

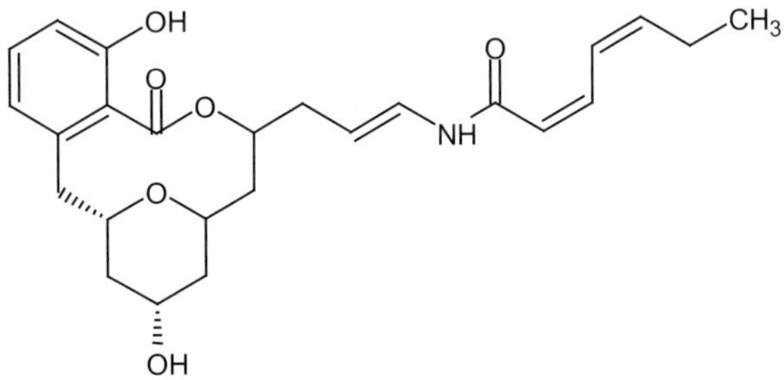

Figure 46 Structure of apicularen A.

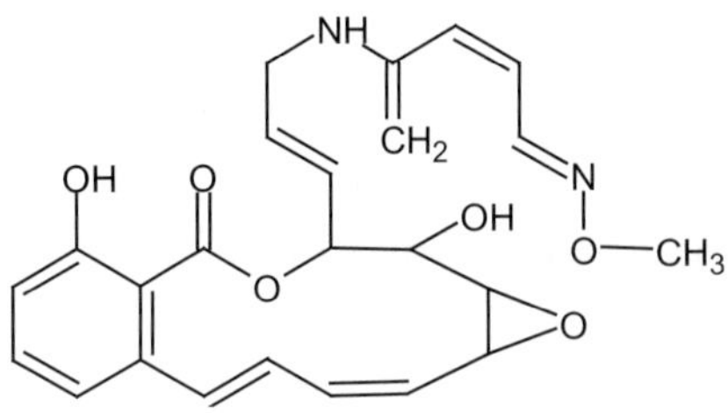

Figure 47 Structure of oximidine I.

N-acetyl glucose derivate of apicularen A, and showed less activity than its precursor.[157] Benzolactone enamides differ in their inhibitory activity towards ATPases of different species. This was demonstrated for apicularen A with preparations of vacuolar membranes from *S. cerevisiae*.[158]

Other benzolactone enamides produced by terrestrial bacteria are the oximidines. Oximidines I (Fig. 47) and II were found in *Pseudonomas* sp. They selectively inhibit the growth of 3Y1 cells transformed with E1A, Ras, and Src oncogenes.[158]

Cruentaren (Fig. 48) has been isolated from active metabolites of the myxobacterium *Byssovorax cruenta* and shows high cytotoxicity against mammalian and fungal cells. Because of its structural relationship to apicularen, it is a promising new drug candidate. However, it inhibits F-ATPase in nanomolar concentration ranges. Therefore, its inhibitory activity is not specific to V-ATPases.[158]

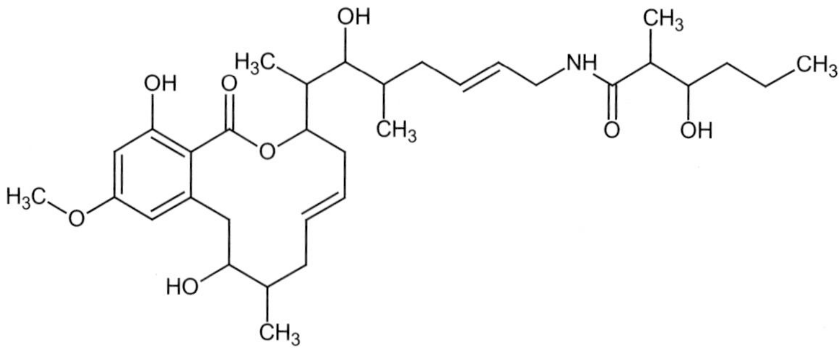

Figure 48 Structure of cruentaren.

14. CONCLUSIONS

Products from terrestrial microbials offer new possibilities for cancer therapies. The compounds mentioned in this chapter were found by high throughput screenings or are already known for other effects. For example geldanamycin is used as an antibiotic. Their activity against cancer cells was discovered later.

Some have only been tested in cell culture while others have reached a further step in clinical research. It has been recognized that further research is required in the field of cell cultures.

Most initial substances have been chemically modified to optimize the pharmacokinetics and reduce their toxicity. Wortmannin is an unstable PI3K inhibitor while a modification of the furane ring increases stability. Geldanamycin shows less water solubility and a higher toxicity than the analogue Alvespirmicin.

Some compounds found by structure-activity-relationships indicate better effects than the initial agent but there are no studies on side effects. Caused by interactions with other targets like kinases with similar binding site, 5-azaindolocarbazoles as seen in Fig. 12 inhibits Chk1 but the selectivity over e.g. CdKs requires further investigation.

The use of these new anticancer drugs can be used singly or as part of a combination with other therapeutic modalities, for example chemotherapy or immunotherapy. They have shown results in more chemosensitive cells, to alleviate the resistance to other drugs, which can then be targeted by another type of therapy.

Even if some substances have shown selectivity towards cancer cells unwanted side effects due to the influence on uninfected cells still occur and that is another reason for further study.

REFERENCES

1. Whitman, Downes CP, Keeler M, *et al.* (1988) Type I phosphatidylinositol kinase makes a novel inositol phospholipid, phosphatidylinositol-3-phosphate. *Nature* **332**: 644–646.
2. Ihle NT, Williams R, Chow S, *et al.* (2004) Molecular pharmacology and antitumor activity of PX-682, a novel inhibitor of phosphoinositide-3-kinase signalling. *Mol Cancer Ther* **3**: 763–772.
3. Fry MJ. (2001) Phosphoinositide 3-kinase signalling in breast cancer: how big a role might it play? *Breast Cancer Res* **3**, 304–312.
4. Wymann MP, Bulgarelli-Leva G, Zvelebil MJ, *et al.* (1996) Wortmannin inactivates phosphoinositide 3-kinase by covalent modification of Lys-802, a residue involved in the phosphate transfer reaction. *Mol Cell Biol* **16**: 1772–1733.
5. Powis G, Bonjouklian R, Berggren MM, *et al.* (1994) Wortmannin, a potent and selective inhibitor of phosphatidylinositol-3-kinase. *Cancer Res* **54**: 2419–2423.
6. Zask A, Kaplan J, Toral-Barza L, *et al.* (2008) Synthesis and structure-activity relationships of ring-opened 17-hydroxywortmannins: potent phosphoinositide 3-Kinase inhibitors with improved properties and anticancer efficacy. *J Med Chem* **51**: 1319–1323.
7. Yuan H, Pupo MT, Blois J, *et al.* (2009) A stabilized demethoxyviridin derivate inhibits PI3 kinase. *Bioorg Med Chem Lett* **19**: 4223–4227.
8. Schultz RM, Merriman RL, Andis SL, *et al.* (1995) In vitro and in vivo antitumor activity of the phosphatidylinositol-3-kinase inhibitor, wortmannin. *Anticancer Res* **15**: 1135–1139.
9. Norman BH, Shih C, Toth JE, *et al.* (1996) Studies on the mechanism of phosphatidylinositol 3-kinase inhibition by wortmannin and related analogs. *J Med Chem* **39**: 1106–1111.
10. Xiao Z, Xue J, Sowin TJ, Zhang H. (2006) Differential roles of checkpoint kinase 1, checkpoint kinase 2 and mitogen-activated protein kinase-activated protein kinase 2 in mediating DNA damage-induced cell cycle arrest: implications for cancer therapy. *Mol Cancer Ther* **5**: 1935–1943.
11. Zhao B, Bower MJ, McDevitt PJ, *et al.* (2002) Structural basis for Chk1 inhibition by UCN-01. *J Biol Chem* **277**: 46609–46615.

12. Lawrie AM, Noble ME, Tunnah P, *et al.* (1997) Protein kinase inhibition by staurosporine revealed in details of the molecular interaction with CDK2. *Nat Struct Biol* **4**: 796–801.
13. Wang Q, Fan S, Eastman A, *et al.* (1996) UCN-01: a potent abrogator of G2 checkpoint function in cancer cells with disrupted p53. *J Natl Cancer Inst* **88**: 956–965.
14. Bunch RT, Eastman A. (1996) Enhancement of cisplatin-induced cytotoxicity by 7-hydroxystaurosporine (UCN-01), a new G2-checkpoint inhibitor. *Clin Cancer Res* **2**: 791–797.
15. Kamata K, Suetsugu T, Yamamoto Y, *et al.* (2006) Bisindole alkaloids from myxomycetes Arcyria denudata and Arcyria obvelata. *J Nat Prod* **69**: 1252–1254.
16. Lefoix M, Coudert G, Routier S, *et al.* (2008) Novel 5-azaindolocarbazoles as cytotoxic agents and Chk1 inhibitors. *Bioorg Med Chem* **16**: 5303–5321.
17. Otani T, Sugimoto Y, Aoyagi Y, *et al.* (2000) New Cdc25B tyrosine phosphatase inhibitors, nocardiones A and B, produced by Nocardia sp. TP-A0248: taxonomy, fermentation, isolation, structural elucidation and biological properties. *J Antibiot (Tokyo)* **53**: 337–344.
18. Johnston PA, Foster CA, Tierno MB, *et al.* (2009) Cdc25B dual-specificity phosphatase inhibitors identified in a high-throughput screen of the NIH compound library. *Assay Drug Dev Technol* **7**: 250–265.
19. Sodeoka M, Sampe R, Kojima S, *et al.* (2001) Synthesis of a tetronic acid library focused on inhibitors of tyrosine and dual-specificity protein phosphatases and its evaluation regarding VHR and cdc25B inhibition. *J Med Chem* **44**: 3216–3222.
20. Sodeoka M, Sampe R, Kagamizono T, Osada H. (1996) Asymmetric synthesis of a 3-acyltetronic acid derivative, RK-682, and formation of its calcium salt during silica gel column chromatography. *Chem Pharm Bull* (Tokyo) **49**: 206–212.
21. Todd JL, Tanner KG, Denu JM. (1999) Extracellular regulated kinases (ERK) 1 and ERK2 are authentic substrates for the dual-specificity protein-tyrosine phosphatase VHR. A novel role in down-regulating the ERK pathway. *J Biol Chem* **274**: 13271–13280.
22. Reddy RS, Swarup G. (1995) Alternative splicing generates four different forms of a non-transmembrane protein tyrosine phosphatase mRNA. *DNA Cell Biol* **14**: 1007–1015.
23. Lapenna S, Giordano A. (2009) Cell cycle kinases as therapeutic targets for cancer. *Nat Rev Drug Discov* **8**: 547–566.

24. Meijer L. (2000) Cyclin-dependent kinases inhibitors as potential anticancer, antineurodegenerative, antiviral and antiparasitic agents. *Drug Resist Updat* **3**: 83–88.
25. Collins I, Garrett MD. (2005) Targeting the cell division cycle in cancer: CDK and cell cycle checkpoint kinase inhibitors. *Curr Opin Pharmacol* **5**: 366–373.
26. Sanchez-Martinez C, Shih C, Zhu G, *et al.* (2003) Studies on cyclin-dependent kinase inhibitors: indolo-[2,3-a]pyrrolo[3,4-c]carbazoles versus bis-indolylmaleimides. *Bioorg Med Chem Lett* **13**: 3841–3846.
27. Engler TA, Furness K, Malhotra S, *et al.* (2003) Novel, potent and selective cyclin D1/CDK4 inhibitors: indolo[6,7-a]pyrrolo[3,4-c]carbazoles. *Bioorg Med Chem Lett* **13**: 2261–2267.
28. Sanchez-Martinez C, Shih C, Faul MM, *et al.* (2003) Aryl[a]pyrrolo[3,4-c]carbazoles as selective cyclin D1-CDK4 inhibitors. *Bioorg Med Chem Lett* **13**: 3835–3839.
29. Zhu G, Conner S, Zhou X, *et al.* (2003) Synthesis of quinolinyl/isoquinolinyl[a]pyrrolo [3,4-c] carbazoles as cyclin D1/CDK4 inhibitors. *Bioorg Med Chem Lett* **13**: 1231–1235.
30. Mahlknecht U, Will J, Varin A, *et al.* (2004) Histone deacetylase 3, a class I histone deacetylase, suppresses MAPK11-mediated activating transcription factor-2 activation and represses TNF gene expression. *J Immunol* **173**: 3979–3990.
31. Wegener D, Wirsching F, Riester D, *et al.* (2003) A fluorogenic histone deacetylase assay well suited for high-throughput activity screening. *Chem Biol* **10**: 61–68.
32. Buggy JJ, Sideris ML, Mak P, *et al.* (2000) Cloning and characterization of a novel human histone deacetylase, HDAC8. *Biochem J* **350**: Pt 1, 199–205.
33. Dangond F, Gullans SR. (1998) Differential expression of human histone deacetylase mRNAs in response to immune cell apoptosis induction by trichostatin A and butyrate. *Biochem Biophys Res Commun* **247**: 833–837.
34. Kao HY, Lee CH, *et al.* (2002) Isolation and characterization of mammalian HDAC10, a novel histone deacetylase. *J Biol Chem* **277**: 187–193.
35. Vaziri H, Dessain SK, Ng Eaton E, *et al.* (2001) hSIR2(SIRT1) functions as an NAD-dependent p53 deacetylase. *Cell* **107**: 149–159.
36. Finnin MS, Donigian JR, Cohen A, *et al.* (1999) Structures of a histone deacetylase homologue bound to the TSA and SAHA inhibitors. *Nature* **401**: 188–193.
37. Johnstone RW. (2002) Histone-deacetylase inhibitors: novel drugs for the treatment of cancer. *Nat Rev Drug Discov* **1**: 287–299.

38. Bertrand P. (2010) Inside HDAC with HDAC inhibitors. *Eur J Med Chem* **45**: 2095–2116.
39. Elaut G, Török G, Vinken M, *et al.* (2002) Major phase I biotransformation pathways of Trichostatin a in rat hepatocytes and in rat and human liver microsomes. *Drug Metab Dispos* **30**: 1320–1328.
40. Yoshida M, Horinouchi S, Beppu T. (1995) Trichostatin A and trapoxin: novel chemical probes for the role of histone acetylation in chromatin structure and function. *Bioessays* **17**: 423–430.
41. Yoshida M, Hoshikawa Y, Koseki K, *et al.* (1990) Structural specificity for biological activity of trichostatin A, a specific inhibitor of mammalian cell cycle with potent differentiation-inducing activity in Friend leukemia cells. *J Antibiot (Tokyo)* **43**: 1101–1106.
42. Kouzarides T. (1999) Histone acetylases and deacetylases in cell proliferation. *Curr Opin Genet Dev* **9**: 40–48.
43. Doetzlhofer A, Rotheneder H, Lagger G, *et al.* (1999) Histone deacetylase 1 can repress transcription by binding to Sp1. *Mol Cell Biol* **19**: 5504–5511.
44. Marks PA, Richon VM, Breslow R, *et al.* (2001) Histone deacetylase inhibitors as new cancer drugs. *Curr Opin Oncol* **13**: 477–483.
45. Richon VM, Emiliani S, Verdin E, *et al.* (1998) A class of hybrid polar inducers of transformed cell differentiation inhibits histone deacetylases. *Proc Natl Acad Sci USA* **95**: 3003–3007.
46. Takakura M, Kyo S, Sowa Y, *et al.* (2001) Telomerase activation by histone deacetylase inhibitor in normal cells. *Nucleic Acids Res* **29**: 3006–3011.
47. Suzuki T, Kimura A, Nagai R, *et al.* (2000) Regulation of interaction of the acetyltransferase region of p300 and the DNA-binding domain of Sp1 on and through DNA binding. *Genes Cells* **5**: 29–41.
48. Marks PA, Rifkind RA, Richon VM, *et al.* (2001) Inhibitors of histone deacetylase are potentially effective anticancer agents. *Clin Cancer Res* **7**: 759–760.
49. Kijima M, Yoshida M, Sugita K, *et al.* (1993) Trapoxin, an antitumor cyclic tetrapeptide, is an irreversible inhibitor of mammalian histone deacetylase. *J Biol Chem* **268**: 22429–22435.
50. Yoshida M, Furumai R, Nishiyama M, *et al.* (2001) Histone deacetylase as a new target for cancer chemotherapy. *Cancer Chemother Pharmacol* **48**: Suppl 1, S20–26.
51. Nakajima H, Kim YB, Terano H, *et al.* (1998) FR901228, a potent antitumor antibiotic, is a novel histone deacetylase inhibitor. *Exp Cell Res* **241**: 126–133.

52. Barnard JA, Warwick G. (1993) Butyrate rapidly induces growth inhibition and differentiation in HT-29 cells. *Cell Growth Differ* **4**: 495–501.
53. Hinnebusch BF, Meng S, Wu JT, *et al.* (2002) The effects of short-chain fatty acids on human colon cancer cell phenotype are associated with histone hyperacetylation. *J Nutr* **132**: 1012–1017.
54. Wang Q, Li N, Wang X, *et al.* (2002) Augmentation of sodium butyrate-induced apoptosis by phosphatidylinositol 3′-kinase inhibition in the KM20 human colon cancer cell line. *Clin Cancer Res* **8**: 1940–1947.
55. Liu LF, Liu CC, Alberts BM. (1979) T4 DNA topoisomerase: a new ATP-dependent enzyme essential for initiation of T4 bacteriophage DNA replication. *Nature* **281**: 456–461.
56. Berger DS. (1998) Understanding and managing resistance. *Posit Aware* **9**: 21.
57. Wang JC. (1996) DNA topoisomerases. *Annu Rev Biochem* **65**: 635–692.
58. Li TK, Liu LF. (2001) Tumor cell death induced by topoisomerase-targeting drugs. *Annu Rev Pharmacol Toxicol* **41**: 53–77.
59. Drake FH, Zimmerman JP, McCabe FL, *et al.* (1987) Purification of topoisomerase II from amsacrine-resistant P388 leukemia cells. Evidence for two forms of the enzyme. *J Biol Chem* **262**: 16739–16747.
60. Perrin D, van Hille B, Barret JM, *et al.* (2000) F 11782, a novel epipodophylloid non-intercalating dual catalytic inhibitor of topoisomerases I and II with an original mechanism of action. *Biochem Pharmacol* **59**: 807–819.
61. Bailly C. (2000) Topoisomerase I poisons and suppressors as anticancer drugs. *Curr Med Chem* **7**: 39–58.
62. Vosberg HP. (1985) DNA topoisomerases: enzymes that control DNA conformation. *Curr Top Microbiol Immunol* **114**: 19–102.
63. Wang JC. (1985) DNA topoisomerases. *Annu Rev Biochem* **54**: 665–697.
64. Champoux JJ. (2000) Structure-based analysis of the effects of camptothecin on the activities of human topoisomerase I. *Ann N Y Acad Sci* **922**: 56–64.
65. Redinbo MR, Stewart L, Kuhn P, *et al.* (1998) Crystal structures of human topoisomerase I in covalent and noncovalent complexes with DNA. *Science* **279**: 1504–1513.
66. Stewart L, Redinbo MR, Qiu X, *et al.* (1998) A model for the mechanism of human topoisomerase I. *Science* **279**: 1534–1541.
67. Bailly C, Qu X, Chaires JB, *et al.* (1999) Substitution at the F-ring N-imide of the indolocarbazole antitumor drug NB-506 increases the cytotoxicity, DNA binding, and topoisomerase I inhibition activities. *J Med Chem* **42**: 2927–2935.

68. Jin S, Kim JS, Sim SP, *et al.* (2000) Heterocyclic bibenzimidazole derivatives as topoisomerase I inhibitors. *Bioorg Med Chem Lett* **10**: 719–723.
69. Soderlind KJ, Gorodetsky B, Singh AK, *et al.* (1999) Bis-benzimidazole anti-cancer agents: targeting human tumour helicases. *Anticancer Drug Des* **14**: 19–36.
70. Vasiutina EL, Bugreev DV, Riabinin VA, *et al.* Interaction of a complex of human DNA and topoisomerase I with oligo-1,3-thiazolecarboxamides and their conjugates with oligonucleotides. *Mol Biol (Mosk)* **34**: 419–426.
71. Lown JW. (1993) Anthracycline and anthraquinone anticancer agents: current status and recent developments. *Pharmacol Ther* **60**: 185–214.
72. Gewirtz DA. (1999) A critical evaluation of the mechanisms of action proposed for the antitumor effects of the anthracycline antibiotics adriamycin and daunorubicin. *Biochem Pharmacol* **57**: 727–741.
73. Dorr RT. (1996) Cytoprotective agents for anthracyclines. *Semin Oncol* **23**: 23–34.
74. Hande KR. (1998) Etoposide: four decades of development of a topoisomerase II inhibitor. *Eur J Cancer* **34**: 1514–1521.
75. Chen GL, Yang L, Rowe TC, *et al.* (1984) Nonintercalative antitumor drugs interfere with the breakage-reunion reaction of mammalian DNA topoisomerase II. *J Biol Chem* **259**: 13560–13566.
76. Fantl WJ, Johnson DE, Williams LT. (1993) Signalling by receptor tyrosine kinases. *Annu Rev Biochem* **62**: 453–481.
77. Rajkumar T. (2001) Growth factors and growth factor receptors in cancer. *Curr Sci India* **81**: 535–541.
78. Tuli SS, Liu R, Chen C, *et al.* (2006) Immunohistochemical localization of EGF, TGF-alpha, TGF-beta, and their receptors in rat corneas during healing of excimer laser ablation. *Current Eye Research* **31**: 709–719.
79. Hubbard SR, Till JH. (2000) Protein tyrosine kinase structure and function. *Ann Rev Biochem* **69**: 373–398.
80. Blank SV, Chang R, Muggia F. (2005) Epidermal growth factor receptor inhibitors for the treatment of epithelial ovarian cancer. *Oncology* (Williston Park) **19**: 553–559; discussion 560–567.
81. Camardella L, Carratore V, Ciardiello MA, *et al.* (2000) Kiwi protein inhibitor of pectin methylesterase amino-acid sequence and structural importance of two disulfide bridges. *Eur J Biochem* **267**: 4561–4565.
82. Griffin J. (2001) The biology of signal transduction inhibition: basic science to novel therapies. *Semin Oncol* **28**: 3–8.
83. Lynch TJ, Bell DW, Sordella R, *et al.* (2004) Activating mutations in the epidermal growth factor receptor underlying responsiveness of non-small-cell lung cancer to gefitinib. *N Engl J Med* **350**: 2129–2139.

84. Hershko A, Ciechanover A, Varshavsky A. (2000) Basic Medical Research Award. The ubiquitin system. *Nat Med* **6**: 1073–1081.
85. Naujokat C, Hoffmann S. (2002) Role and function of the 26S proteasome in proliferation and apoptosis. *Lab Invest* **82**: 965–980.
86. Adams J, Palombella VJ, Elliott PJ. (2000) Proteasome inhibition: a new strategy in cancer treatment. *Invest New Drugs* **18**: 109–121.
87. Wilson PG, Borisy GG. (1997) Evolution of the multi-tubulin hypothesis. *Bioessays* **19**: 451–454.
88. Wade RH. (2009) On and around microtubules: an overview. *Mol Biotechnol* **43**: 177–191.
89. Mandelkow E, Mandelkow EM. (1995) Microtubules and microtubule-associated proteins. *Curr Opin Cell Biol* **7**: 72–81.
90. Harrison MR, Holen KD, Liu G. (2009) Beyond taxanes: a review of novel agents that target mitotic tubulin and microtubules, kinases, and kinesins. *Clin Adv Hematol Oncol* **7**: 54–64.
91. Geldof AA, Minneboo A, Heimans JJ. (1998) Vinca-alkaloid neurotoxicity measured using an in vitro model. *J Neurooncol* **37**: 109–113.
92. Wienecke A, Bacher G. (2009) Indibulin, a novel microtubule inhibitor, discriminates between mature neuronal and nonneuronal tubulin. *Cancer Res* **69**: 171–177.
93. Mayer TU, Kapoor TM, Haggarty SJ, *et al.* (1999) Small molecule inhibitor of mitotic spindle bipolarity identified in a phenotype-based screen. *Science* **286**: 971–974.
94. Tao W, South VJ, Diehl RE, *et al.* (2007) An inhibitor of the kinesin spindle protein activates the intrinsic apoptotic pathway independently of p53 and de novo protein synthesis. *Mol Cell Biol* **27**, 689–698.
95. Blangy A, Lane HA, d'Hérin P, *et al.* (1995) Phosphorylation by p34cdc2 regulates spindle association of human Eg5, a kinesin-related motor essential for bipolar spindle formation in vivo. *Cell* **83**: 1159–1169.
96. Kapoor TM, Mayer TU, Coughlin ML, *et al.* (2000) Probing spindle assembly mechanisms with monastrol, a small molecule inhibitor of the mitotic kinesin, Eg5. *J Cell Biol* **150**: 975–988.
97. Maliga Z, Kapoor TM, Mitchison TJ. (2002) Evidence that monastrol is an allosteric inhibitor of the mitotic kinesin Eg5. *Chem Biol* **9**: 989–996.
98. Schmidt M, Bastians H. (2007) Mitotic drug targets and the development of novel anti-mitotic anticancer drugs. *Drug Resist Updat* **10**: 162–181.
99. Purcell JW, Davis J, Reddy M, *et al.* (2010) Activity of the kinesin spindle protein inhibitor ispinesib (SB-715992) in models of breast cancer. *Clin Cancer Res* **16**: 566–576.

100. Holen K, DiPaola R, Liu G, *et al.* (2011) A phase I trial of MK-0731, a Kinesin Spindle Protein (KSP) inhibitor, in patients with solid tumors. *Invest New Drugs* **30**: 1088–1095.
101. Woessner R, Tunquist B, Lemieux C, *et al.* (2009) ARRY-520, a novel KSP inhibitor with potent activity in hematological and taxane-resistant tumor models. *Anticancer Res* **29**: 4373–4380.
102. Fumoleau P, Coudert B, Isambert N, *et al.* (2007) Novel tubulin-targeting agents: anticancer activity and pharmacologic profile of epothilones and related analogues. *Ann Oncol* **18**: Suppl 5, v9–15.
103. Hussain M, Tangen CM, Lara PN Jr, *et al.* (2005) Ixabepilone (epothilone B analogue BMS-247550) is active in chemotherapy-naive patients with hormone-refractory prostate cancer: a Southwest Oncology Group trial S0111. *J Clin Oncol* **23**: 8724–8729.
104. Rustin G, Reed N, Jayson GC, *et al.* (2011) A phase II trial evaluating two schedules of sagopilone (ZK-EPO), a novel epothilone, in patients with platinum-resistant ovarian cancer. *Ann Oncol* **22**: 2411–2416.
105. Shah SA, Potter MW, McDade TP, *et al.* (2001) 26S proteasome inhibition induces apoptosis and limits growth of human pancreatic cancer. *J Cell Biochem* **82**: 110–122.
106. Seeger M, Ferrell K, Dubiel W. (1997) The 26S proteasome: a dynamic structure. *Mol Biol Rep* **24**: 83–88.
107. Coux O, Tanaka K, Goldberg AL. (1996) Structure and functions of the 20S and 26S proteasomes. *Ann Rev Biochem* **65**: 801–847.
108. Newton K, Vucic D. (2007) Ubiquitin ligases in cancer: ushers for degradation. *Cancer Invest* **25**: 502–513.
109. Glickman MH, Rubin DM, Fu H, *et al.* (1999) Functional analysis of the proteasome regulatory particle. *Mol Biol Rep* **26**: 21–28.
110. Blagosklonny MV, Wu GS, Omura S, el-Deiry WS. (1996) Proteasome-dependent regulation of p21WAF1/CIP1 expression. *Biochem Biophys Res Commun* **227**: 564–569.
111. Brown K, Park S, Kanno T, *et al.* (1993) Mutual regulation of the transcriptional activator NF-kappa B and its inhibitor, I kappa B-alpha. *Proc Natl Acad Sci USA* **90**: 2532–2536.
112. Montagut C, Rovira A, Albanell J. (2006) The proteasome: a novel target for anticancer therapy. *Clinical & translational oncology*: official publication of the Federation of Spanish Oncology Societies and of the National Cancer Institute of Mexico **8**: 313–317.

113. Cusack JC Jr, Liu R, Baldwin AS Jr. (2000) Inducible chemoresistance to 7-ethyl-10-[4-(1-piperidino)-1-piperidino]-carbonyloxycamptothecin (CPT-11) in colorectal cancer cells and a xenograft model is overcome by inhibition of nuclear factor-kappaB activation. *Cancer Res* **60**: 2323–2330.
114. Hanada M, Sugawara K, Kaneta K, *et al.* (1992) Epoxomicin, a new antitumor agent of microbial origin. *J Antibiot (Tokyo)* **45**: 1746–1752.
115. Meng L, Mohan R, Kwok BH, *et al.* (1999) Epoxomicin, a potent and selective proteasome inhibitor, exhibits in vivo anti-inflammatory activity. *Proc Natl Acad Sci USA* **96**: 10403–10408.
116. Parlati F, Lee SJ, Aujay M, *et al.* (2009) Carfilzomib can induce tumor cell death through selective inhibition of the chymotrypsin-like activity of the proteasome. *Blood* **114**: 3439–3447.
117. Stadler M, Bitzer J, Mayer-Bartschmid A, *et al.* (2007) Cinnabaramides A-G: analogues of lactacystin and salinosporamide from a terrestrial streptomycete. *J Nat Prod* **70**: 246–252.
118. Ramel C, Tobler M, Meyer M, *et al.* (2009) Biosynthesis of the proteasome inhibitor Syringolin A: the ureido group joining two amino acids originates from bicarbonate. *BMC Biochemistry* **10**: 26.
119. Clerc J, Florea BI, Kraus M, *et al.* (2009) Syringolin A selectively labels the 20S proteasome in murine EL4 and wild-type and bortezomib-adapted leukaemic cell lines. *Chem Bio Chem* **10**: 2638–2643.
120. Jego G, Hazoumé A, Seigneuric R, Garrido C. (2010) Targeting heat shock proteins in cancer. *Cancer Letters* **332**: 275–285.
121. Young JC, Agashe VR, Siegers K, Hartl FU. (2004) Pathways of chaperone-mediated protein folding in the cytosol. *Nat Rev Mol Cell Biol* **5**: 781–791.
122. Calderwood SK, Khaleque A, Sawyer DB, Ciocca DR. (2006) Heat shock proteins in cancer: chaperones of tumorigenesis. *Trends Biochem Sci* **31**: 164–172.
123. Stebbins CE, Russo AA, Schneider C, *et al.* (1997) Crystal structure of an Hsp90-geldanamycin complex: targeting of a protein chaperone by an anti-tumor agent. *Cell* **89**: 239–250.
124. Neckers L, Ivy SP. (2003) Heat shock protein 90. *Curr Opin Oncol* **15**: 419–424.
125. Pearl LH, Prodromou C. (2006) Structure and mechanism of the Hsp90 molecular chaperone machinery. *Annu Rev Biochem* **75**: 271–294.
126. Pick E, Kluger Y, Giltnane JM, *et al.* (2007) High HSP90 expression is associated with decreased survival in breast cancer. *Cancer Res* **67**: 2932–2937.

127. Shimamura T, Shapiro GI. (2008) Heat shock protein 90 inhibition in lung cancer. *J Thorac Oncol* **3**: S152–159.
128. Campisi J. (2005) Senescent cells, tumor suppression, and organismal aging: good citizens, bad neighbors. *Cell* **120**: 513–522.
129. Neckers L. (2002) Hsp90 inhibitors as novel cancer chemotherapeutic agents. *Trends Mol Med* **8**: S55–S61.
130. Fukuyo Y, Hunt CR, Horikoshi N. (2010) Geldanamycin and its anti-cancer activities. *Cancer Lett.* **290**: 24–35.
131. Agatsuma T, Ogawa H, Akasaka K, *et al.* (2002) Halohydrin and oxime derivatives of radicicol: synthesis and antitumor activities. *Bioorg Med Chem* **10**: 3445–3454.
132. Blagosklonny MV. (2002) Hsp-90-associated oncoproteins: multiple targets of geldanamycin and its analogs. *Leukemia* **16**: 455–462.
133. Smith V, Sausville EA, Camalier RF, *et al.* (2005) Comparison of 17 dimethylaminoethylamino-17-demethoxygeldanamycin (17DMAG) and 17-allylamino-17 demethoxygeldanamycin (17AAG) in vitro: effects on Hsp90 and client proteins in melanoma models. *Cancer Chemother Pharmacol* **56**: 126–137.
134. Uehara Y. (2003) Natural Product origins of Hsp90 Inhibitors. *Current Cancer Drug Targets* **3**: 325–330.
135. Srethapakdi M, Liu F, Tavorath R, Rosen N. (2000) Inhibition of Hsp90 function by ansamycins causes retinoblastoma gene product dependent G1 arrest. *Cancer Res* **60**: 3940–3946.
136. Marcu MG, Schulte TW, Neckers L. (2000) Novobiocin and related coumarins and depletion of heat shock protein 90-dependent signaling proteins. *J Natl Cancer Inst* **92**: 242–248.
137. Kudo N, Matsumori N, Taoka H, *et al.* (1999) Leptomycin B inactivates CRM1/exportin 1 by covalent modification at a cysteine residue in the central conserved region. *Proc Natl Acad Sci USA* **96**: 9112–9117.
138. Newlands ES, Rustin GJ, Brampton MH. (1996) Phase I trial of elactocin. *Br J Cancer* **74**: 648–649.
139. Vigneri P, Wang JY. (2001) Induction of apoptosis in chronic myelogenous leukemia cells through nuclear entrapment of BCR–ABL tyrosine kinase. *Nat Med* **7**: 228–234.
140. Kau TR, Silver PA. (2003) Nuclear transport as a target for cell growth. *Drug Discovery Today* **8**: 78–85.
141. Karet FE, Finberg KE, Nelson RD, *et al.* (1999) Mutations in the gene encoding B1 subunit of H+-ATPase cause renal tubular acidosis with sensorineural deafness. *Nat Genet* **21**: 84–90.

142. Frattini A, Orchard PJ, Sobacchi C, *et al.* (2000) Defects in TCIRG1 subunit of the vacuolar proton pump are responsible for a subset of human autosomal recessive osteopetrosis. *Nat Genet* **25**: 343–346.
143. Nishi T, Forgac M. (2002) The vacuolar (H+)-ATPases — nature's most versatile proton pumps. *Nat Rev Mol Cell Biol* **3**: 94–103.
144. Beyenbach KW, Wieczorek H. (2006) The V-type H+-ATPase: molecular structure and function, physiological roles and regulation. *J Exp Biol* **209**: 577–589.
145. Martinez-Zaguilan R, Lynch RM, Martinez GM, Gillies RJ. (1993) Vacuolar-type H(+)- ATPases are functionally expressed in plasma membranes of human tumor cells. *Am J Physiol Cell Physiol* **265**: C1015–C1029.
146. Bowman EJ, Siebers A, Altendorf K. (1988) Bafilomycins: a class of inhibitors of membrane ATPases from microorganisms, animal cells, and plant cells. *Proc Natl Acad Sci USA* **85**: 7972–7976.
147. Werner G, Hagenmaier H, Drautz H, *et al.* (1984) Metabolic products of microorganisms. 224. Bafilomycins, a new group of macrolide antibiotics. Production, isolation, chemical structure and biological activity. *J Antibiot (Tokyo)* **37**: 110–117.
148. Dröse S, Bindseil KU, Bowman EJ, *et al.* (1993) Inhibitory effect of modified bafilomycins and concanamycins on P- and V-type adenosinetriphosphatases. *Biochemistry* **32**: 3902–3906.
149. Bowman EJ, Bowman BJ. (2000) Cellular role of the V-ATPase in Neurospora crassa: analysis of mutants resistant to concanamycin or lacking the catalytic subunit A. *J Exp Biol* **203**: 97–106.
150. Bowman EJ, Siebers A, Altendorf K. (1988) Bafilomycins: a class of inhibitors of membrane ATPases from microorganisms, animal cells, and plant cells. *Proc Natl Acad Sci USA* **85**: 7972–7976.
151. Huss M, Wieczorek H. (2009) Inhibitors of V-ATPases: old and new players. *J Exp Biol* **212**: 341–346.
152. Bowman BJ, Bowman EJ. (2002) Mutations in subunit c of the vacuolar ATPase confer resistance to bafilomycin and identify a conserved antibiotic binding site. *J Biol Chem* **277**: 3965–3972.
153. Huss M, Ingenhorst G, König S, *et al.* (2002) Concanamycin A, the specific inhibitor of V-ATPases, binds to the V(o) subunit c. *J Biol Chem* **277**: 40544–40548.
154. Bowman EJ, Graham LA, Stevens TH, Bowman BJ. (2004) The bafilomycin/concanamycin binding site in subunit c of the V-ATPases from Neurospora crassa and Saccharomyces cerevisiae. *J Biol Chem* **279**: 33131–33138.

155. Beutler JA, McKee TC. (2003) Novel marine and microbial natural product inhibitors of vacuolar ATPase. *Curr Med Chem* **10**: 787–796.
156. Bockelmann S, Menche D, Rudolph S, *et al.* (2010) Archazolid A binds to the equatorial region of the c-Ring of the vacuolar H+-ATPase. *J Biol Chem* **285**: 38304–38314.
157. Hong J, Yokomakura A, Nakano Y, *et al.* (2006) Inhibition of vacuolar-type (H+)-ATPase by the cytostatic macrolide Apicularen A and its role in Apicularen A-induced apoptosis in RAW 264.7 cells. *FEBS Lett* **580**: 2723–2730.
158. Pérez-Sayáns M, Somoza-Martín JM, Barros-Angueira F, *et al.* (2009) V-ATPase inhibitors and implication in cancer treatment. *Cancer Treat Rev* **35**: 707–713.

CHAPTER 6

Marine Compounds

Jennifer Honek and Thomas Efferth†*

ABSTRACT

The world's oceans harbour a vast diversity of organisms. During the past decade, interest in marine-derived compounds with potential use in anti-cancer treatment has grown, and investigation of these compounds has intensified. As a result, the number of marine-derived anti-cancer lead compounds that are in clinical or pre-clinical studies has increased considerably. The coexistence of species in a spatially limited habitat coincides with the evolution of various mechanisms of defense and competition. Sessile species such as algae, corals and sponges are at high risk of predation and in constant competition for suitable attachment space. Consequently, they have developed highly sophisticated chemical defence mechanisms based on the synthesis of toxic secondary metabolites. Marine natural products in clinical trials or on the market as anti-cancer drugs act on a variety of cellular and molecular targets. Some compounds directly inhibit DNA synthesis by interfering with DNA polymerase; others inhibit histone deacetylases (HDACs), which are required for rendering the DNA accessible for transcription. Some compounds hinder tumor metastasis by serving as matrix metalloprotease inhibitors. Other groups of substances interfere with Rho-GTPase function and thereby oppose tumor progression. Another mechanism of anti-tumor activity is the induction of apoptosis or necrosis in cancer cells. The ability to inhibit NF-κB has been attributed to a number of marine natural compounds. This chapter provides an overview on promising marine compounds and their modes of action against tumors.

*Institute of Pharmacy and Molecular Biotechnology, University of Heidelberg, Im Neuenheimer feld 364, 69120 Heidelberg, Germany.

†Institute of Pharmacy and Biochemistry, Johannes Gutenberg University, Staudinger Weg 5, 55128 Mainz, Germany.

1. INTRODUCTION

The current biological diversity is a result of billions of years of evolution. Natural processes and selection have forced plants and animals to adapt to their habitat and optimize their defence mechanisms. According to the World Conservation Monitoring Centre, about 1.75 million species have been identified to date.[1] However, some scientists estimate that the number of species may be 50 times more.

The term *biodiversity* describes habitat diversity, genetic diversity and species diversity, each of which is tightly linked to the other two. The term further represents the variability among all organisms and ecological systems.

The world's oceans harbour a vast diversity of organisms, many of which lack physical defence mechanisms due to restricted mobility. Consequently, they have highly sophisticated chemical defence mechanisms based on the synthesis of toxic secondary metabolites.

The therapeutic potential of natural products has been acknowledged for hundreds of years and plant-derived biologically active compounds have been a hallmark of traditional medicine for even longer.

The majority of marine species inhabit the ocean fringe and the deep ocean thermal vent communities, which are amongst the most biodiverse environments. Together, the ocean fringe and deep sea vents cover <1 percent of the earth's surface, yet they harbour the majority of the world's species. Due to the great diversity of species living in such a restricted area, marine organisms must be highly competitive in order to survive and spread. Sessile organisms especially, such as sponges, corals and algae, find themselves in a constant struggle for attachment space, nutrients and survival. Chemical defence strategies are based on the production of toxic secondary metabolites such as terpenoids, alkaloids, steroids, sugar derivatives and peptides. Apart from these common forms of bioactive secondary metabolites, marine-derived compounds often contain covalently bound halogen atoms such as chlorine and bromine.

During the past decade, interest in marine-derived compounds with potential use in anti-cancer treatment has grown, and investigation of these compounds has intensified. As a result, the number of marine-derived anti-cancer lead compounds that are in clinical or preclinical studies has considerably increased.

2. NATURAL PRODUCTS IN CANCER THERAPY

Natural products from plants, bacteria and fungi have been a traditional source of drugs for centuries, and these compounds have found wide-ranging applications in biology, pharmacy and medicine.[2] Regarding cancer chemotherapy, various important drugs are derived from nature Structural modification of natural compounds as well as the production of semi-synthetic and synthetic compounds based on natural compounds has yielded potent antitumor drugs such as paclitaxel.

However, it was not until the late 1950s that the therapeutic value of marine natural products was noticed.[3] Due to rather poor accessibility of these compounds their potential was not realized for many years. Nowadays, extensive screening of marine natural products has resulted in the identification of marine-derived compounds that affect the cell cycle and cellular metabolism.[3]

Since drug resistance is a major issue in chemotherapy, seeking novel lead compounds for chemotherapy is crucial to improve the cure rates of this multi-faceted disease. Semi-synthetic modification of natural lead compounds may increase bioactivity while lowering the therapeutic index, thereby reducing side effects.[2] Recent achievements in bioinformatics and biotechnology, including high-throughput screening, computational chemistry and the possibility of genetic modification, have opened new avenues for the discovery and design of novel anticancer drugs as well as new methods for large-scale production.[2]

Chemotherapy relies strongly on the biological differences between cancer cells and normal cells. Consequently, strategies to improve therapeutic efficacy while lowering toxic side effects focus on targeting molecules and pathways that are biologically essential for neoplastic cells but not crucial for the survival of healthy cells. However, since these differences are frequently marginal, cancer patients have to cope with severe adverse effects, which in many cases temporarily aggravate their condition. Therefore, increasing therapeutic selectivity is — apart from circumventing drug resistance — one of the most urgent issues in cancer therapy.

3. ACCESS AND SUPPLY OF MARINE NATURAL PRODUCTS

The number of marine-derived natural products in anti-cancer preclinical and clinical trials has increased during the past decade.[4]

As a potential drug candidate, a compound is required to meet several criteria. There must exist a sustainable, industrially feasible supply; the compound must exhibit moderate toxicity; and there must be a possibility to develop a formulation that is applicable in the clinic. Furthermore, analytical methods are required to determine characteristics such as the compound's pharmacokinetic parameters. The compound must be able to be metabolized in a way that maximizes efficacy and limits adverse effects to a tolerable level. In this respect, pharmacogenetic differences between individuals have to be considered and the therapeutic index should be as high as possible.[4]

To guarantee a sustainable supply, several possibilities are available for the retrieval of natural compounds. One option is the controlled use of natural resources. Alternatives include mariculture or aquaculture. For less easily accessible compounds or compounds that are available only in limited amounts, hemisynthesis and total syntheses are promising options. Recent advances in molecular biology and biotechnology allow for the generation of transgenic bacteria, which are easy to cultivate and can produce many desired compounds upon genetic modification.[4]

4. MARINE ORGANISMS AS SOURCE OF CANDIDATE DRUGS FOR CANCER THERAPY

The marine environment harbors immeasurable biological and chemical diversity. It is estimated that up to two million different species live in the world's oceans, especially in biodiversity hotspots like the Palaun Islands.[4–6]

The coexistence of such a large number of species in a spatially limited habitat (as mentioned, the majority of sea life is confined to the ocean fringe and hydrothermal vent areas) coincides with the evolution of various mechanisms of defense and competition. In particular, sessile species such as algae, corals and sponges are at high risk of predation and in constant competition for suitable attachment space. Growth-limiting factors such as nutrient supply, water current and temperature enhance the competition between different individuals and species.

Consequently, a large diversity of chemical compounds has evolved. Many of these compounds constitute secondary metabolites such as alkaloids, peptides, etc., but halogenated compounds are also common among marine-derived natural products.[4,6]

In the 1960s, marine biodiversity was heralded as a promising source for novel pharmaceuticals raising interesting therapeutic options.

Studies have revealed a broad range of potential applications of several marine-derived compounds due to their anti-infectious, anti-proliferative, cytotoxic or anti-tumor activity.[4,6]

Research conducted in the past years has increasingly provided evidence that marine microalgae, cyanobacteria and heterotrophic bacteria living in symbiotic relationships with marine invertebrates are the best sources of biologically active compounds in a large number of cases[7] (Fig. 1).

4.1. Algae

The term *algae* is used to describe a variety of different organisms belonging to different phylogenetic groups. They are plant-like (commonly photosynthetic and aquatic) organisms lacking roots, leaves and vascular tissue. Algae are distributed globally in the sea and fresh water as well as in moist terrestrial areas. Although most algae are microscopic, some marine seaweed exceed 50 m in length. Most algae are eukaryotic organisms containing chloroplasts, and are thus able to conduct photosynthesis.

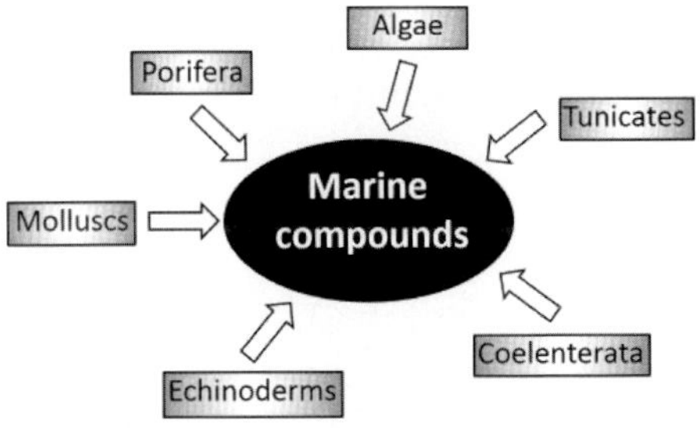

Figure 1 Sources of marine compunds with cytotoxic and anti-cancer activity.

Cyanobacteria represent a specific, prokaryotic phylogenetic group of algae. These autotrophic organisms have photosynthetic ability; however, they lack chloroplasts and conduct this energy converting process in the cytoplasm. Cyanobacteria can be either single-celled or colonial.

Dinoflagellata are flagellated, hetero- or autotrophic algae. Many species belonging to this group of protists are endosymbionts of marine invertebrates.

4.2. Porifera

Sponges belong to the *phylum porifera* and are characterized by a hollow outer part that is held together by a jelly-like collagen-containing mesophyll. They have neither a digestive, nor a circulatory system, and rely heavily on maintaining a constant water flow through their tissue in order to receive nutrients and oxygen and get rid of waste products. Sponges filter food particles out of the water that is flowing through their form. Oxygen is taken up by diffusion. These sessile metazoans populate tidal areas as well as the depths of the ocean. They attach preferentially to firm surfaces provided by rocks, although some can attach to soft sediment. The majority of sponges feed on bacteria; however, some of them harbour photosynthesizing organisms such as cyanobacteria as endosymbionts. Marine sponges are invertebrates and — due to their restricted motility — lack physical defences. Therefore, in order to protect themselves from predators and be able to compete for space and nutrients, they have developed chemical defence mechanisms by producing highly toxic secondary metabolites. Amongst all marine invertebrates, sponges produce the highest quantity of these metabolites. Interestingly, these compounds are effective at extremely low doses and exhibit unique modes of action.

It should be noted that the bioactive compounds are not always produced by the sponges themselves, but rather by the great variety of bacteria they harbour.

4.3. Coelenterates

Coelenterata comprises two animal phyla, namely Ctenophora (comb jellies) and Cnidaria (coral animals) containing corals, jellyfish, hydra and sea anemones. Soft corals commonly live in a symbiotic relationship with dinoflagellate algae, which provide the corals with food. Coelenterates are characterized by a rather simple physiology. Their jelly-like body encloses the internal organs. Some species feed by stunning or paralysing prey with their so-called nematocysts while others feed mainly on plankton. Due to their fleshy and often colourful body, predators such as fish and molluscs are attracted to these marine invertebrates. The coelenterates's defence mechanisms are based on stinging cells as well as on toxic secondary metabolites. These metabolites also act as a potent weapon in the struggle for territory and attachment space by inhibiting the growth of neighboring animals and algae.

4.4. Echinoderms

Echinoderms constitute a phylum of marine animals that live at every ocean depth. The phylum comprises starfish, brittle stars, crinoids, sea urchins and sea cucumbers. The echinoderm body is radially symmetrical and many echinoderms are characterized by their slow mobility. However, sessile species have been reported. Perhaps due to their low or non-mobility and inability to escape predators by physical means, echinoderms have not disposed of the ability to defend themselves using chemical means. Consequently echinoderms rely on chemical defence mechanisms, as do the majority of marine invertebrates.

A great variety of biologically active substances such as carotenoids, ether lipids, glycolipids, porphyrins and naphthoquinones have been isolated from these marine invertebrates. Some of these compounds contain an acetylenic unit. As described in detail later, acetylenic compounds have a great therapeutic potential and are potential drug candidates due to their anti-bacterial, anti-fungal, anti-inflammatory and also anti-tumor properties.

4.5. Tunicates

Tunicates (urochordates) constitute a subphylum of underwater filter-feeding animals possessing a box-like anatomy. The have incurrent and excurrent siphons and feed by filtering water.

4.6. Molluscs

Molluscs are members of the phylum Mollusca, which includes Caudofoveata (deep-sea worm-like creatures), Aplacophora (solenogasters), Polypalcophora (chitons), Scaphopoda (tusk shells), Cephalopoda (squids, octopuses, nautilus, cuttlefish), Bivalvia (clams, oysters, scallops, mussels) and Gastropoda (nudibranchs, snails, slugs, limpets, sea hares). This diverse group of animals constitutes the largest marine phylum, containing approximately one-fourth of all marine organisms that have been described thus far. Although anatomy differs among the various mollusk species, a common feature is a mantle with a cavity for breathing and excretion.

4.7. Targets of Cancer Therapy

During the past years, remarkable progress has been made in elucidating the molecular mechanisms underlying cancer formation and progression. Molecular alterations observed in neoplastic lesions provide potential targets for cancer chemotherapy. Exploiting the overexpression of certain cellular components can render tumors more vulnerable to treatment than normal tissue. Interfering with biological pathways involved in proliferation, angiogenesis and cell-cell-adhesion, all of which play a critical role in tumor progression, provides the basis for the anti-tumor effect of many chemotherapeutic drugs.

Marine natural products in clinical trials or on the market as anticancer drugs act on a variety of cellular and molecular targets. Some compounds, such as Ara-C, directly inhibit DNA synthesis by interfering with DNA polymerase; others inhibit histone deacetylases (HDACs), which are required for rendering the DNA accessible for transcription.[8] Some compounds hinder tumor metastasis by serving as matrix metalloprotease inhibitors.[9] Other groups of substances interfere with Rho-GTPase function and thereby oppose tumor progression.[10] Another mechanism of anti-tumor activity is the induction of apoptosis or necrosis in cancer cells. The ability to inhibit NF-κB has been attributed to a small number of marine natural compounds. These compounds exert their function by either targeting IKK-dependent degradation of IκB, interfering with the proteolytic activity of the 26S proteasome or by blocking binding of NF-κB to its corresponding DNA binding site. However, the mode of action of many marine derived anti-cancer drugs has not yet been elucidated.[11]

5. GENERAL PROBLEMS IN CANCER THERAPY

5.1. Efficiency vs. Specificity (Targeting of Tumor Cells vs. "Normal Cells")

A major challenge that remains in the field of cancer chemotherapy is the delicate balance between the specificity of a drug and its selectivity. In the course of chemotherapeutic as well as radiotherapeutic treatment of cancer, both normal and neoplastic cells are affected. On the one hand, chemotherapeutics aim at efficiently damaging and ultimately killing cancer cells. However, the drugs must simultaneously be highly specific in order to avoid causing devastating damage to normal, non-neoplastic cells. Damage to normal cells frequently evokes severe side effects in chemotherapy due to the restricted specificity of conventional anti-tumor drugs. These undesired side effects can be attributed to the interference of chemotherapeutics with basic cellular pathways that are involved in the correct functioning of every cell.

Cancer therapy is often based on non-targeted drugs that act downstream of signaling pathways such as DNA synthesis or microtubule assembly rather than targeting specific cellular structures that are only present or modified in malignant cells.[12] Chemotherapy also frequently exploits the differential proliferative potential between normal and malignant cells. Due to the anti-proliferative nature of anti-neoplastic agents, rapidly dividing cell types in normal tissues, such as bone marrow, hair follicles and skin, are subject to the adverse effects of the drug as well. Healthy tissue is thus affected by the drug, frequently leading to the development of severe adverse effects such as alopecia, leucopenia and neutropenia, mucositis, nausea and vomiting, diarrhea and infections. Although these side effects are only temporary and can be reversed upon cessation of chemotherapy, they aggravate the patient's condition and impose a significant physical as well as psychological burden on the patient.

Targeted chemotherapy is also tightly linked to severe side effects, since the target molecule is most commonly present in normal, healthy tissue as well as in cancer cells. However, the target molecules are often overexpressed or modified in some way in neoplastic cells in comparison to normal cells. Consequently, the dosage required to achieve an antitumor effect does not affect normal cells to the same extent. Furthermore, during non-dosage intervals, normal tissue regenerates faster than neoplastic tissue. For example, anti-proliferative drugs preferentially target cancer cells due to their high rate of proliferation. Common targets in cancer therapy are growth factor receptors, signaling molecules, cell cycle regulators and molecules involved invasion and angiogensis.

5.2. Drug Resistance

Resistance of cancer cell subpopulations to chemotherapeutic treatment is one of the most challenging issues in tumor therapy. Such resistance can be accomplished by several different mechanisms that are explained in the following sections. If only a small number of cells resist chemotherapeutic treatment, the tumor decreases in size suggesting improvement in the patient's condition. However, the survival of a few chemotherapy resistant tumor cells is sufficient to cause a relapse resulting in the formation of a highly aggressive tumor. Treatment of such secondary tumors is complicated by the fact that they feature resistance to a number of conventional chemotherapeutic drugs.

5.2.1. *Multidrug resistance*

The phenomenon of multidrug resistance (MDR) is an important issue in cancer therapy. MDR describes the fact that tumor cells possess intrinsic or acquired cross-resistance to various chemotherapeutic agents. MDR complicates effective treatment of cancer and aggravates the prognosis of this disease. Different proteins such as P-glycoprotein (P-gp), multidrug resistance associated protein (MRP1) and breast cancer resistance protein (BCRP) are involved in MDR. These proteins are members of the Adenosine triphosphate (ATP)-binding cassette (ABC) transporter protein superfamily, and actively expel drug from the intracellular space in an ATP-dependent manner, thereby hindering the intracellular accumulation of the therapeutic agent. Such efflux pumps prevent the drugs from reaching their target organelles in sufficient quantities. Overexpression of MDR-associated proteins increases a tumor's resistance to a broad variety of chemotherapeutic agents currently used in cancer treatment.

5.2.2. *Anti-oxidative defense*

Reactive oxygen species (ROS) are constantly generated under physiological conditions via several different pathways. The most important source of ROS generation is the respiratory chain in the mitochondria of every cell. During oxidative phosphorylation, electrons are delivered to the respiratory chain. A proton gradient across the inner mitochondrial membrane is established and serves as an energy source for ATP generation. Reactive oxygen species such as superoxide and hydrogen peroxide are oxygen containing chemical species with highly reactive properties that are generated as a byproduct of cellular respiration. These molecules can bind to DNA or other cellular macromolecules such as lipids or proteins and thereby cause harm to the cell. In the normal physiological environment, ROS are scavenged by intrinsic cellular defense mechanisms such as the enzymes catalase and superoxide dismutase (SOD) as well as the GSH and TRX system.

However, when this delicate equilibrium between ROS generation and detoxification is shifted towards generation, the cell can no longer balance the oxidative stress caused by these reactive molecules and is therefore subjected to oxidative insults leading to mutations, increased proliferation and/or apoptosis.

Some cancer drugs attempt to eradicate tumor cells by generating ROS. However, in many tumor cells, antioxidant enzymes are overexpressed and can effectively scavenge therapy-induced ROS and protect the tumor cell from oxidative insult.[13]

Numerous studies indicate that high SOD levels in tumors are associated with poor prognosis, resistance to chemotherapeutic and irradiation treatment, and aggressive malignant transformation of cells.[14,15]

Even though the mechanisms leading to overexpression of antioxidant enzymes in cancer cells remain to be clarified, there is evidence that anticancer drugs induce the activity of glutathione-related enzymes.[16–18]

5.2.3. *Anti-apoptotic mechanisms*

Apoptotic cell death is a crucial mechanism in maintaining a balance between cell death and proliferation, and is therefore essential for tissue homeostasis. Apoptosis, or programmed cell death, is a feature of various physiological as well as pathological processes. This tightly regulated mechanism of cell demise involves both transcriptional and post-transcriptional gene activation. Apoptosis is closely linked to the machinery regulating cell proliferation and differentiation through oncogenes and tumor suppressor genes. The inability of cells to undergo apoptosis following certain stimuli is implicated in tumor promotion and progression. It has been demonstrated that malfunctions in the apoptotic pathway are a common feature of human tumor cells.

Apoptotic cell demise in normal cells, triggered by excessive oxidative stress and chemical or physical insult to the DNA, is mediated via two distinct pathways: an intrinsic, mitochondrial pathway and an extrinsic, death receptor mediated pathway. Both pathways eventually result in the activation of cysteine asparagine proteases (caspases), which cleave a number of nuclear and cytoplasmic substrates such as DNA. Caspases act as death effector molecules and are subdivided into activator and effector caspases.

The extrinsic pathway is triggered by the binding of a substrate to death receptors of the tumor necrosis factor (TNF) superfamily. The TNF receptors feature an extracellular cysteine-rich domain, a membrane spanning domain and an intracellular death domain with a common sequence. Upon ligand binding and trimerization of the receptors, accessory proteins are recruited and eventually the death inducing signaling complex (DISC) is assembled. Formation of DISC leads to the activation of caspases 2, 8 and 10, thereby initiating the caspase-mediated cell death cascade. These proteases in turn cleave and activate downstream effector caspases such as caspase 3, 6 and 7.

The intrinsic apoptotic pathway is independent from receptor activity, and is initiated by permeabilization of the outer mitochondrial membrane. Mitochondria sequester pro-apoptotic proteins such as Apoptotic inducing factor 1 (AIP1) and Second mitochondria-derived activator of caspase/Direct IAP-binding protein with low PI (Smac/DIABOLO), which play a pivotal role in programmed cell death. A crucial step in the intrinsic apoptotic pathway is the release of cytochrome c from the mitochondria into the cytosol. Cytochrome c is a member of the electron transport chain and — under normal conditions — is attached to the inner mitochondrial membrane. Release of cytochrome c from the mitochondria is achieved by the formation of the mitochondrial permeability transition (mPT) pore. The mPT pore is formed from both the outer and inner mitochondrial membrane and consists of a large number of protein complexes. However, only molecules below 5 kDa are able to pass through this pore. It is a matter of debate, whether cytochrome c is expelled from the mitochondria through specific pores in the outer mitochondrial membrane or whether it is released upon rupturing of the membrane, which is induced by osmotic swelling.

Pro-apopototic proteins such as Bax and the cytosolic protein Bid are also involved in the opening of the mPT pore. Bid is present in an inactive state in the cytosol and is activated upon cleavage by caspase-8 yielding truncated-Bid (tBid), which translocates into the mitochondria and induces the release of cytochrome c.

Members of the Bcl-2 protein family tightly regulate the opening of the mPT pore and a delicate balance of pro- and anti-apoptotic proteins determines the fate of the cell. Bcl-2 proteins including pro-apoptotic Bax and anti-apoptotic Bcl-2 form dimers and the relative abundance of each protein dictates whether a cell will progress into cell demise or rescue pathways. The rescue molecule Bcl-2 prevents the release of cytochrome c from the mitochondria under normal conditions by functioning as a gatekeeper of the mPT pore, while Bax promotes the opening of this pore.

Cells subjected to genotoxic insult leading to DNA damage usually undergo apoptosis. However, it has been demonstrated that the apoptotic pathway is frequently disrupted in cancer cells. Consequently, DNA alterations can be established and are passed on during cell division, resulting in the formation of preneoplastic foci with a growth selection advantage that benefits their proliferation and tumor promotion and progression (quelle: Toxicology book).

Resistance to apoptosis in human cancers can be caused by several distinct mechanisms. Alterations in the death-receptor pathway of apoptosis can result from impaired death receptor expression or function as well as from aberrant expression of anti-apoptotic proteins that oppose the activation of caspase 8. Down regulated surface expression or complete absence of death receptors limits the transduction of the initial death signal to the intracellular signaling pathways that initiate apoptotic cell death. In chemotherapy-resistant neuroblastoma and leukemia cells, diminished expression of the CD95 receptor has been reported. These findings hint at a pivotal role for the CD95 receptor in drug sensitivity. Dysfunctions on a post-transcriptional level, such as in the transport of receptors from intracellular storage to the cell surface also hinder the response of tumor cells to chemotherapeutics. This phenomenon has been demonstrated for the tumor necrosis factor-related apoptosis-inducing ligand (TRAIL) receptors TRAIL-R1 and TRAIL-R2. The genes coding for both CD95 and TRAIL-R are located on chromosome 8p, and this genetic locus is frequently missing in human cancers. Furthermore, mutations in the CD95 gene have been detected in hematological cancers and various solid tumors. However, it is not only genetic changes but also epigenetic modifications (such as CpG-island hypermethylation in promoter regions) that can lower surface expression of death receptors and thereby impair death-receptor signaling and apoptotic cell demise (quelle: Tumor resistance to apoptosis paper).

Furthermore, aberrant expression of anti-apoptotic genes such as the cellular FLICE inhibitory protein (cFLIP) can interfere with the apoptotic pathway by preventing recruitment of caspase 8 to activated death receptors and thereby hampering activation of the caspase cascade. The underlying mechanism of inhibiting caspase 8 recruitment is based on the high sequence similarity of these anti-apoptotic proteins with pro-caspase 8. Due to this homology, they are recruited to the DISC instead of pro-caspase 8.

Another factor influencing the transmission of the death signal is the expression level of caspase 8 itself. Mutations in the gene coding for caspase 8 result in the expression of aberrant proteins that prevent the recruitment of wild type caspase 8 to activated death receptors. Thus, this mutation acts in a dominant negative manner. However, such mutations are only rarely observed in human cancers. Moreover, alternative splicing resulting in the generation of caspase 8L, which lacks a catalytic domain, can also act as a dominant-negative inhibitor of apoptosis. Epigenetic modifications, namely, hypermethylation of regulatory sequences of the gene coding for caspase 8, have been reported in malignant brain tumors, small cell lung carcinoma and neuroblastoma. Hypermethylation results in silencing of the respective gene and thereby contributes to down regulation of caspase 8.

Regarding the mitochondrial pathway of apoptosis, alterations on both the genetic as well as epigenetic level have been reported. Tumor cell survival and chemotherapy resistance are enhanced by a shift mutation of the pro- and anti-apoptotic proteins of the Bcl-2 family. Translocation of the *bcl-2* locus into the immunoglobulin heavy chain locus has been reported in human follicular lymphoma, resulting in an overexpression of anti-apoptotic Bcl-2.

6. MARINE-ORIGIN ESTABLISHED DRUGS AND THOSE IN DEVELOPMENT

6.1. ET-743

Ecteiascidin (ET)-743 is a compound isolated from the marine tunicate *Ectenascidia turbinate.*[19] (Fig. 2) After its discovery, it took almost 40 years for it to be released into the market as the first marine-derived anti-cancer drug. As early as 1969, extracts from the marine tunicates *Ectenascidia* were recorded as having antitumor activity. However, further analysis of the compounds responsible for this effect was difficult to accomplish since only minute amounts of the active components, the alkaloids named ecteinascidins, are present in the tunicate. As described by Rinehart in 1990, the most abundant active components in tunicate extract are ET-743 and its N-demethyl analogue ET-729; both possess a similar potency.[20] Chemically, these substances comprise three fused tetrahydroisoquinolone rings; the third tetrahydroisoquinolone ring is connected to the base structure by a thioether bridge.[21,22] This connection completes a 10-membered lactone, which is a characteristic feature of ecteinascidins. Early studies reported the cytotoxic effect of these alkaloids against L1210 leukemia cells.[21] Subsequent *in vivo* studies demonstrated a potent anti-tumor activity in mouse models. The anti-tumor effect was observed in mouse models of P388 lymphoma, B16 melanoma, M5076 ovarian sarcoma, and human lung carcinoma as well as mammary carcinoma xenografts.

ET-743, but not ET-729, was chosen for further investigation simply due to the fact that ET-729 is less abundant in tunicate extract and therefore

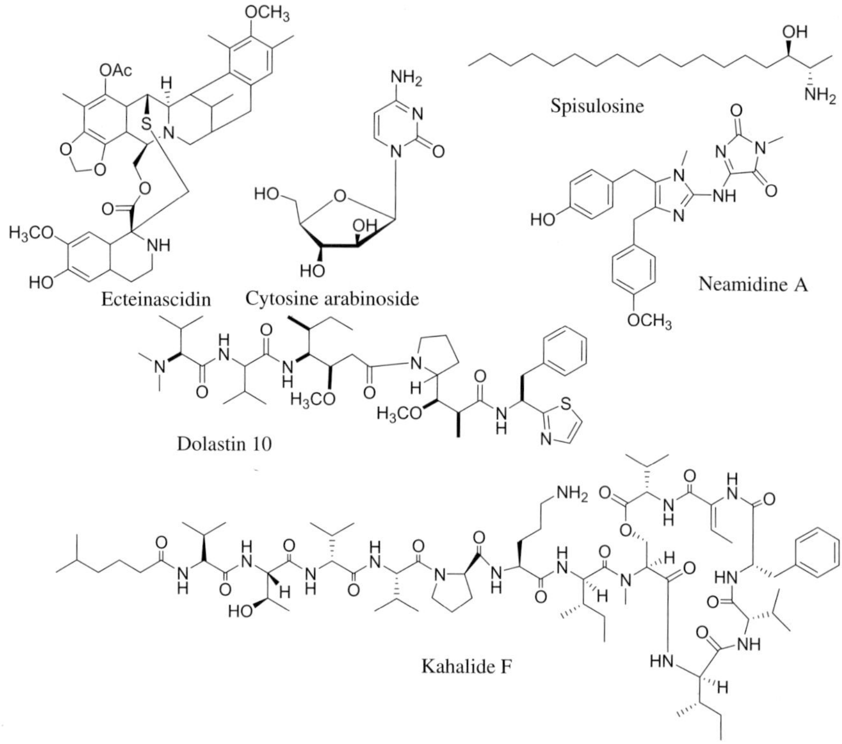

Figure 2 Chemical structures of selected marine compounds.

it is more difficult to conduct comprehensive studies on this compound. However, accessibility of ET-743 has been a critical issue in conducting further studies because the yield when isolating it from tunicates is very low. Over the years, a semi-synthesis process for ET-743 has been optimized, providing sufficient material for *in vitro* as well as *in vivo* studies. Following wide-ranging studies and clinical trials, ET-743 was approved for the treatment of refractory soft-tissue sarcomas by the European Commission in July 2007 and is currently in the market (Yondelis).[23,24]

Studies investigating combinatory effects of ET-743 with doxorubicin, paclitaxel and irinotecan, respectively, suggested synergistic effects.[19]

Dose-limiting toxicities associated with ET-743 were found to be neutropaenia and thrombocytopaenia.[25]

ET-743 mediates its antitumor effect by covalently binding to the minor groove of the DNA. The compound alkylates guanine nucleotides specifically at the N2 position.[21,26] ET-743 selectively binds to CpG islands in the promoter region forming a complex. Adduct formation with the double stranded DNA causes the helix to bend towards the major groove. Ring 3 of ET-743 has been shown to protrude into the minor groove and interferes with DNA binding factors in the promoter region of certain genes, *e.g.* cell cycle control genes, thereby modifying their transcription. As opposed to other common alkylating agents that are used in cancer therapy, binding of ET-743 to the DNA is reversible upon denaturation.[19]

Interestingly, only promoters containing a CCAAT box are affected. *Mdr1* (multi drug resistance gene 1), the gene encoding for P-glycoprotein that is associated with the development of drug resistance, is regulated by a promoter containing a CCAAT box. Therefore, ET-743 is efficient against tumor cell lines that are normally resistant to treatment due to *mdr1* overexpression. ET-743 inhibits only cancer-activated but not basal transcription.[3,27]

Furthermore, ET-743 affects transcription-coupled nucleotide excision repair ((TC)-NER) and induces cell death. Interference of ET-743 with nucleotide excision repair has been investigated by Takebayashi *et al.*[19] They demonstrated that adducts formed by ET-743 with the DNA are recognized by the cellular nucleotide excision repair (NER) system. Their findings imply that (TC)-NER plays an essential role in the sensitivity of cancer cells to treatment with ET-743. The details of how (TC)-NER mediates the cytotoxicity of ET-743 have not been fully elucidated yet.[19] Takebayashi *et al.* suggested that repair complexes might produce irreversible single strand breaks (SSBs) in the DNA while ineffectively excising ET-743-alkylated DNA. According to their findings, the formation of SSBs is tightly correlated with NER activity and cytotoxicity, thereby implying that ET-743 induces cell death via induction of TC-NER-mediated SSBs in actively transcribed genes. The effect may be mediated by transcription-associated DNA-damage resulting in SSBs flanking the lesion and eventually leading to the demise of affected cells.[19] This effect was observed at the same concentrations of ET-743 that have been used in Phase I clinical trials.

6.2. Cytosine Arabinoside (Ara-C)

Nucleoside analogues comprise a pharmacologically diverse family of cytotoxic, antiviral and immunosuppressive agents. In general, nucleoside analogues used in anti-cancer therapy include analogues of physiologically occurring purine and pyrimidine nucleosides and nucleobases.[28] The analogues are antimetabolites that interfere with the synthesis of nucleic acids in processes such as DNA replication. Their cytotoxic effect is exerted by incorporation into the RNA or DNA, or by interference with enzymes that are involved in nucleic acid synthesis. Furthermore, alteration of the metabolism of physiologically occurring nucleosides has been reported.[28]

Due to their mostly hydrophilic nature, nucleoside analogues require specialized nucleoside transporters in order to cross the outer cell membrane and enter the cell. It has been suggested that the abundance and tissue distribution of these transporters determines the sensitivity and specificity of nucleoside analogs in a given tissue.[28]

The pyrimidine analogue cytosine arabinoside (Ara-C) has been isolated from the Carribean sponge *Cryptothethya crypta*[25] (Fig. 2). The compound is a desoxycytidine analogue and apart from its cytotoxic activity inhibits DNA-polymerase competitively, thereby interfering with DNA replication. Ara-C selectively targets rapidly dividing cells and is used in the treatment of hematologic malignancies.[28,29] However, the compound has not shown efficacy in the treatment of solid tumors.[28,30] Intravenous administration of Ara-C for seven days yielded complete remission of acute myeloid leukaemia in about one-third of cases. Combining Ara-C treatment with administration of an anthracycline improves the remission rate to 65–75 percent in previously untreated patients.[28]

A key element in the prognosis of patients treated with Ara-C is the expression level of the human equilibrative nucleoside-transport-facilitating protein 1 (hENT1), the enzyme catalyzing the rate-limiting step in the uptake of the compound into the cell. Low transport rates have been shown to be correlated with poor clinical response to Ara-C treatment.[31] However, this problem can be circumvented by administering high doses of Ara-C. In high-dose treatment (10 μM), diffusion rates overcome pump-mediated transport of the compound.[32]

Inside the cell, the drug is converted by deoxycytidine kinase to arabinosyl CMP, which in turn is ultimately phosphorylated to the triphosphate Ara-CTP by deoxycytidylate kinase.[33] Subsequent to incorporation of Ara-CTP into the DNA, the nucleoside analogue blocks initiation and elongation of the replicating DNA strand. Replication is also interrupted by inhibition of DNA polymerases α and β, which are responsible for *de novo* synthesis of DNA and DNA repair, respectively.[34] The major adverse effects of Ara-C treatment are leucopenia, thrombocytopenia, nausea, vomiting, mucositis and alopecia. In high-dose treatment, neurotoxicity and pericarditis have been reported as additional side effects.[28,35]

Ara-CTP is rapidly detoxified within the body via deamination to arabinosyluridine (ara-U) in a reaction catalyzed by deoxycytidine deaminase. Metabolic deamination as well as ara-C's rather low affinity for the activating enzyme, desoxycytidine kinase, markedly limit the cytotoxic activity of this compound.[36] The design of desoxycytidine analogues yielded the compound gemcitabine. This novel drug is more lipophilic than ara-C and therefore has a higher affinity for nucleoside membrane transporters, which facilitates its uptake into the cell. Gemcitabine further has a remarkably high affinity for desoxycytidine kinase, which is required for metabolic activation. Therefore, this compound has beneficial properties regarding accumulation and retention within the cell, and thus an enhanced cytotoxic potential.[28,37] Gemcitabine has shown consistent activity against various solid tumors. Its anti-tumor activity is derived from several mechanisms. Incorporation of gemcitabine into DNA inhibits DNA synthesis via inhibition of ribonucleotide reductase.[28,38] Furthermore, gemcitabine activity is self-potentiating: the formation of gemcitabine-derived triphosphates reduces the intracellular concentration of normal deoxynucleotide triphosphates, especially dCTP. Decreased availability of dCTP in turn results in increased incorporation of gemcitabine nucleotides into newly synthesized DNA. The activating enzyme deoxycytidine kinase is normally down-regulated by high levels of dCTP; however, since dCTP concentrations decrease upon gemcitabine treatment, deoxycytidine kinase concentrations increase. Deoxycytidine kinase catalyzes the reactions that produce active gemcitabine di- and triphosphates, resulting in increased levels of these compounds.[28,37,39–41]

Due to the fact that gemcitabine can also be incorporated into RNA, the compound also interferes with translation processes.[28,42]

The most common adverse effects of gemcitabine treatment include mild myelosuppression, mild nausea and vomiting, influenza-like syndrome and rash.[28] Treatment protocols are in the process of being optimized; the compound has not been released into the market yet.[43]

6.3. Keyhole Limpet Hemocyanin (KLH)

Hemocyanin is a copper-containing protein with a high molecular mass that is isolated from the marine mollusk giant keyhole limpet *Megathura crenulata*. In the 1960s, it was discovered that hemocyanin possesses immunostimulatory properties in humans and animals.[44–46] The compound activates both the humoral and cellular immune response. Immunologists took great interest in this marine-derived protein, but the compound may hold potential in tumor biology as well as in the treatment of parasitic diseases. As opposed to other marine-derived compounds, large amounts of KLH can be isolated from the hemocyanin-rich hemolymph of *M. crenulata*. It was discovered only in the 1990s that two different oligomeric isoforms of KLH, namely KLH1 and KLH2, exist. Naturally occurring hemocyanin is synthesized into so-called pore cells in the connective tissue of certain pulmonate gastropods. Pore cells are large cells with a high percentage of endoplasmatic reticulum (ER) and a characteristic fenestrated surface. Hemocyanin accumulates in ER of these pore cells as intact cylindrical macromolecules; the formation of intracellular crystalline arrays has been reported by Sminia *et al.* (1977).[47] Each functional unit within the molluscan hemocyanin macromolecule has a binuclear copper binding site through which molecular oxygen can be bound. Oxygenated hemocyanins demonstrate a characteristic blue color caused by light absorbance around 340 nm after oxygen binds the copper.[45]

Since the 1980s, KLH has been implicated in a number of clinical studies and it has been suggested that it is beneficial in tumor immunotherapy. Intra-vesicular administration of KLH to patients with bladder carcinoma after surgical excision of the tumor yielded successful outcomes in many cases.[48,49] KLH's beneficial effect has been attributed to the disaccharide epitope Gal(b1-3)GalNAc. It was hypothesized that this epitope cross reacts with an equivalent epitope on the tumor cell surface. The compound causes an antibody response in parallel with an increase in the activity of natural killer cells.[45,49,50] Taken together, the humoral and cell-mediated response to KLH leads to a cytolytic reduction of tumor growth. Clinical studies have demonstrated that the efficacy of KLH is comparable to non-specific immunotherapy and chemotherapy with compounds such as mitomycin C or ethoglucin.[49] Clinical use of KLH has proven to be extremely safe; adverse effects are rare and less severe than those associated with conventional treatment. In addition to bladder carcinomas, KLH treatment has also been studied in other tumors. Mucin-like epitopes of pancarcinoma antigens that are expressed on epthelially derived adenocarcinomas such as colorectal, breast and ovarian carcinomas and ganglioside epitopes on malignant melanomas constitute potential targets for immunotherapeutic treatment with KLH. Gangliosides are widely distributed throughout vertebrate tissues. However, gangliosides expressed on melanomas show slight structural differences from those on normal cells, thereby providing specific targets for KLH-based therapy. Currently, the generation of melanoma-targeted antibodies provides one method for specifically targeting cancer cells. Conjugation of synthetic oligosaccharides to KLH may open a new avenue for provoking a specific cytotoxic antibody response in parallel with the nonspecific stimulation of the immune system via activation of cytotoxic T-cells and NK cells. A number of other epitopes present on tumor cells hold similar potential for designing specific and targeted treatments.[49] The uniqueness of these epitopes is sufficient to allow for specific targeting of neoplastic cells thereby supporting the concept of Active Specific Immunotherapy (ASI).[49,51–53]

6.4. Dolastatin 10

The pentapeptide dolastatin (DOLA)-10, isolated from the mollusk *Dolabella auricularia*, has been identified as possessing anti-tumor properties (Fig. 2). The anti-tumor activity of dolastatins was first revealed during a screening in 1972. Several members of this substance class were tested and DOLA-10 was the most potent cell proliferation inhibitor in the class.[6,54–56] Dolastatin-10 exhibited highly cytotoxic activity against cancer cells. The anti-proliferative effect of DOLA-10 exceeds that of established chemotherapy drugs such as vinblastine and paclitaxel.[28,56,57] The compound revealed inhibitory activity *in vitro* against several different human cancer cell lines such as melanoma, sarcoma and ovarian cancer cells. Phase I clinical trials identified the maximum tolerable dose as 300–400 μg per m^2. The major dose-limiting toxicity is granulocytopaenia.[57] However, although DOLA-10 has exhibited great anti-proliferative potency against ovarian carcinoma xenografts in tumor models, in a phase I clinical trial the administration of dolastatin-10 could not prevent progression in advanced solid tumors.[57]

DOLA-10 is a non-linear depsipeptide containing four unique amino acids (N,N-dimethylvaline, dolaisoleuince, dolaphenine and doleproine) C-terminally conjugated to the primary amine dolaphenine.[6] DOLA-10 inhibits tubulin polymerization as well as tubulin-dependent GTP-hydrolysis. It binds to tubulin at a distinct site close to the vinca peptide binding site, and thereby interferes with the formation of the mitotic spindle.[58] Consequently, cells lack sufficient microtubules to progress through mitosis, and accumulate in the S-phase and G2/M-phase.[59,60] Additionally, DOLA-10 induces apoptosis by causing the down-regulation of the anti-apoptotic gene *bcl2*.[57,61,62] It also up-regulates p53, which is involved in the transactivation of several target genes inducing cell-cycle arrest, apoptosis and DNA-repair.[57,62]

The tumor suppressor protein p53 is among the most frequently altered proteins in human cancer. Approximately 50 percent of all human tumors harbor mutations in the gene coding for p53.[63] In many cases, p53 alterations arise from point mutations, deletions of the short arm of chromosome 17, or amplification of its regulator MDM2. As a result of these types of alterations, high levels of altered p53 are present in the cell or alternatively p53 protein may be completely absent.[64–67] A lack of functional p53 is often associated with metastatic cancer. Such a shortage negatively influences patient prognosis because p53 is involved in many normal cellular processes.[68] The p53 protein is involved in regulating senescence, differentiation, cell death and response to genotoxic stress. Consequently, loss of p53 results in a disregulation of these processes, ultimately leading to resistance to death stimuli, metastasis, enhanced proliferation and genomic instability.[69] Under physiological conditions in normal cells, only minute amounts of p53 are detectable in the cytoplasm. Constant low levels of p53 are maintained in unstressed cells by the E3 ubiquitin ligase MDM2, which targets p53 for ubiquitin-mediated degradation by the proteasome.[69–71] However, upon DNA damage or oncogenic activation, p53 is activated and stabilized through post-translational modifications mediated by various pathways.[69,72,73] Subsequently, p53 binds to downstream response elements and transactivates different target genes involved in DNA damage repair, cell cycle regulation, senescence and apoptosis. In addition to inducing gene expression, p53 also acts as a repressor for certain genes. However, the mechanism underlying its transcriptional repressor activity remains to be clarified.[69,74]

6.5. Acetylenic Lipids

Naturally occurring acetylenic lipids and their derivatives have been found in numerous plants, fungi and marine invertebrates. Interesting biological activities have been attributed to some of the acetylenic lipids with respect to anti-tumor, antibacterial, antimicrobial and antifungal therapy.[75,76] Marine sponges are characterized by limited mobility and therefore lack an important element of physical defense.[77,78] Consequently, sponges have evolved chemical-based mechanisms to protect themselves from predators and to compete for space and nutrients. In particular, they produce secondary metabolites, many of which feature complex structures and unique modes of action. Many of these metabolites exert potent effects at very low concentrations and are of great interest for developing novel valuable pharmacological compounds.

More than 100 metabolites produced by marine sponges possess markedly antitumor activity. The antitumor potential of these compounds is most frequently based on cytotoxicty. In the Brown Bowl Sponge (*Cribrochalina vasculum*), a number of closely related acetylenic alcohols with cytotoxic potential have been discovered. *In vitro* studies and cytotoxicity assays have demonstrated their antitumor activity against various cancer cell lines such as H-522 non-small cell lung line. Five of these metabolites also feature immunosuppressant activity.[78,79]

Apart from cytotoxicity, a rather unspecific way of eliminating tumor cells, by using compounds with a more specific mode of action has also been reported. Lembehyne A, isolated from the Indonesian marine sponge *Haliclona* sp., is a novel long-chain polyacetylene that specifically stops the cell cycle of neuroblastoma cells at the G1 phase.[78,80,81]

Different carotenoids such as carotene, cynthiaxanthin, astaxanthin and halocynthiaxanthin have been isolated from marine molluscs and tunicates.[78] These substances demonstrated anti-mutagenic and anticancer effects against various human cancer cell lines and have been shown to prevent genotoxicity.[82] Genotoxicity is characterized by sister chromatid exchanges that are provoked by cellular insults such as reactive oxygen species. It was demonstrated in a cell culture model that beta-carotene offers significant protection against the generation of sister chromatid exchange (SCE) after treatment of cells with different tumor promoters such as phorbol myristyl acetate.[82,83] Beta-caroten also induces the release of reactive oxygen species from polymorphonuclear leukocytes (PMNL). However, these *in vitro* results were not able to be correlated with results obtained from *in vivo* studies. Furthermore, it has been shown in CHO cells treated with the carcinogen methyl methanesulfonate, which induces genotoxic symptoms such as chromosomal aberrations and translocations that the addition of beta-carotene inhibits genotoxicity in a dose-dependent manner.[82,84]

A protective effect of beta-carotene was also described by Banerjee *et al.*, who treated mouse mammary cell organ cultures with the carcinogens dimethylbenzanthracene (DMBA), N-nitrosodiethylamine and N-methylnitrosourea. Upon addition of the anti-oxidant beta-carotene, the number of SCE events decreased significantly.[78,85]

Carotenes have also been shown to decrease the extent of malignant transformation. They are thought to act in the promotional stage of this process.[82,86]

Callysponginol sulphate A is a C24 acetylenic fatty acid isolated from the marine sponge *Callyspongia truncate.* It is a membrane type 1 matrix metalloprotease (MT1-MMP) inhibitor, and therefore plays a role in the prevention of metastasis.[78,87,88]

Matrix metalloproteases (MMP) are involved in the degradation of extracellular matrix (ECM) macromolecules and are implicated in tumor invasion and metastasis.[89] Physiologically, these proteins play a key-role in ECM re-modelling during development and morphogenesis.[90,91] Their activity is tightly regulated at a transcriptional level by a delicate balance between activation and inhibition.[89,92] It has been found that matrix-metalloprotease 2 (MMP-2) is constitutively active in many tumors.[93–95] The MMP-2 activation ratio has further been shown to correlate with the level of MT1-MMP expression. This finding, along with other studies, implies that MT1-MMP activates MMP-2 in neoplastic tissue.[89,96] Although MT1-MMP is present in normal tissue, its expression is significantly upregulated during oncogenic transformation, for example by v-src or erbB2.[89,97]

The ECM is involved in regulating proliferation, morphogenesis and cell migration and constitutes a physiological protection mechanism against tumor cell invasion by functioning as a physical barrier. ECM degradation by MMPs removes this barrier, and thereby is a critical step in enabling metastasis. Furthermore, MMPs have also been implicated in angiogenesis, another critical process in tumor growth.

6.6. ES-285

Spisulosine (ES-285) is isolated from the marine mollusc *Spisula polynyma* (Fig. 2). The compound's 2S,3R-2-amino-3-octanodecanol structure shares significant analogies with sphingosine-related lipids.[98]

Two different hypotheses regarding the mechanism by which it exerts its cytotoxic effect on cancer cells have been suggested. The first hypothesis proposes that ES-285 acts at the extracellular level as an antagonist of the phospholipid growth factors (PLGF) S1P/LPA.[98] PLGFs are pleiotropic molecules that provoke a myriad of cellular responses affecting proliferation, survival, morphological reorganization, adherence, chemotactic movement and ionic conductance.[98–107] ES-285s effect PLGFs via the endothelial differentiation gene (EDG) receptors, which constitute a subgroup of G-protein coupled receptors (GPCRs). However, it was found that these receptors do not play a crucial role in ES-285 mediated cell demise.[98]

Extensive studies indicate that ES-285 acts as an activator of some protein kinase C (PKC) isoforms.[98] PKCs are serine/threonine kinases and are involved in signaling pathways that regulate growth, differentiation and apoptosis. PKC is named for its calcium-dependent activity and is physiologically activated via the phosphoinositol 3-kinase (PI3K) pathway which, in turn, is initiated by G-protein coupled receptors (GPCR). When substrate binds a GPCR, GDP is exchanged for GTP, resulting in activation of the trimeric G-protein, which consists of α, β and γ subunits. Activation of the G-protein corresponds to a change in its conformation, which subsequently induces dissociation of the α subunit of the G-protein from the GPCR. The activated G-protein then induces phospholipase C-β (PLC-β) activity. This enzyme hydrolyses phosphatidylinositol-4,5-biphosphate ($PI(4,5)P_2$) thereby generating intracellular secondary messengers, namely inositol-1,4,5-triphosphate (IP_3) and diacyl glycerol (DAG). While IP_3 induces release of calcium ions from the endoplasmic reticulum (ER) into the cytosol, DAG activates PKC in a concerted action with phosphatidyl serine and Ca^{2+}.[108] Activation of PKC results changes in its subcellular localization and the phosphorylation of various target molecules. However, little is known about the downstream events of PKC activation. The major PKC-activated pathway is the MEK-ERK pathway, which is implicated in tumor progression and promotion.[109,110] There is evidence that PKC isoforms α, δ and ε phosphorylate and thereby activates the serine/theronine kinase *Raf1*.[110–112] Activated *Raf1* induces the mitogen-activated protein kinase pathway via phosphorylation of MEK1 and MEK2. This ultimately results in the expression of genes involved in cell proliferation.[110,113] The PKC isoforms α, $\beta 1$ and γ have been reported to specifically inactivate glycogen synthetase kinase 3β (GSK3β) preventing repression of the signal transducing transcription factor *c-jun*, which is implicated in cell cycle progression and cell transformation.[110,114,115]

Tumor promoters such as phorbol esters also induce PKC activation comparably to the endogenous activator DAG.[110,116] Studies have demonstrated that increased activation and overexpression PKC induces malignant transformation of cells and that inhibition of PKC is associated with the induction of apoptosis and reduced tumor growth.[98,117–119]

6.7. Kahalalide F

The secondary metabolite Kahalalide F (KF) has been isolated from the mollusk *Elysia rufescens* and preclinical experiments have shown promising results for KF treatment *in vitro* as well as *in vivo* (Fig. 2). The depsipeptide KF is a C_{75} cyclic tripeptide containing unusual amino acids such as Z-dehydroaminobutyric acid.[3] It was later found that the compound is not produced by *E. rufescens* itself, but rather by the alga *Bropsis* spp. on which the mollusk feeds. Both organisms exploit the compound for defense against predators. KF concentrations are about four orders of magnitude lower in *Bryopsis* spp. than in *E. rufescens*, and therefore are not toxic to the latter.[3]

Solid-phase synthesis allows for large-scale synthesis of Kahalalide F. In the early 1990s, the compound was licensed.

KF is active against various types of solid tumors such as prostate, breast, non-small cell lung, ovarian and colon carcinomas.[120,121] As reported by Suarez *et al.*, non-tumoral cell lines were 5 to 40 times less sensitive to KF than tumoral cell lines.[122] Gomez *et al.* demonstrated that bone marrow progenitor cells were not affected by treatment with concentrations of KF exceeding those that showed pharmacological effects.[123] Therefore, severe adverse effects such as leukopaenia should not be expected. Thus, KF holds promise as an effective anti-tumor agent that will not show serious side effects.

Interestingly, the efficiency of Kahalalide F is neither affected by *mdr1* nor by *her2/neu* over expression, allowing it to be used as a second line drug as well as being used to treat highly malignant tumors.[4,122] To date, the mechanism underlying the antitumor effect of Kahalalide F has not been fully elucidated. However, it was reported that the active compound mainly targets lysosomes and mediates the induction of vacuolization. In some cancer cell lines, KF was shown to cause necrosis-like processes; cytotoxicity is mediated by down-regulation of the ERBB3 protein and inhibition of the phosphatidylinositol 3-kinase (PI3K)-AKT signaling pathway.[120]

The ErbB3 protein is a member of the ErbB family of receptor tyrosine kinases (RTK). The ErbB family consists of four different receptors: EGFR (ErbB1/HER1), ErbB2 (neu, HER2), ErbB3 (HER3) and ErbB4 (HER4).[124,125] ErbB receptors are activated upon binding of their respective ligands, all of which are growth factors of the EGF family.[126–128] Three different groups of EGFs have been described: the first group comprises, among others, EGF and transforming growth factor α; these ligands bind specifically to the EGFR. Ligands of the second group, such as epiregulin (EPR) and heparin-binding growth factor (HB-EGF), bind to both EGFR and ErbB4. Neuregulins (NRG), members of the third group, bind to ErbB3 and/or ErbB4.[129,130]

Activation of ErbB recpetors is mediated via autocrine or paracrine secretion of the ligands, which then activate the receptors by binding to extracellular domains. Ligand binding induces dimerization and eventually trans-autophosphorylation of the receptors via their tyrosine kinase domains.[126] Upon phosphorylation of a specific tyrosine residue within the cytoplasmic tail proteins containing an Src homology 2 (SH2) and a phosphotyrosine binding (PTB) domain, enzymes are recruited to the receptors, resulting in the induction of intracellular signaling cascades. It has been asserted by different groups that all ErbB ligands and receptors induce the activation of the ras/raf/MEK/MAPK pathway, which is involved in tumorigenesis. Furthermore, ErbB receptors activate Phosphatidylinositol-3-kinase (PI3K). The activity of PI3K is required for Ras-induced cellular transformation that causes malignancies. Studies in transgenic mice indicate that PI3K and AKT play a direct role in tumorigenesis. This oncogenic potential comes from the downstream effects mediated by the PI3K/Akt signaling pathway, which is involved in survival, angiogenesis and growth processes. PI3K is a heterodimer consisting of a p85 regulatory domain and a p110 catalytic domain.[131,132] Upon activation, PI3K phosphorylates phosphatidylinositol-4,5-bisphosphate (PIP2) to PIP3, which recruits further downstream molecules such as the serine-threonine kinase Akt. Subsequently, Akt is activated by being phosphorylated at Ser-473. As a regulator of multiple cellular target proteins involved in cell proliferation, survival, growth, angiogenesis and other processes, Akt also plays a role in carcinogenesis, and aberrant PI3K/Akt signaling due to mutations or overexpression may result in tumor formation.[132,133] Cancer cell subpopulations with hyperactive PI3K/Akt signaling are frequently genetically selected in the course of tumorigenesis. The activated Akt protein partly balances its anti-apoptotic effect by phosphorylation of the pro-apoptotic Bcl2 family member, Bad. Phosphorylated Bad is sequestered in the mitochondrial membrane and its activation via cleavage to truncated Bad (t-Bad) is prevented.[132,134] Furthermore, cell survival is promoted via phosphorylation of IκB kinase (IKK), indirectly activating the anti-apoptotic transcription factor nuclear factor κB (NFκB).[132,135,136] The PI3K-Akt signaling pathway has also been implicated in promoting the degradation of the tumor suppressor protein p53 by phosphorylating and thereby activating its negative regulator Mdm2.[132,137,138] It has been demonstrated that degradation of p53 results in the circumvention of normal apoptosis pathways.[132,139]

Another tumor promoting feature of the PI3K-Akt signaling pathway is its regulation of cell cycle progression, especially at the G1/S transition. Although the detailed mechanism has not yet been fully elucidated, it has been found that Akt contributes to the stabilization of cyclin D1 and Myc.[140,141] molecules that are frequently overexpressed in cancer cells.[132,142–144]

Overexpression of ErbB receptors, especially the HER2/Neu receptor, is associated with resistance to anticancer drugs that target these receptors.[145] Due to the fact that ErbB3 itself lacks catalytic function and its activation relies on formation of heterodimers, it has been suggested that ErbB3-associated drug resistance is mediated by paracrine or autocrine signaling. ErbB3 might enable certain ligands to activate catalytic RTKs.[145]

Frequently, activation of one receptor of the ErbB-family coincides with transmission of the signal to other receptors of this family, thereby amplifying and diversifying the signal. This phenomenon is associated with the transformation of cells that might lead to malignancy.

Physiologically, ErbB receptors are involved in many developmental processes and are required for the formation of various organs such as the heart and mammary gland. Furthermore, they play a role in renewal of stem cells in normal tissues such as skin, liver and gut.[146,147]

Despite the essential role ErbB receptors play during normal development, the receptor family is also associated with pathological conditions. Various studies have demonstrated that overexpression of ErbB receptors may cause neoplastic transformation, and that increased expression of EFGR is closely linked to enhanced proliferation of carcinoma cells. High expression levels of EGFR and ErbB3 have been described in the majority of human carcinomas: 50 percent to 70 percent of lung, colon and breast carcinomas overexpress these receptors.[121]

Another mechanism by which KF elicits its antitumor effect is the induction of oncosis, a form of passive cell death closely related to necrosis.[3,4,122]

Early stages of oncosis are characterized by significant alterations in cell shape and volume. Affected cells show blebbing along luminal borders and vascular space. Oncotic cells are swollen and the ER and Golgi apparatus appear dilated.

6.8. Naamidine A

The dibenzylated 2-aminoimidazole alkaloid Naamidine A was first isolated from the sponge *Leucett* sp. and acts as an inhibitor of the epidermal growth factor receptor (EGFR)[148] (Fig. 2).

As described earlier, the EFGR signaling pathway plays a pivotal role in the development of some human tumors. Overexpression of this receptor is associated with cell growth and tumorigenicity in some tumor types. Therefore, compounds interfering at some level of this signaling pathway can be used as potential anti-proliferative agents in cancer therapy. A screening of extracts of different marine organisms revealed that the ethanol extract of a yellow sponge of the *Leucetta* species possesses the ability to inhibit EGF-dependent DNA synthesis and proliferation.[148] Further investigations resulted in the isolation of three alkaloids from the extract, namely naamidine A, isonaamidine B and isonaamidine C. However, it was demonstrated that the anti-proliferative activity of the crude extract can be solely attributed to naamidine A.[148]

Naamidine A selectively inhibits EGF-mediated mitogenesis.[148] *In vivo* studies have demonstrated that the compound inhibits the growth of squamous cell carcinoma cell lines overexpressing the EGF receptor by 85 percent at a maximal tolerable dose of 25 mg/kg. It has been found that naamidine A inhibits tumor growth by acting downstream of the ligand binding and tyrosine kinase domains to inhibit EFG-mediated DNA synthesis and cell proliferation. This mechanism was elucidated by a study that showed that neither the extracellular nor the tyrosine kinase domain of the EGF receptor appeared to be affected by naamidine A treatment.[148]

Tumor growth is frequently associated with mutations in growth factor signaling pathways such as the EGFR signaling pathway. Interactions between EGF and its membrane-bound receptor trigger activation of the MAPK pathway, which leads to a cascade of intracellular signals culminating in cellular responses including transcription, cell division, adhesion, and cell death.[127,149–151] Triggering of the MAPK cascade results in activation of ERKs and subsequently in activation of transcription factors are involved in proliferation.[151,152] The MAPK pathway has been identified as a link between the membrane-bound oncogene Ras and certain nuclear events. It has been hypothesized that the mode of action of naamidine A is based on changes in the phosphorylation state and phosphotransferase activity of ERKs. In the human epidemoid carcinoma cell line A-431, which overexpresses EGF receptors, Naamidine A treatment was shown to induce phosphotransferase activity of ERK enzymes thereby altering downstream signaling of ERKs, ultimately resulting in complete abrogation of DNA synthesis 30 h after treatment.[151] Cells were arrested in the G1 phase of the cell cycle. ERK enzymes propagate signals that then activate several proteins that can trigger proliferation, differentiation and cell death. It can be concluded that ERKs function as primary targets of naamidine A, since it has been demonstrated that sustained ERK signaling induces G1 arrest.[151] Due to the fact that Naamidine A selectively targets EGF-mediated mitogenesis, tumor cells overexpressing the EGF receptor are more sensitive to this compound than are healthy cells. Therefore, the intrinsic overexpression of this receptor in many tumor cells accounts for the high selectivity of naamidine A.

6.9. NFκB Modulators

NF-κB inhibiting properties have been attributed to a small number of marine natural compounds. These compounds exert their function by either targeting IKK-dependent degradation of IκB, interfering with the proteolytic activity of the 26S proteasome, or by blocking the binding of NF-κB to its DNA binding site. However, the mode of action of many marine derived anti-cancer drugs has not yet been elucidated.[11]

Nuclear factor κB (NF-κB) acts as a transcription factor exerting a regulatory function in gene expression by binding to a specific sequence within a regulatory region of the DNA. This inducible transcription factor is involved in the expression of a large number of genes and can be activated rapidly. Many of the genes regulated by NF-κB play a role in diseases, carcinogenesis and inflammation processes.[11,153–155] It has been recently reported that NF-κB plays a key role in linking inflammation and cancer.[11,156] NF-κB dimerizes via its N-terminal Rel homology domain (RHD), interacts with a cytoplasmic inhibitory protein, IκB, and binds to the DNA.[11,157] At the C-terminal end of this RHD, the protein contains a nuclear localization signal (NLS) enabling the activated NF-κB to be translocated into the nucleus.

Normally, NF-κB is localized in the cytoplasm in its inactive form that is bound to IκB. Upon cellular damage such as ionizing radiation, oxidative stress, cytokines or toxins the IκB kinase complex (IKK) is activated. The IKK then phosphorylates IκB, which results in polyubiquitinylation of IκB, targeting it for proteasomal degradation.[11,158] Upon degradation of IκB, NF-κB translocates into the nucleus where it exerts its regulatory function.[11,157,159]

REFERENCES

1. Groombridge B, Jenkins MD. (2000) *Global Biodiversity: Earth's Living Resources in the 21st Century.* World Conversation Press, Cambridge, UK.
2. Gordaliza M. (2007) Natural products as leads to anticancer drugs. *Clin Transl Oncol* **9**: 767–776.
3. Molinski TF, Dalisay DS, Lievens SL, Saludes JP. (2009) Drug development from marine natural products. *Nat Rev Drug Discov* **8**: 69–85.
4. Jimeno J, Lopez-Martin JA, Ruiz-Casado A, *et al.* (2004) Progress in the clinical development of new marine-derived anticancer compounds. *Anticancer Drugs* **15**: 321–329.
5. John Faulkner D, Newman DJ, Cragg GM. (2004) Investigations of the marine flora and fauna of the Islands of Palau. *Nat Prod Rep* **21**: 50–76.
6. Simmons TL, Andrianasolo E, McPhail K, *et al.* (2005) Marine natural products as anticancer drugs. *Mol Cancer Ther* **4**: 333–342.
7. Konig GM, Kehraus S, Seibert SF, *et al.* (2006) Natural products from marine organisms and their associated microbes. *Chembiochem* **7**: 229–238.
8. Bruserud O, Stapnes C, Ersvaer E, *et al.* (2007) Histone deacetylase inhibitors in cancer treatment: a review of the clinical toxicity and the modulation of gene expression in cancer cell. *Curr Pharm Biotechnol* **8**: 388–400.
9. Tu G, Xu W, Huang H, Li S. (2008) Progress in the development of matrix metalloproteinase inhibitors. *Curr Med Chem* **15**: 1388–1395.
10. Welsh CF. (2004) Rho GTPases as key transducers of proliferative signals in g1 cell cycle regulation. *Breast Cancer Res Treat* **84**: 33–42.
11. Folmer F, Jaspars M, Dicato M, Diederich M. (2008) Marine natural products as targeted modulators of the transcription factor NF-kappaB. *Biochem Pharmacol* **75**: 603–617.
12. Hait WN, Hambley TW. (2009) Targeted cancer therapeutics. *Cancer Res* **69**: 1263–1267.

13. Sun J, Chen Y, Li M, Ge Z. (1998) Role of antioxidant enzymes on ionizing radiation resistance. *Free Radic Biol Med* **24**: 586–593.
14. Kinnula VL, Crapo JD. (2004) Superoxide dismutases in malignant cells and human tumors. *Free Radic Biol Med* **36**: 718–744.
15. Sinha BK, Mimnaugh EG. (1990) Free radicals and anticancer drug resistance: oxygen free radicals in the mechanisms of drug cytotoxicity and resistance by certain tumors. *Free Radic Biol Med* **8**: 567–581.
16. de Vries EG, Meijer C, Timmer-Bosscha H, *et al.* (1989) Resistance mechanisms in three human small cell lung cancer cell lines established from one patient during clinical follow-up. *Cancer Res* **49**: 4175–4178.
17. Hao XY, Bergh J, Brodin O, *et al.* (1994) Acquired resistance to cisplatin and doxorubicin in a small cell lung cancer cell line is correlated to elevated expression of glutathione-linked detoxification enzymes. *Carcinogenesis* **15**: 1167–1173.
18. Kahlos K. (1999) The expression and possible role of manganese superoxide dismutase in malignant pleural mesothelioma. Dissertation, University of Oulu.
19. Takebayashi Y, Pourquier P, Zimonjic DB, *et al.* (2001) Antiproliferative activity of ecteinascidin 743 is dependent upon transcription-coupled nucleotide-excision repair. *Nat Med* 7: 961–966.
20. Rinehart KL, Holt TG, Fregeau NL, *et al.* (1990) Bioactive compounds from aquatic and terrestrial sources. *J Nat Prod* **53**: 771–792.
21. Simmons TL, Andrianasolo E, McPhail K, *et al.* (2005) Marine natural products as anticancer drugs. *Mol Cancer Ther* **4**: 333–342.
22. Wright A, Forleo D, Gunawardana G, *et al.* (1990) Antitumor tetrahydroisoquinoline alkaloids from the colonial ascidian Ecteinascidia turbinata. *J Org Chem* **55**: 4508–4512.
23. Le Cesne A, Domont J, Cioffi A. (2008) The new era of trabectedin in soft tissue sarcomas. *EJHP* **14**: 72–75.
24. (2006) Trabectedin: Ecteinascidin 743, Ecteinascidin-743, ET 743, ET-743, NSC 684766. *Drugs R D* 7: 317–328.
25. Chabner BA, Longo DL. (2005) *Cancer Chemotherapy and Biotherapy: Principles and Practice*, 4th edn. Lippincott Williams & Wilkins, p. 367.
26. Zewail-Foote M, Hurley LH. (1999) Ecteinascidin 743: a minor groove alkylator that bends DNA toward the major groove. *J Med Chem* **42**: 2493–2497.
27. Fayette J, Coquard IR, Alberti L, *et al.* (2005) ET-743: a novel agent with activity in soft tissue sarcomas. *Oncologist* **10**: 827–832.
28. Galmarini CM, Mackey JR, Dumontet C. (2002) Nucleoside analogues and nucleobases in cancer treatment. *Lancet Oncol* **3**: 415–424.

29. Chabner BA LD. (2005) *Cancer Chemotherapy & Biotherapy. Principles and Practice*, 4th edn. Lippincott Williams & Wilkins, p. 184.
30. Czaykowski PM, Samuels T, Oza A. (1997) A durable response to cytarabine in advanced breast cancer. *Clin Oncol (R Coll Radiol)* **9**: 181–183.
31. Galmarini CM, Thomas X, Calvo F, *et al.* (2002) Potential mechanisms of resistance to cytarabine in AML patients. *Leuk Res* **26**: 621–629.
32. Capizzi RL, Yang JL, Cheng E, *et al.* (1983) Alteration of the pharmacokinetics of high-dose ara-C by its metabolite, high ara-U in patients with acute leukemia. *J Clin Oncol* **1**: 763–771.
33. Hubeek I, Litvinova E, Peters GJ, *et al.* (2004) The effect of G-CSF on the in vitro cytotoxicity of cytarabine and fludarabine in the FLAG combination in pediatric acute myeloid leukemia. *Int J Oncol* **25**: 1823–1829.
34. Snyder DL, Roberts J, Friedman E. (1996) *Handbook of Pharmacology of Aging*, 2nd edn. CRC Press, Boca Raton (online version), p. 324.
35. Vaickus L, Letendre L. (1984) Pericarditis induced by high-dose cytarabine therapy. *Arch Intern Med* **144**: 1868–1869.
36. (HSDB) HSDB (2009) Toxnet. National Library of Medicine.
37. Plunkett W, Huang P, Xu YZ, *et al.* (1995) Gemcitabine: metabolism, mechanisms of action, and self-potentiation. *Semin Oncol* **22**: 3–10.
38. Huang P, Plunkett W. (1995) Induction of apoptosis by gemcitabine. *Semin Oncol* **22**: 19–25.
39. Heinemann V, Schulz L, Issels RD, Plunkett W. (1995) Gemcitabine: a modulator of intracellular nucleotide and deoxynucleotide metabolism. *Semin Oncol* **22**: 11–18.
40. Plunkett W, Huang P, Searcy CE, Gandhi V. (1996) Gemcitabine: preclinical pharmacology and mechanisms of action. *Semin Oncol* **23**: 3–15.
41. Rieger J, Durka S, Streffer J, *et al.* (1999) Gemcitabine cytotoxicity of human malignant glioma cells: modulation by antioxidants, BCL-2 and dexamethasone. *Eur J Pharmacol* **365**: 301–308.
42. Ruiz van Haperen VW, Veerman G, Vermorken JB, Peters GJ. (1993) 2′,2′-Difluoro-deoxycytidine (gemcitabine) incorporation into RNA and DNA of tumour cell lines. *Biochem Pharmacol* **46**: 762–766.
43. Galmarini CM, Clarke ML, Falette N, *et al.* (2002) Expression of a non-functional p53 affects the sensitivity of cancer cells to gemcitabine. *Int J Cancer* **97**: 439–445.
44. Curtis JE, Hersh EM, Harris JE, *et al.* (1970) The human primary immune response to keyhole limpet haemocyanin: interrelationships of delayed hypersensitivity, antibody response and in vitro blast transformation. *Clin Exp Immunol* **6**: 473–491.

45. Harris JR, Gebauer W, Sohngen SM, *et al.* (1997) Keyhole limpet hemocyanin (KLH), II: characteristic reassociation properties of purified KLH1 and KLH2. *Micron* **28**: 43–56.
46. Weigle WO. (1964) Immunochemical properties of hemocyanin. *Immunochemistry* **1**: 295–302.
47. Sminia T, Vlugh-van Dallen JE. (1977) Haemocyanin synthesis in pore cells of the terrestrial snail Helix aspersa. *Cell Tissue Res* **183**: 299–301.
48. Jurincic-Winkler CD, Metz KA, Beuth J, Klippel KF. (2000) Keyhole limpet hemocyanin for carcinoma in situ of the bladder: a long-term follow-up study. *Eur Urol* 37 Suppl **3**: 45–49.
49. Harris JR, Markl J. (1999) Keyhole limpet hemocyanin (KLH): a biomedical review. *Micron* **30**: 597–623.
50. Wirguin I, Suturkova-Milosevic L, Briani C, Latov N. (1995) Keyhole limpet hemocyanin contains Gal(beta 1-3)-GalNAc determinants that are cross-reactive with the T antigen. *Cancer Immunol Immunother* **40**: 307–310.
51. Kudryashov V, Kim HM, Ragupathi G, *et al.* (1998) Immunogenicity of synthetic conjugates of Lewis(y) oligosaccharide with proteins in mice: towards the design of anticancer vaccines. *Cancer Immunol Immunother* **45**: 281–286.
52. Kudryashov V, Ragupathi G, Kim IJ, *et al.* (1998) Characterization of a mouse monoclonal IgG3 antibody to the tumor-associated globo H structure produced by immunization with a synthetic glycoconjugate. *Glycoconj J* **15**: 243–249.
53. Miles DW, Towlson KE, Graham R, *et al.* (1996) A randomised phase II study of sialyl-Tn and DETOX-B adjuvant with or without cyclophosphamide pretreatment for the active specific immunotherapy of breast cancer. *Br J Cancer* **74**: 1292–1296.
54. Pettit GR, Srirangam JK, Barkoczy J, *et al.* (1995) Antineoplastic agents 337. Synthesis of dolastatin 10 structural modifications. *Anticancer Drug Des* **10**: 529–544.
55. Amador ML, Jimeno J, Paz-Ares L, *et al.* (2003) Progress in the development and acquisition of anticancer agents from marine sources. *Ann Oncol* **14**: 1607–1615.
56. Aherne GW, Hardcastle A, Valenti M, *et al.* (1996) Antitumour evaluation of dolastatins 10 and 15 and their measurement in plasma by radioimmunoassay. *Cancer Chemother Pharmacol* **38**: 225–232.
57. Madden T, Tran HT, Beck D, *et al.* (2000) Novel marine-derived anticancer agents: a phase I clinical, pharmacological, and pharmacodynamic study of

dolastatin 10 (NSC 376128) in patients with advanced solid tumors. *Clin Cancer Res* **6**: 1293–1301.

58. Bai R, Pettit GR, Hamel E. (1990) Dolastatin 10, a powerful cytostatic peptide derived from a marine animal. Inhibition of tubulin polymerization mediated through the vinca alkaloid binding domain. *Biochem Pharmacol* **39**: 1941–1949.
59. Turner T, Jackson WH, Pettit GR, *et al.* (1998) Treatment of human prostate cancer cells with dolastatin 10, a peptide isolated from a marine shell-less mollusc. *Prostate* **34**: 175–181.
60. Vaishampayan U, Glode M, Du W, *et al.* (2000) Phase II study of dolastatin-10 in patients with hormone-refractory metastatic prostate adenocarcinoma. *Clin Cancer Res* **6**: 4205–4208.
61. Kalemkerian GP, Ou X, Adil MR, *et al.* (1999) Activity of dolastatin 10 against small-cell lung cancer in vitro and in vivo: induction of apoptosis and bcl-2 modification. *Cancer Chemother Pharmacol* **43**: 507–515.
62. Maki A, Diwakaran H, Redman B, *et al.* (1995) The bcl-2 and p53 oncoproteins can be modulated by bryostatin 1 and dolastatins in human diffuse large cell lymphoma. *Anticancer Drugs* **6**: 392–397.
63. Hollstein M, Rice K, Greenblatt MS, *et al.* (1994) Database of p53 gene somatic mutations in human tumors and cell lines. *Nucleic Acids Res* **22**: 3551–3555.
64. Baker SJ, Fearon ER, Nigro JM, *et al.* (1989) Chromosome 17 deletions and p53 gene mutations in colorectal carcinomas. *Science* **244**: 217–221.
65. Bond GL, Hu W, Bond EE, *et al.* (2004) A single nucleotide polymorphism in the MDM2 promoter attenuates the p53 tumor suppressor pathway and accelerates tumor formation in humans. *Cell* **119**: 591–602.
66. Marin MC, Jost CA, Brooks LA, *et al.* (2000) A common polymorphism acts as an intragenic modifier of mutant p53 behaviour. *Nat Genet* **25**: 47–54.
67. Momand J, Jung D, Wilczynski S, Niland J. (1998) The MDM2 gene amplification database. *Nucleic Acids Res* **26**: 3453–3459.
68. Navone NM, Labate ME, Troncoso P, *et al.* (1999) p53 mutations in prostate cancer bone metastases suggest that selected p53 mutants in the primary site define foci with metastatic potential. *J Urol* **161**: 304–308.
69. Chari NS, Pinaire NL, Thorpe L, *et al.* (2009) The p53 tumor suppressor network in cancer and the therapeutic modulation of cell death. *Apoptosis* **14**: 336–347.
70. Wu X, Bayle JH, Olson D, Levine AJ. (1993) The p53-mdm-2 autoregulatory feedback loop. *Genes Dev* **7**: 1126–1132.
71. Barak Y, Juven T, Haffner R, Oren M. (1993) mdm2 expression is induced by wild type p53 activity. *EMBO J* **12**: 461–468.

72. Vousden KH, Lu X. (2002) Live or let die: the cell's response to p53. *Nat Rev Cancer* **2**: 594–604.
73. Harris SL, Levine AJ. (2005) The p53 pathway: positive and negative feedback loops. *Oncogene* **24**: 2899–2908.
74. Oren M. (1994) Relationship of p53 to the control of apoptotic cell death. *Semin Cancer Biol* **5**: 221–227.
75. Heinrich M, Robles M, West JE, *et al.* (1998) Ethnopharmacology of Mexican asteraceae (Compositae). *Annu Rev Pharmacol Toxicol* **38**: 539–565.
76. Ebermann R, Alth G, Kreitner M, Kubin A. (1996) Natural products derived from plants as potential drugs for the photodynamic destruction of tumor cells. *J Photochem Photobiol B* **36**: 95–97.
77. Halstead BW. (1969) Marine biotoxins: a new source of medicinals. *Lloydia* **32**: 484–488.
78. Dembitsky VM. (2006) Anticancer activity of natural and synthetic acetylenic lipids. *Lipids* **41**: 883–924.
79. Hallock YF, Cardellina JH, 2nd, Balaschak MS, *et al.* (1995) Antitumor activity and stereochemistry of acetylenic alcohols from the sponge Cribrochalina vasculum. *J Nat Prod* **58**: 1801–1807.
80. Aoki S, Matsui K, Takata T, *et al.* (2001) Lembehyne A, a spongean polyacetylene, induces neuronal differentiation in neuroblastoma cell. *Biochem Biophys Res Commun* **289**: 558–563.
81. Aoki S, Matsui K, Takata T, Kobayashi M. (2003) In situ photoaffinity labeling of the target protein for lembehyne A, a neuronal differentiation inducer. *FEBS Lett* **544**: 223–227.
82. Krinsky NI. (1993) Effects of carotenoids on cells. *Mol Aspects Med* **14**: 241–246.
83. Stich HF, Dunn BP. (1986) Relationship between cellular levels of beta-carotene and sensitivity to genotoxic agents. *Int J Cancer* **38**: 713–717.
84. Stich HF, Tsang SS, Palcic B. (1990) The effect of retinoids, carotenoids and phenolics on chromosomal instability of bovine papillomavirus DNA-carrying cells. *Mutat Res* **241**: 387–393.
85. Manoharan K, Banerjee MR. (1985) beta-Carotene reduces sister chromatid exchanges induced by chemical carcinogens in mouse mammary cells in organ culture. *Cell Biol Int Rep* **9**: 783–789.
86. Pung A, Rundhaug JE, Yoshizawa CN, Bertram JS. (1988) Beta-carotene and canthaxanthin inhibit chemically- and physically-induced neoplastic transformation in 10T1/2 cells. *Carcinogenesis* **9**: 1533–1539.
87. Fujita M, Nakao Y, Matsunaga S, *et al.* (2002) Sodium 1-(12-hydroxy)-octadecanyl sulfate, an MMP2 inhibitor, isolated from a tunicate of the family Polyclinidae. *J Nat Prod* **65**: 1936–1938.

88. Fujita M, Nakao Y, Matsunaga S, *et al.* (2003) Callysponginol sulfate A, an MT1-MMP inhibitor isolated from the marine sponge Callyspongia truncata. *J Nat Prod* **66**: 569–571.
89. Sato H, Takino T, Miyamori H. (2005) Roles of membrane-type matrix metalloproteinase-1 in tumor invasion and metastasis. *Cancer Sci* **96**: 212–217.
90. Seiki M, Yana I. (2003) Roles of pericellular proteolysis by membrane type-1 matrix metalloproteinase in cancer invasion and angiogenesis. *Cancer Sci* **94**: 569–574.
91. Woessner JF, Jr. (1991) Matrix metalloproteinases and their inhibitors in connective tissue remodeling. *FASEB J* **5**: 2145–2154.
92. Matrisian LM. (1990) Metalloproteinases and their inhibitors in matrix remodeling. *Trends Genet* **6**: 121–125.
93. Nakada M, Nakamura H, Ikeda E, *et al.* (1999) Expression and tissue localization of membrane-type 1, 2, and 3 matrix metalloproteinases in human astrocytic tumors. *Am J Pathol* **154**: 417–428.
94. Nomura H, Sato H, Seiki M, *et al.* (1995) Expression of membrane-type matrix metalloproteinase in human gastric carcinomas. *Cancer Res* **55**: 3263–3266.
95. Sato H, Takino T, Okada Y, *et al.* (1994) A matrix metalloproteinase expressed on the surface of invasive tumour cells. *Nature* **370**: 61–65.
96. Holmbeck K, Bianco P, Caterina J, *et al.* (1999) MT1-MMP-deficient mice develop dwarfism, osteopenia, arthritis, and connective tissue disease due to inadequate collagen turnover. *Cell* **99**: 81–92.
97. Kadono Y, Okada Y, Namiki M, *et al.* (1998) Transformation of epithelial Madin-Darby canine kidney cells with p60(v-src) induces expression of membrane-type 1 matrix metalloproteinase and invasiveness. *Cancer Res* **58**: 2240–2244.
98. Salcedo M, Cuevas C, Alonso JL, *et al.* (2007) The marine sphingolipid-derived compound ES 285 triggers an atypical cell death pathway. *Apoptosis* **12**: 395–409.
99. Fukushima N, Weiner JA, Kaushal D, *et al.* (2002) Lysophosphatidic acid influences the morphology and motility of young, postmitotic cortical neurons. *Mol Cell Neurosci* **20**: 271–282.
100. Hisano N, Yatomi Y, Satoh K, *et al.* (1999) Induction and suppression of endothelial cell apoptosis by sphingolipids: a possible in vitro model for cell-cell interactions between platelets and endothelial cells. *Blood* **93**: 4293–4299.
101. Hooks SB, Santos WL, Im DS, *et al.* (2001) Lysophosphatidic acid-induced mitogenesis is regulated by lipid phosphate phosphatases and is Edg-receptor independent. *J Biol Chem* **276**: 4611–4621.

102. Kranenburg O, Poland M, Gebbink M, *et al.* (1997) Dissociation of LPA-induced cytoskeletal contraction from stress fiber formation by differential localization of RhoA. *J Cell Sci* **110**: 2417–2427.
103. Postma FR, Jalink K, Hengeveld T, *et al.* (1996) Serum-induced membrane depolarization in quiescent fibroblasts: activation of a chloride conductance through the G protein-coupled LPA receptor. *EMBO J* **15**: 63–72.
104. Tigyi G, Fischer DJ, Sebok A, *et al.* (1996) Lysophosphatidic acid-induced neurite retraction in PC12 cells: neurite-protective effects of cyclic AMP signaling. *J Neurochem* **66**: 549–558.
105. van Corven EJ, Groenink A, Jalink K, *et al.* (1989) Lysophosphatidate-induced cell proliferation: identification and dissection of signaling pathways mediated by G proteins. *Cell* **59**: 45–54.
106. Weiner JA, Chun J. (1999) Schwann cell survival mediated by the signaling phospholipid lysophosphatidic acid. *Proc Natl Acad Sci U S A* **96**: 5233–5238.
107. Zhang Q, Peyruchaud O, French KJ, *et al.* (1999) Sphingosine 1-phosphate stimulates fibronectin matrix assembly through a Rho-dependent signal pathway. *Blood* **93**: 2984–2990.
108. Alberts B, Johnson A, Lewis J, *et al.* (2003) Molekularbiologie der Zelle. Wiley-VCH 4. Auflage: pp. 1001.
109. Kolch W, Heidecker G, Kochs G, *et al.* (1993) Protein kinase C alpha activates RAF-1 by direct phosphorylation. *Nature* **364**: 249–252.
110. Mackay HJ, Twelves CJ. (2003) Protein kinase C: a target for anticancer drugs? *Endocr Relat Cancer* **10**: 389–396.
111. Cai H, Smola U, Wixler V, *et al.* (1997) Role of diacylglycerol-regulated protein kinase C isotypes in growth factor activation of the Raf-1 protein kinase. *Mol Cell Biol* **17**: 732–741.
112. Ueda Y, Hirai S, Osada S, *et al.* (1996) Protein kinase C activates the MEK-ERK pathway in a manner independent of Ras and dependent on Raf. *J Biol Chem* **271**: 23512–23519.
113. Marshall CJ. (1996) Cell signalling. Raf gets it together. *Nature* **383**: 127–128.
114. Goode N, Hughes K, Woodgett JR, Parker PJ. (1992) Differential regulation of glycogen synthase kinase-3 beta by protein kinase C isotypes. *J Biol Chem* **267**: 16878–16882.
115. Bossy-Wetzel E, Bakiri L, Yaniv M. (1997) Induction of apoptosis by the transcription factor c-Jun. *EMBO J* **16**: 1695–1709.
116. Castagna M, Takai Y, Kaibuchi K, *et al.* (1982) Direct activation of calcium-activated, phospholipid-dependent protein kinase by tumor-promoting phorbol esters. *J Biol Chem* **257**: 7847–7851.
117. Fujii T, Garcia-Bermejo ML, Bernabo JL, *et al.* (2000) Involvement of protein kinase C delta (PKCdelta) in phorbol ester-induced apoptosis in LNCaP

prostate cancer cells. Lack of proteolytic cleavage of PKCdelta. *J Biol Chem* **275**: 7574–7582.
118. O'Brian C, Vogel VG, Singletary SE, Ward NE. (1989) Elevated protein kinase C expression in human breast tumor biopsies relative to normal breast tissue. *Cancer Res* **49**: 3215–3217.
119. Takenaga K, Takahashi K. (1986) Effects of 12-O-tetradecanoylphorbol-13-acetate on adhesiveness and lung-colonizing ability of Lewis lung carcinoma cells. *Cancer Res* **46**: 375–380.
120. Janmaat ML, Rodriguez JA, Jimeno J, *et al.* (2005) Kahalalide F induces necrosis-like cell death that involves depletion of ErbB3 and inhibition of Akt signaling. *Mol Pharmacol* **68**: 502–510.
121. Normanno N, De Luca A, Bianco C, *et al.* (2006) Epidermal growth factor receptor (EGFR) signaling in cancer. *Gene* **366**: 2–16.
122. Suarez Y, Gonzalez L, Cuadrado A, *et al.* (2003) Kahalalide F, a new marine-derived compound, induces oncosis in human prostate and breast cancer cells. *Mol Cancer Ther* **2**: 863–872.
123. Gomez SG, Bueren JA, Faircloth GT, *et al.* (2003) In vitro toxicity of three new antitumoral drugs (trabectedin, aplidin, and kahalalide F) on hematopoietic progenitors and stem cells. *Exp Hematol* **31**: 1104–1111.
124. Yarden Y. (2001) Biology of HER2 and its importance in breast cancer. *Oncology* **61**: Suppl 2: 1–13.
125. Ferguson KM, Berger MB, Mendrola JM, *et al.* (2003) EGF activates its receptor by removing interactions that autoinhibit ectodomain dimerization. *Mol Cell* **11**: 507–517.
126. Olayioye MA, Neve RM, Lane HA, Hynes NE. (2000) The ErbB signaling network: receptor heterodimerization in development and cancer. *EMBO J* **19**: 3159–3167.
127. Yarden Y, Sliwkowski MX. (2001) Untangling the ErbB signalling network. *Nat Rev Mol Cell Biol* **2**: 127–137.
128. Harari D, Tzahar E, Romano J, *et al.* (1999) Neuregulin-4: a novel growth factor that acts through the ErbB-4 receptor tyrosine kinase. *Oncogene* **18**: 2681–2689.
129. Carraway KL, 3rd, Weber JL, Unger MJ, *et al.* (1997) Neuregulin-2, a new ligand of ErbB3/ErbB4-receptor tyrosine kinases. *Nature* **387**: 512–516.
130. Carraway KL, Carraway CA, Carraway KL, 3rd (1997) Roles of ErbB-3 and ErbB-4 in the physiology and pathology of the mammary gland. *J Mammary Gland Biol Neoplasia* **2**: 187–198.
131. Cantley LC. (2002) The phosphoinositide 3-kinase pathway. *Science* **296**: 1655–1657.

132. Luo J, Manning BD, Cantley LC. (2003) Targeting the PI3K-Akt pathway in human cancer: rationale and promise. *Cancer Cell* **4**: 257–262.
133. Vivanco I, Sawyers CL. (2002) The phosphatidylinositol 3-Kinase AKT pathway in human cancer. *Nat Rev Cancer* **2**: 489–501.
134. Datta SR, Brunet A, Greenberg ME. (1999) Cellular survival: a play in three Akts. *Genes Dev* **13**: 2905–2927.
135. Romashkova JA, Makarov SS. (1999) NF-kappaB is a target of AKT in anti-apoptotic PDGF signalling. *Nature* **401**: 86–90.
136. Ozes ON, Mayo LD, Gustin JA, *et al.* (1999) NF-kappaB activation by tumour necrosis factor requires the Akt serine-threonine kinase. *Nature* **401**: 82–85.
137. Mayo LD, Donner DB. (2001) A phosphatidylinositol 3-kinase/Akt pathway promotes translocation of Mdm2 from the cytoplasm to the nucleus. *Proc Natl Acad Sci U S A* **98**: 11598–11603.
138. Zhou BP, Liao Y, Xia W, *et al.* (2001) HER-2/neu induces p53 ubiquitination via Akt-mediated MDM2 phosphorylation. *Nat Cell Biol* **3**: 973–982.
139. Schmitt CA. (2003) Senescence, apoptosis and therapy — cutting the lifelines of cancer. *Nat Rev Cancer* **3**: 286–295.
140. Diehl JA, Cheng M, Roussel MF, Sherr CJ. (1998) Glycogen synthase kinase-3beta regulates cyclin D1 proteolysis and subcellular localization. *Genes Dev* **12**: 3499–3511.
141. Sears R, Nuckolls F, Haura E, *et al.* (2000) Multiple Ras-dependent phosphorylation pathways regulate Myc protein stability. *Genes Dev* **14**: 2501–2514.
142. Barnes DM, Gillett CE. (1998) Cyclin D1 in breast cancer. *Breast Cancer Res Treat* **52**: 1–15.
143. Pelengaris S, Khan M, Evan G. (2002) c-MYC: more than just a matter of life and death. *Nat Rev Cancer* **2**: 764–776.
144. Sherr CJ, Roberts JM. (1999) CDK inhibitors: positive and negative regulators of G1-phase progression. *Genes Dev* **13**: 1501–1512.
145. Stern DF. (2008) ERBB3/HER3 and ERBB2/HER2 duet in mammary development and breast cancer. *J Mammary Gland Biol Neoplasia* **13**: 215–223.
146. Campbell ID, Bork P. (1993) Epidermal growth factor-like modules. *Curr Opin Struct Biol* **3**: 385–392.
147. Salomon DS, Kim N, Saeki T, Ciardiello F. (1990) Transforming growth factor-alpha: an oncodevelopmental growth factor. *Cancer Cells* **2**: 389–397.
148. Copp BR, Fairchild CR, Cornell L, *et al.* (1998) Naamidine A is an antagonist of the epidermal growth factor receptor and an in vivo active antitumor agent. *J Med Chem* **41**: 3909–3911.
149. Freeman M. (1998) Complexity of EGF receptor signalling revealed in Drosophila. *Curr Opin Genet Dev* **8**: 407–411.

150. Hai T, Hartman MG. (2001) The molecular biology and nomenclature of the activating transcription factor/cAMP responsive element binding family of transcription factors: activating transcription factor proteins and homeostasis. *Gene* **273**: 1–11.
151. James RD, Jones DA, Aalbersberg W, Ireland CM. (2003) Naamidine A intensifies the phosphotransferase activity of extracellular signal-regulated kinases causing A-431 cells to arrest in G1. *Mol Cancer Ther* **2**: 747–751.
152. Aaronson SA. (1991) Growth factors and cancer. *Science* **254**: 1146–1153.
153. Darnell JE, Jr. (2002) Transcription factors as targets for cancer therapy. *Nat Rev Cancer* **2**: 740–749.
154. Pikarsky E, Porat RM, Stein I, *et al.* (2004) NF-kappaB functions as a tumour promoter in inflammation-associated cancer. *Nature* **431**: 461–466.
155. Sebolt-Leopold JS, English JM. (2006) Mechanisms of drug inhibition of signalling molecules. *Nature* **441**: 457–462.
156. Karin M, Greten FR. (2005) NF-kappaB: linking inflammation and immunity to cancer development and progression. *Nat Rev Immunol* **5**: 749–759.
157. Keutgens A, Robert I, Viatour P, Chariot A. (2006) Deregulated NF-kappaB activity in haematological malignancies. *Biochem Pharmacol* **72**: 1069–1080.
158. Evans PC. (2005) Regulation of pro-inflammatory signalling networks by ubiquitin: identification of novel targets for anti-inflammatory drugs. *Expert Rev Mol Med* **7**: 1–19.
159. Karin M. (2006) Nuclear factor-kappaB in cancer development and progression. *Nature* **441**: 431–436.

CHAPTER 7

Ethnopharmacology and Phytotherapy

Ariane Löhnert, Susanne Löhnert, Viktoriya Mogilevskaya and Sandra Schick

ABSTRACT

For the development of novel drugs, it may be helpful to look at traditional medicines because of their time-tested value. All human societies have developed their own medical concepts that are rooted deeply in their "knowledge, skills and practices based on theories, beliefs, and experiences." (WHO, 2000). Ethnopharmacologists focus particularly on the biological and pharmacological effects of indigenous medicines of past and present cultures. Since plants have been extensively used in traditional medicine since prehistoric times, ethnopharmacology largely deals with ethnobotany and phytotherapy. Using cancer as an example, we give an overview of the impact of ancient European medicine, European medieval cloister medicine, traditional medicine of Native Americans, traditional Chinese medicine, Kampo and Ayurveda for the treatment of this disease.

1. INTRODUCTION

Modern medicine developed rapidly in the nineteenth century based on the current scientific knowledge of the time. Pharmaceuticals are prime tools of this evidence-based medicine. Although in some cases existing pharmaceuticals are very effective, many diseases (e.g. malignant neoplasia) still cannot be sufficiently healed with the drugs available. For the development of novel drugs, it may be helpful to look at traditional medicines because of their time-tested value. All human societies have developed their own medical concepts that are rooted deeply in their "knowledge, skills and practices

Institute of Pharmacy and Biochemistry, Johannes Gutenberg University, Staudinger Weg 5, 55128 Mainz, Germany

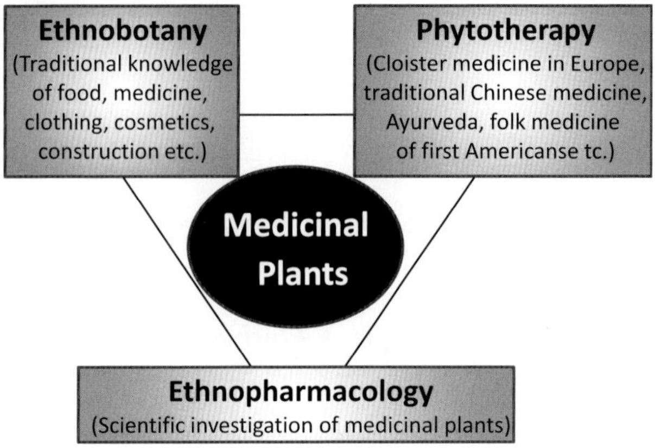

Figure 1 Medicinal plants at the interface of ethnobotany, phytotherapy, and ethnopharmacology.

based on theories, beliefs, and experiences."[1] Traditional medicines use different kinds of biological drugs for healing including plants, fungi, animals, microorganisms, and minerals.

Ethnopharmacologists focus particularly on the biological and pharmacological effects of indigenous medicines of past and present cultures. Since plants have been extensively used in traditional medicine since prehistoric times, ethnopharmacology largely deals with ethnobotany and phytotherapy (Fig. 1). The relationships between humans and plants in all their complexity are at the centre of finding a cure for patients. However, less than 15 percent of vascular plants have been systematically investigated.[2] Hence, there is much potential in this field of research.

2. THE HISTORY OF A DISEASE

Tumors are not a new phenomenon of the modern world and lifestyle. The first documented bone changes attributable to cancer were found on the upper arm bone of a warrior in a grave from the early Iron Age (500 BC) in Münsingen, Switzerland.[3] Further evidence for the existence of cancer in ancient times comes from the upper jaw bones of an Australopithecus from Kenya whose cancer was likely caused by Burkitt's lymphoma after Eppstein-Barr-Virus infection.[4]

Hippocrates of Kos (460–370 BC) first coined the term “cancer” to describe malignant diseases. He described large tumors in female breasts that were nourished by blood vessels reminding him of the legs and claws of crustaceans. For this resemblance, he termed this malignant tissue formation “karkinos” (gr. cancer). Tumors of the breast, cervix and bladder were also described in the “Medical Papyrus of Kahun” and ancient Egyptian “Papyrus Ebers.”[3] Treatments such as surgery to remove proliferating malignancies together with the surrounding healthy areas are recorded in the “Edwin Smith Papyrus.” The first known chemotherapies trace back to ancient India and ancient China. In these regions, tumors were coated with pastes of arsenic and mercury, and medicinal herbs were used for cancer therapy.[5]

3. PLANTS IN CANCER THERAPY TODAY

Due to the rising incidence of cancer and the fact that current options for cancer therapy are frequently not successful, it is imperative to find new strategies for cancer therapy and to improve chances of cancer prevention. Phytochemicals represent a promising option in achieving improved therapies.

In the past few years, cancer patients have been attracted by complementary and alternative medicine (CAM), which can be combined with allopathic chemotherapy or employed as a replacement for the usual primary therapy.[6] Jonathan Hartwell began to pursue the use of therapeutic phytochemicals as early as the 1960s and 1970s. He described “over 3,000 species of plants which [had] been reported to have alleged anti-cancer properties.”[7] Graham *et al.* published an additional list of more than 350 plant species used against cancer in the year 2000. The National Cancer Institute (NCI) screened more than 114,000 extracts of more than 35,000 plant samples for anticancer activity between 1960 and 1981.[8] In fact, over 70 percent of anticancer compounds are either natural products or substances derived from natural products.[2] Prominent examples are paclitaxel, vinblastine, and camptothecin.

Increased understanding of the molecular mechanisms of cancer development led to improved screenings of plants for anticancer activity. For example, NF-κB plays a critical role in the pathogenesis of a number of cancer types. Therefore, screening of plant constituents which suppress NF-κB activation represents a promising approach for anticancer drug development.[9,10]

The following section describes selected examples of ethnobotanical and ethnopharmacological concepts and phytotherapies for cancer treatment.

4. ANCIENT EUROPEAN MEDICINE

One of the most famous doctors of European classical antiquity was Hippocrates (460-370 B.C., Ancient Greece). He was the first to create a written record of medical information. This represented a significant progress, because oral communication was the common way to hand down knowledge at that time. His best known work is the "*Corpus Hippocraticum.*"[11]

Hippocratic medicine is based on four central points: (1) exact and differentiated observation of the patient, (2) profound formulation of a prognosis, and (3) consequent treatment by drugs or surgery. These three important points have been supplemented by a fourth core idea, empirical research. One important development pioneered by Hippocratic medicine was the demystification of the former belief that diseases are the result of an individual's fate.

Hippocrates established the doctrine that a healthy body reflects inner balance and harmony. He also further developed the basic principles for the concept of humoral pathology, which had been important for many centuries.

According to Hippocrates, diseases are a logical consequence of disturbed harmony in the body, especially as caused by an unfavourable mixture of corporal humours. This unbalanced mixture was termed dyscrasia, in contrast to a harmonious mixture, termed eucrasia. Restoration of eucrasia is the central aim of Hippocratic medicine. Hippocratic medicine dictates that a balance must be reached between sleeping and being awake, between working and relaxing, between thinking and dreaming, and so on.[11] In this context, dietary foods and dietetic treatments played an important role in treatments in order to promote a healthy balance in all spheres of life.

Another key personality in ancient medicine was Cornelius Celsus, who lived during the 1st century in the Roman Empire. There he published the book "*de Medicina*" describing the early origins of medicine and its progression up to his time. His work dealt with the prognosis of disease and the influences of climate, age, and other conditions on health and disease.[11] Thus, Celsus had already recognized a relationship between disease incidence and lifestyle. Today, thousands of years later, the link between lifestyle and health has been documented in many epidemiological and clinical studies. For example, today we know that a healthy diet consisting of fish, olive oil and few fats — the so-called Mediterranean diet — protects against diseases like heart-attacks, diabetes and metabolic syndrome.

Pedanius Dioskurides also played an important role in ancient times, and his influence extends into the twentieth and twenty-first centuries. Dioskurides published "*Materia Medica*" describing the effect of various medicines, foods, drugs, ointments and minerals as well as magic treatments. Medical knowledge fixed in written documents allowed more people to participate in medical healing practices than oral tradition did. Thus, Dioskurides' work became a basis for subsequent books on herbal medicine published later in the Middle Ages.[11]

Another important figure in medical history is Galenos of Pergamon, who thoroughly and precisely studied the humoral pathology ideology developed by Hippocrates. Like Hippocrates, he believed that imbalance among different humours of the body provoke diseases whereas well-balanced humours bring health.

Galenos' doctrine was based on the existence of four different humours in the body: blood, yellow gall, black gall, and mucus.[12] Humoral pathology remained the most important doctrine in medicine for many centuries. Later, urine gained relevance as an additional humour.

Specific methods to maintain or restore eucrasia included vomiting, fleecing and blowing of the nose. A general principle of humoral pathology was "contraria contraies, which means "acting against." Specific character traits were also seen as consequences of humoral pathology. For example:

- An excess of yellow gall leads to aggressiveness with disobedient behaviour
- An excess of black gall leads to sad and depressive behaviour and melancholy
- An excess of blood leads to overreactions, agitations or alternately joyous gaiety and creates a sanguine personality

- An excess of mucus leads to the introverted and hesitant personality of phlegmatic persons.[11]

4.1. European Cloister Medicine

In the Middle Ages, living conditions of the general population deteriorated. In these times, the cloisters were cultural centers in Europe.

One of the most important Christian monks, Benedict of Nursia, founded the cloister Monte Cassino in 480-547. This cloister became famous for its benediction rules *ora et labora* — pray and work. Benedict of Nursia taught the monks to read and write. In doing so, he ensured that the art of writing and reading was handed down from generation to generation. He also cared deeply about the people living around the cloisters. He established the precedent that every cloister should have a hospital. It was the monks' duty to care for the health of the neighbouring population. Thereby, the fundamentals of cloister medicine were created.

Another monk called Cassiodor later founded the Vivarium cloister near Squillace, a village in Calabria in the south of Italy. He encouraged his monks to read the works of Dioskurides, Hippocrates and Galen to learn about the characteristics of herbs and herbal mixtures.

These monasteries became enclaves of medical knowledge. It was through these monasteries that the term "cloister medicine" found its way into the medical progress of the Middle Ages. The so called *Lorscher Arzneibuch* (ger. Pharmacopoia of Lorsch) represents an exquisite document of cloister medicine. It is a scripture from the Lorsch monastery situated close to Worms (Germany). The *Lorscher Arzneibuch* is the oldest available book of cloister medicine.

Interestingly, costs played a similar role in the Middle Ages as they do today. Therefore, it is not surprising that the herbs of the *Lorscher Arzneibuch* that grew locally around Lorsch were more frequently used than the expensive herbs found in oriental countries.[13]

Many other famous works, e.g. the *Hortulus* by the monk Walafrid Strabo, who lived at the cloister Reichenau (Bodensee, Germany), followed the Lorscher medical book. One work on cloister medicine written by Odo Magdunensis was called *de Viribus Herbarum*, but was later renamed *Macer Floridus*. The works by Odo and Strabo were didactic poems. Odo's book became one of the favorite scriptures on cloister medicine at that time. In contrast to Strabo, who simply described the plants, Odo was more interested in the medical effects of the plants. Remarkably, *Macer Floridus* was the very first work on cloister medicine translated into the German language.[14]

Among the authors who wrote about cloister medicine, one person holds an outstanding position: Hildegard of Bingen (Germany). Being a nun, she published two books on medicine. *Causae et curae* dealt with theological views on the planet and its relationship to nature and medicine. *Physica* described medical treatments, e.g. the use of grain and herbs, different elements and of trees and scrubs. Additionally, she considered certain animals for medical purposes. Some parts of *Physica* were translated into the German language. After the invention of book printing by Johannes Gutenberg, books describing herbs of the Middle Ages experienced a great revival. *Gart der Gesundheit* (ger. garden of health) was the first German book about herbs and became very popular.[13]

4.1.1. *Plants of the cloister medicine — applied then and now*

Phytotherapy is extraordinarily popular today. Many laymen regard it as harmless, and believe it promises a mild therapy.

Nevertheless, it must be said that this is not the total truth!

Natural floras contain many of the most dangerous poisons in the world. Plants have evolved these poisons to defend themselves from the attacks of predators and herbivores. This phenomena has been termed "animal plant warfare" and is described in detail elsewhere.[15]

Well-known, severely poisonous plants include *Datura stramonium* (Solanaceae) containing the alkaloids hyoscyamine and scopolamine, *Atropa belladonna* (Solanaceae) containing atropine, and *Digitalis lanata* (Plantaginaceae) with cardiac glycoside constituents. Cardiac glycosides are currently used as drugs, but the monitoring of these drugs is of utmost importance because they have a small therapeutic window and overdosing is lethal.

Thus, phytotherapy cannot be regarded as a harmless alternative to classical medicine. Remarkably, many medicinal plants used in the Middle Ages are still being used today. Rather than providing an overview on these plants, which would be out of the scope of this chapter, we present a few, impressing examples.

4.1.2. *Mistletoe (Viscum album, Loranthaceae)*

In the times when Celtic and German tribes settled in the European continent, mistletoe was used by special cults. In the Middle Ages, mistletoe was used against epileptical diseases. It was also applied to women during childbirth.[16] Together with resin and wax, mistletoe-containing plasters were thought to heal boils. Hildegard of Bingen used mistletoe settlings grown on pear trees to prevent against bronchitis and other lung diseases.

At the beginning of the nineteenth century, mistletoe was introduced as a possible therapy for cancer. A fermented extract mixed with metals was produced. Today it is known that the viscotoxins A_2, A_3, B and Ps1 and lectins are responsible for mistletoe's cytostatic effects. These toxic polypeptides inhibit the growth of human tumor cells. Mistletoe induces necrosis when administered via subcutaneous injection. It can also be used to treat degenerative bone diseases, a property which was recognized as early as the Middle Ages.[17]

In the past two decades, numerous clinical trials on the effects of mistletoe extracts have been performed. Ostermann and colleagues concluded that patients treated with mistletoe extracts had a better prognosis to overcome cancer.[18] In addition, a better quality of life has been observed for patients treated with mistletoe extracts during chemotherapy.[19] In animal experiments, mistletoe induced tumor remissions in mice, but not in rats.[20] Further investigations are necessary to fully explore the potential of mistletoe in cancer therapy.

4.1.3. *Hop (Humulus lupulus, Cannabaceae)*

Since ancient times, hop has been used for brewing beer. This plant was not only used in breweries, but also for medical treatment against agitation.

Even today, hop is frequently used in combination with valerian in pharmacy. Regarding its cytostatic effects, the compound xanthohumol is important due to its antiestrogenic and antioxidative characteristics. Xanthohumol inhibits proliferation and induces differentiation of cells.[17] As xanthohumol inhibits the activity of the enzyme aromatase, it may possess activity against estrogen-dependent tumors, e.g. breast cancer.[21] Furthermore, xanthohumol kills B-chronic lymphocytic leukemia cells *in vitro*.[22] Considering the broad spectrum of cancer-inhibiting mechanisms demonstrated, *Humulus lupulus* L. or some of its constituents may be promising in cancer treatment.

4.1.4. *Barberry (Berberis vulgaris, Berberidaceae)*

The ingredients of barberry are antimycotic and inhibit acetylcholinesterase. Barberry is known as a medicinal plant in Europe and Asia. In former times, this plant was known as acid punch (German: *Berberitze, Sauerdorn, Essigbeere*).[23] It has been assumed that there is an interaction between berberine, a benzyl-isochinoline alkaloid, and the mitochondria of cells, which contributes to the cytotoxic activity of this compound towards tumor cells.[24]

4.1.5. *Yew (Taxus baccata, Taxaceae)*

The yew belongs to the typical vegetation of the European continent and was mentioned by Galen. The toxicity of this plant is well-known and has often been used to evoke abortion and to kill worms and other parasites. The toxic ingredients in the needles, seeds and bark are taxanes. The production of mush from the non-toxic aril used to be a common practice in Europe (www.botanik.univie.ca.at/hbv/download/ib_taxus baccata.pdf).[25] The species *Taxus baccata* is cultivated in Europe, while *Taxus brevifolia*, the Pacific Yew, is cultivated in America. The cytostatic ingredients in yew are paclitaxel and its precursors. Paclitaxel hinders mitosis by inhibiting the depolymerisation of the microtubules. The synthetic derivative, docetaxel has higher binding affinity to microtubules than paclitaxel.[17] Taxanes are used for the treatment of ovarian, breast, and lung cancer.

4.1.6. *Meadow saffron (Colchicum autumnale, Colchicaceae)*

Meadow saffron was known as *Hermodactylus* in the Middle Ages.[13] Hildegard of Bingen used the name *Heylheubt*,[16] because it was used to kill head lice by washing one's hair with extracts of the blossoms and bulbs.[13]

The German name, *Herbstzeitlose* (late bloomers) was used because the plant blooms in the autumn and shows only green leaves in the spring. In cloister medicine, this plant was traditionally used against gout in the feet, hands and bowels. Meadow saffron was boiled with honey and water and afterwards applied together with cinnamon and fennel as a plaster. To treat nasal adenoids, a cotton ball was filled with meadow saffron powder and then placed into the nose.[13]

Even today meadow saffron is used against gout. Colchicine, the main alkaloid, inhibits mitosis. It inhibits the polymerization of microtubules by binding to ß-tubulin, leading to cell death or polyploid cells.[17] Additionally, it has antiphoglistic and antichemotactic effects.[26] Colchicine inhibits the phagocytosis of urea crystals and the consequent release of lactic acid. This prevents further decrease of pH and further crystallisation of urea.[17] Hence, acute gout attacks cannot take place. Upon acute gout attacks, 1 mg colchicine is given orally and followed by oral administration of 0.5–1.5 mg colchicine every 2 h. The total daily dose should not be more than 8 mg, and treatment for long time periods should be avoided due to the potent toxicity of colchicin.[26]

As colchicine has a similar mode of action to the *Vinca* alkaloids, it has been considered as an anti-mitotic spindle poison in cancer therapy. However, its use in cancer treatment is obsolete, because of non-tolerable high neurotoxicity. However, at lower concentrations, colchicines can be used for rheumatic diseases.[17]

4.1.7. *Milk thistle (Silybum marianum, Asteraceae)*

Silybum marianum [L.] Gaertn. (Asteraceae; milk thistle) is a traditional, relatively safe and well-tolerated herb that has been used for almost 2000 years in a variety of settings. The extract of milk thistle mainly consists of diverse flavonolignans of which silymarin, a complex of silybinin (silybin A and B), silydianin, isosilybin (A and B) and silychristin, is the main active constituent.[27,28]

Many *in vitro* and *in vivo* studies and human clinical studies indicate the use of silymarin or one of its compounds in oncology. It adversely affects cancer cells in a number of ways and is nontoxic to normal cells.[29]

Another benefit recommending its use is its biologically availability and non-invasive oral administration.

One valuable application of the milk thistle in cancer therapy is its ability to reduce the iatrogenic toxicity associated with classical chemotherapy, because of its hepatoprotective, antitoxic, antioxidant, anticancer, antidiabetic and cardioprotective effects.[28,30] It can also normalize immunoregulatory defects and shows a variety of different protective effects. For instance, it can be used for detoxification after chemotherapy; can protect liver, kidney, and heart; prevent and treat hepatotoxicity during or after chemotherapy; reduce the adverse effects of anesthesia and ameliorate long-term effects of cancer treatment, especially diabetes mellitus and disorders related to hepatic and cardiovascular function.[27]

As recently summarized, milk thistle demonstrates anticancer and chemopreventive effects against various types of cancer.[29,31] A range of possible anti-cancer activities of silymarin or one of its compounds is well-demonstrated against the hallmarks of cancer.[32] It inhibits the cancer cell proliferation of various cancer cells,[28,33] induces apoptotic death in cancer cells,[28,29,33] possesses anti-angiogenic efficacy,[29,33] shows anti-invasive and antimetastatic efficacy,[29,33] has anti-inflammatory and antioxidative properties[29,33] and inhibits cellular glucose uptake.[34]

Since high-quality studies are limited and negative reports on the compound also exist, extensive further explorations with standardized samples are necessary. Nevertheless, this multi-functional and multi-target drug is promising for cancer prevention, for adjuvant cancer treatment alongside chemo- and radiation therapy and even for direct anticancer therapy[27] due to its ability to block all stages of carcinogenesis, initiation, promotion and progression.[33]

4.2. Mediterranean Food

It has been long known that in Mediterranean countries such as Italy, France and Greece the risk of heart related insults is less than in northern European countries. This fact may be related to the Mediterranean diet, which is high in vegetarian components, such as various fruits and vegetables, nuts, and whole grain products as well as olive oil and red wine. This food is characterized by a high proportion of unsaturated fatty acids, e.g. in olive oil. Furthermore, red wine contains many polyphenols, which counteract oxidative stress. Resveratrol in red wine activates nitric oxide and inhibits endothelin-1. Anthocyanes in wine also reveal antioxidative effects.[17] In addition to the preventive function towards heart-related diseases this diet may also play a role in prevention of cancer and other diseases.

In Scandinavian countries, the USA, and Great Britain, the incidence of breast cancer, prostate cancer and endometrial cancer is strikingly higher than in Mediterranean countries. This trend may be related to Mediterranean nourishment. It is evident that the population in northern countries eats less fruits and vegetables and more meat.[35] Furthermore, it is hypothesized that fish and olive oil may prevent breast cancer, prostate cancer and bowel cancer.[36] Full grain products may also reduce the risk for cancer.[37]

5. NATIVE AMERICAN KNOWLEDGE OF HEALING

The folk medicine of the Native Americans has its origins in prehistoric times.[38] The knowledge of healing was derived from natural principles. The basis for this unique natural medicine is their ideology that the Earth does not belong to men, but rather that men belong to Earth. The natives believe that every plant and every animal has a spirit. Thus, treating animals and plants respectfully is one of the most important aspects of their doctrine of well-being. The shamans and native healers divide diseases into different categories dependent on evil spirits. For example, wind infection may be the cause of asthma. The Native American body of medical knowledge assumes that all diseases result from a dysfunctional relationship between men and magical or supernatural powers.[39] Since medicine is so closely associated with magic, there are many magical rituals to cure diseases. Often the Indians of South America perform a ritual of "drawing out," in which the healer sucks on the site of the suspected disease and extracts an object from the human body that was the cause of the disease. This object can be a thorn, a dead lizard or a splinter of wood.[38] In this ritual, the disease becomes more concrete and the patients are strengthened by the belief that the disease has been removed from their body. Apart from application of scientifically tested and proven drugs, the key difference between modern medicine and Native American medicine is the fact that the process of healing and the environment are central aspects of treatment among the latter. The optimistic mental attitude and strong focus on the healing procedure positively influences the health of the patient. It is scientifically documented that psychological aspects play a crucial role in the treatment of diseases with alternative methods.[40–42] A patient who strongly believes in his recovery may have better chances to be cured. In modern medicine this effect is known as the placebo (lat. I shall please) effect. It has repeatedly been reported that between 5 to 20 percent of positive treatment effects are due to the patient's own conviction.[17,41] With the help of various amulets and talismans in traditional American Indian medicine, the patient is strengthened through the faith that he will escape the disease. Similar to traditional Chinese medicine, acupuncture at different points in the feet is known in Native American medicine to treat various diseases.[38,41] However, in contrast to the Chinese, the North American Indians use Agave leaves instead of metal needles for the acupuncture procedure. Another similarity to Chinese medicine is the use of massage in Native American medicine. Sweat lodges comparable to the Scandinavian saunas support the healing process.[38] In the modern world it is well-known that regular sauna visits stimulate blood circulation and general well-being. Various herbs are also used in sweat lodges of the North American natives. These inhalations are analogous to aroma therapies applied in industrialized countries.

6. OLD TREASURES WITH MODERN PERSPECTIVES

The achievements of indigenous peoples with respect to the healing power of plants date back to ancient intuitive and magical knowledge that is combined with observational experience.[38,39] All the more amazing is the fact that the medicines and drugs used by Native Americans may be equal to or in some cases superior to those of Western medicine. Modern pharmacological research has confirmed the activity of several of these drugs. A well known example is the use of willow bark for the production of salicylic acid, which is the raw material for aspirin.[17,38] Several tribes such as Choctaw, Alabama and Shikasaw used boiled willow bark and roots for the treatment of headaches and fever.[38]

In fact, puerperal fever, maternal mortality and premature births were largely unknown amongst the Native Americans.[38,39] The use of natural oral contraceptives such as yam roots, which contain steroid-like substances, reduced birth rates.[38] Shoshone women chewed stoneseeds (*Lithospermum ruderale*) for contraception. Chemists have isolated an estrogen-like substance from this plant, diosgenin, which has contraceptive features.[38,39] Another example is the root of *Rauvolfia serpentina*, known as snakeroot, which were chewed by Native Americans to reassure. Experimental investigations led to the discovery of the active compound, reserpine, which was used for many years in psychiatry. Today, reserpine is no longer used because of its undesirable side effects.[38,43]

At this point, it is important to emphasize that traditional medicine should not be viewed as a library of chemical substances. The traditional knowledge of healers also includes aspects such as strategies to minimize side effects by special processing of the drug or using it in combination with other herbs. The phytochemical isolation of bioactive components may in fact be disadvantageous because pronounced side effects may emerge without the mitigating activity of other plant ingredients.

7. FROM FOLK MEDICINE TO MODERN DRUGS

One of the best known plants from indigenous medicine is tobacco (*Nicotiana tabacum*). Many rituals include the smoking of tobacco as an essential component. In contrast to tobacco smoking in the Western world, the indigenous usage of tobacco incorporated the addition of other herbs with healing and intoxicating effects.[38] Besides the magical significance of tobacco in the rituals, the plant also boasts some traditional medical applications. Compresses of tobacco leaves were used by different tribes to disinfect wounds and as a remedy for tick bites.[38,39]

Taking into account that the biological function of nicotine is the protection of plants against herbivores, its healing effect is not surprising. Mixed with other medicinal plants, tobacco was also taken as an analgesic.[38,39] It is well known that nicotine binds to specific receptors and stimulates the release of neurotransmitters such as dopamine, serotonin, norepinephrine, and endorphins.[17,44] In contrast to Indians, who hold the smoke of tobacco in their mouths and exhale it, smokers of the modern world inhale the smoke.[38]

Apart from the fact that nicotine is one of the most highly addictive compounds, carcinogenic substances such as carbon monoxide and benzpyrenes are inhaled during smoking. About 30 percent of all cancer cases are attributed to lung tumors, of which of 87 percent are caused by smoking.[45] Thus, cancer incidence would decrease by one quarter if smokers would stop smoking.

Tobacco is not the only drug that is abused. The Native Americans have used the Coca shrub containing the well-known compound cocaine for medicinal purposes for 5,000 years. Due to their heavy usage of this plant, coca has been termed "aspirin of the Andes."[39] It was adopted to treat any kind of pain, neuralgia, rheumatism, cold, or flu as well as various digestive disorders.[38,39] Albert Nieman Gottingen, a chemist discovered in 1859 that this substance can be used as local anesthetic in dentistry. This was a milestone for the medical use of cocaine.[38] However, its use was discontinued after a short time because of its euphoric side effects.

An important crop plant of the ancient and medieval Europe that also was used as medical plant is Cannabis. Later, Cannabis gained popularity as a highly addictive drug. The use of Indian hemp (*Cannabis sativa*) for medical purposes was first mentioned 2737 BC in *Pen Tsao*, the Chinese pharmacopoeia of Emperor Shen Nung.[46] It described more than 120 medical applications of hemp. After the drug's ban in 1925, the drug has experienced a comeback in the last decade for symptomatic treatment of pain.

Especially in patients suffering from severe cancer pain, the application of cannabinoids was promising.[47,48] Despite the potential risk of addiction, the use of cannabinoids may be a good alternative in advanced disease stages, when other drugs become ineffective. After the elucidation of the exact interaction of cannabinoid molecules with specific receptors located mainly in the tissues of the nervous system in past years, there have been some attempts to establish synthetic substances such as δ-9-tetrahydrocannabitol (THC).[47–49] Current results look promising. Using previous data on receptors and the effects of individual substances, production of a synthetic derivative is a possible way to maximize the analgesic effect of a natural substance such as Cannabis while minimizing the addictive potential. The carcinogenic effects of aromatic carbon hydrogen arising during smoking can be reduced by intravenous application of cannabinoids.[48]

8. TRADITIONAL MEDICINES FROM ASIA BASED IN ACADEMIC SCHOOL SYSTEMS

Many patients seek alternative methods for disease treatment in Asian medicines, such as Traditional Chinese Medicine, Kampo, or Ayurveda.

8.1. Traditional Chinese Medicine (TCM)

Because of its 5000 year-old tradition, Traditional Chinese Medicine (TCM) holds a strong position both in rural as well as developed areas of China. The country has made enormous efforts to modernize TCM. Even in the western world, there is a continuously increasing interest in TCM, as it has much potential to supplement Western medicine. Medicinal herbs have always played and continue to play an important role in western medicine, although they have lost prominence due to the progress in organic chemistry during the twentieth century. Nowadays, a revival of interest in TCM in western countries can be observed. TCM, with its holistic approach of analyzing the patient in his or her entirety, provides an attractive alternative to "high-tech" western medicine. However, there is some skepticism as to the quality control of TCM products. In many cases, it is still difficult to guarantee constant quality of herbal remedies.[50]

8.1.1. *History of TCM*

The historical origins of phytotherapy in China are unknown. It likely went beyond the beginning of human culture, since even animals prefer special herbs or plants when they feel sick. It is known that the Emperor, Shen Nong, in 2700 BC was a founder of agriculture and phytotherapy. His work, *Shen Nong Ben Cao*, described 365 medical plants, most of which are still used today.[51] Shen Nong's work can be divided into three parts. The first part describes 120 drugs, which are classified as nontoxic for the human organism and which strengthen human life. Here, plants such as *Panax ginseng* and *Glycyrrhiza glabra* are listed. The second part describes another

120 drugs that affect the human body and may also be toxic. *Zingiber officinale* and *Paeonia lactiflora* are examples of this part. The last volume of the book focuses on 125 drugs called "earthly" that should not be continuously taken, as they show definite toxic activity. *Rheum rhabarbarum* and *Aconitum napellum* are found in this category.[52]

Around the same time, another book was published by Huangdi Neijing. It represented an encyclopedic work that assembled all medical knowledge of that time. Even today, it is part of the basic education at TCM universities.[51] Another basic work of the TCM emerged, the *Shang Han Lunh* ("essay about damaging cold"). The traditional Japanese medicine (Kampo medicine) is mainly founded on this work.[51] The most famous doctor of the post-classical period was Li Shi Shen, who lived in the sixteenth century. He summarized all knowledge of medicine up to this time and organized it in his textbook *Compendium of Materia Medica.* It includes more than 4,000 TCM formulations and 2,000 single herbs that were already partially mentioned by Shen Nong. The continued use of these plants over a thousand years is a strong case for their effectiveness.

Since the middle of the last century, there has been a renaissance of TCM in China. Twenty-five new universities were founded and for decades TCM remedies have been a popular export item to western countries. The interest of foreign countries has brought a new reputation to TCM.[51] The scientific interest in western countries continues to grow.

8.1.2. *Philosophy of TCM*

More than 2,500 years ago the first philosophers appeared in China. Two philosophers in particular, Laotse and Confucius, strongly influenced the development of philosophy in China. The roots of Chinese culture go far back into the past. The most famous philosophical text is called Yijing ("Book of changes"), which is considered to be the origin of Chinese philosophy.

Confucius (551–479 BC) created a new way of viewing man. Before his time, an old belief in demons prevailed as a motivation for proper behavior. In contrast, Confucius emphasized the importance of ethical behavior of an individual for the welfare of the entire society. In the sixth century BC, Laotse established Daoism, which had a strong focus on nature and natural laws. In Daoism, nature becomes the centre of life. Daoism had a strong influence on TCM. The Daoistic idea of returning to a simple and natural life is reflected in the basic theories of TCM. An important area in which Daoism influenced TCM is in the prevention of diseases, not only in their treatment.[51,52]

8.1.3. *Basic theories, treatment and diagnostics of TCM*

TCM is based on the theory of yin and yang that represents balance between two contrasting principles, e.g. cold and heat, up and down, male and female, summer and winter, black and white, or heaven and earth. The idea of yin and yang has its origin in the philosophical macrocosmos. The relation between yin and yang is responsible for the state of one's health. If the two elements are in balance, the human body feels good, while an imbalance leads to illness.

Another basic theory of TCM is based on the five elements: wood, fire, earth, metal and water. These elements are assigned to special organ systems and emotions. In a healthy body, the five organ systems are in harmony with each other.[52] Medicinal plants are classified as specific tastes, e.g. sour, butter, sweet, pungent, or salty.

Other basic principles of TCM encompass the special five substances: *qi, xue, jiing, shen* and *jin-ye. Qi* can be translated as life-energy, *xue* is associated with blood, *jiing* with fluid-like essences and *shen* with spirit. *Jin-ye* represents body fluids.

We find an alternate classification of the organ systems in TCM into *zang* and *fu.* Lung, heart, spleen, liver and kidney belong to *zang*, which represents filled organs preserving *qi* and *xue.* Stomach, intestine, colon, gall bladder and (triple warmer (non-anatomical) organ) belong to *fu*, hollow organs that collect *qi* and *xue.* The TCM theory holds that all organ systems are interconnected by nonvisible tracks (channels or meridians) and luo (collaterals). These connections maintain the constant circulation of *qi* and *xue* and the balance of *yin* and *yang.*[52]

TCM diagnostics uses four main methods: general observation of the patient, listening to the patient's voice and breath, smelling breath and other body fluids, and monitoring the patient's pulse. Treatment is based on the anamnesis of these diagnostic methods. External interferences of the meridians are one mode of treatment, and methods include acupuncture, acupressure, cupping and moxibustion. These methods use external manipulations to trigger internal changes in the body. Internal modes of interference important in TCM include proper nutrition, herbal medicines, meditation.[52]

8.1.4. *TCM-based phytotherapy*

Rather than listing hundreds of herbs, here we focus on a few select examples to highlight the general relevance of TCM-based phytotherapy. One of the most popular plants in TCM is ginseng (*Panax ginseng*, Araliaceae). Ginseng roots have been used extensively in Asia as a tonic and panacea for thousands of years. It may also be effective in cancer prevention and therapy, especially in maintaining as individual's constitution.[53] The active compounds in *Panax* species are the ginsenosides, which are classified as either panaxadiol saponins, panaxatriols saponins or oleanolic acid saponins. However, many other chemical constituents have also been isolated from *P. ginseng*. The costly plant is also popular in Europe and North America. The toxicity of ginseng appears to be low for most individuals. However, overuse or herb-drug interactions may lead to adverse symptoms such as overstimulation, headaches, insomnia, hypertension, palpitations and gastrointestinal problems. Ginseng's clinical potential in cancer treatment lies in its numerous, cancer-related pharmacological activities. *P. ginseng* stimulates the immune system, and act as antioxidant and an antihyperglycemic.[53] It also shows activity against various stressors, exhibits endocrine, hormonal (potentially estrogenic) and blood coagulation effects, and functions as anti-fatigue and antidepressant substance.[54] Furthermore *P. ginseng* reveals cytotoxic activity towards cancer cells.[53] Ginseng's *in vitro* and *in vivo* anticancer activities include induction of apoptosis, inhibition of cell cycle progression, inhibition of invasion and metastasis and antiangiogenic activity. However, no clinical trials have yet confirmed these experimental results.[53] *Panax ginseng* may also induce differentiation of neoplastic cells into normal tissue.[55] Many of the studies are controversial and further studies are needed to prove the efficacy of this plant.

In standard tumor therapy, many natural products are used. Some of them are derived directly from TCM (e.g. camptothecin), demonstrating the great potential TCM holds.

8.1.5. *Classical targets for TCM-derived drugs applied in cancer therapy*

Camptothecin, isolated from *Camptotheca accuminata*, is one of the most popular examples of TCM-derived drugs currently used in western medicine. Its semi-synthetic derivatives, topotecan and irinotecan, are used for the treatment of ovarian and colon cancer. All three of these drugs target topoisomerase I. Inhibition of topoisomerase I leads to single strand breaks in DNA during replication and ultimately to cell death.

Other examples of TCM-derived anticancer drugs are the topoisomerase II inhibitors, etoposid and teniposide. These drugs are based on podophyllotoxins, which were initially identified in the North-American plant, *Podophyllum peltatum*, but are also present in *Podophyllum emodi var.chinensis.* Topoisomerase II inhibitors induce double strand breaks in DNA that ultimately lead to apoptosis.

The alkaloids vincristine and vinblastine from *Catharantus roseus* inhibit polymerization of the microtubules of the mitotic spindle. Thus, they inhibit the separation and distribution of chromosomes to daughter cells. The taxanes are another class of microtubule inhibitors. One taxane, Paclitaxel, is isolated from *Taxus brevifolia* and is also found in *Taxus chinensis* and inhibits depolymerization of microtubules.

The TCM-derived substances described above have been thoroughly investigated and their efficacy is well established. In contrast, several novel targets for TCM-derived natural compounds for cancer therapy have demonstrated potential. Emodine, from *Rheum palmatum*, inhibits casein kinase 2, which is implicated in cell cycle control. Another example of a novel target is telomerase, which plays a critical role in the immortalization of tumor cells. Telomerase inhibitors include verbascoside, from *Pedicularis striata*, and gambogic, from *Garcinia hanburyi.*

The formation of blood vessels is essential for tumor cell nutrition and tumor survival. Disturbing angiogenesis of a tumor can thus inhibit its progression. Capsaicin (*Capsicum* sp.) and sinomemine (*Sinomenium acutum*) are anti-angiogenic substances from TCM.

Artemisinin, which comes from the Chinese medical plant, *Artemisia annua*, and its semi-synthetic derivatives, Artesunate and Artemether, are well known in malaria therapy. In the past two decades, their activity against tumor cells has been recorded and systematically analyzed. This analysis has yielded the conclusion that genes regulating apoptosis and/or angiogenesis are involved in the anti-cancer effects of artemisinin-type drugs. One advantage of artemisinin and its derivatives is that they cannot act (or can only act weakly) as substrates for multidrug resistance transporters. Hence, these drugs could represent a new possibility for the therapy of multidrug-resistant tumor cells.

Another example of a novel drug is homoharringtonine, obtained from *Cephalotaxus harringtonia*, a popular compound in TCM. Homoharringtonine is currently under investigation as a cancer treatment, and has been described as an inhibitor of protein synthesis.[50] Homoharringtonine is highly cytotoxic and shows elevated activity against leukemia cells.

Arsenic-based compounds represent yet another interesting class of substances to show anticancer activity. Arsenic trioxide was used in the eighteenth and nineteenth centuries to treat cancer and syphilis in Greek, Roman and Chinese medicine. In the twentieth century, Paul Ehrlich, the founder of chemotherapy, discovered the arsenic compound arsphenamine, which was used to treat syphilis for decades. In a microarray-based approach, predictors for the basis of tumor cells' sensitivity to arsenic trioxide have been found. Several candidate genes have been identified that show apoptosis as a mechanism of arsenic trioxide's cytotoxicity.[50]

Cantharidin, isolated from the blister beetle *Mylabris*, is another therapeutic agent derived from TCM. It acts by inhibiting the protein phosphatases 1 and 2a. This activity is necessary for growth inhibition or cytotoxicity.[50]

8.1.6. *TCM-based nutrition for prevention of cancer*

In China, the balance between yin and yang plays an eminent role. Hence, nutrition shall keep the yin-yang balance in each individual organism for the maintenance of health. Here, foods are classified according to their yin and yang qualities and energetic conditions.

In the classical texts of TCM, there does not seem to exist a term for a tumor, but there are many descriptions of cancerous symptoms. TCM teaches that these cancer-like disorders are caused by a deficiency of life energy, qi, and a lack of organ function. TCM therapies involve many foods with chemo-preventive properties, e.g. asparagus, cauliflower, white cabbage, celery, carrots, tomatoes, onions and garlic. Soybeans have been reported not only to inhibit the growth of tumor cells, but also to prohibit metastasis.[56]

Green tea plays a great role in Chinese tradition as well. In 2737 BC, green tea was introduced by Emperor Chen Nong. It is said to enhance an individual's capability for seeing and concentration. Furthermore, green tea possesses demonstrated activity against cancer.[56]

Important substances isolated from green tea (*Camellia sinensis*, Theaceae) include the catechines, which belong to the chemical group of polyphenoles. Together with vitamins C and E, catechines show antioxidant activity. They are even able to reduce vascular damage.[56] Japanese scientists isolated one active ingredient of green tea called EGCG (epigallocatechine-3-galate). This substance binds to the surface of tumor cells and inhibits their growth. This effect has been demonstrated in lung cancer patients drinking three cups of tea per day.[56] Some investigation as to whether EGCG has an inhibitory effect towards the growth of breast cancer has also been conducted, but statistically significant results remain lacking.[57] Nevertheless, consumption of green tea might prevent recurrence of breast cancer. Because of the lack of sufficient clinical studies, a definite answer remains elusive.[57] It also remains to be clarified whether peroral application of green tea produces comparably favorable effects.[17]

As one of the oldest known systems of medicine, TCM also represents the origin of other traditional medicines. One such system is Kampo, traditional Japanese medicine, which is described in the following section.

8.2. Traditional Japanese Medicine (Kampo)

The origin of Kampo traces back to China and is, therefore, strongly influenced by TCM. Over 1,500 years, Japanese medicine has changed and adapted a great deal. Treatment with botanicals has always been a central theme of medicine in the eastern world.[58]

The oldest Chinese Pharmacopoeia (Shen Nong Ben Cao Jing of the fifth century) arrived in Japan with the Chinese and Korean monks. The medical recipes it contained spread throughout Japan and developed independently.

In 1984, Yasunori Tanba wrote the medical textbook *Ishinho* (Essence of healing) based on the work *Shang Han Lun* by Chang Zhong-jung. This book is regarded as the starting point for the development of Japanese medicine.

In the eighteenth century, the first Kampo school, the Koho School, was built. This was a milestone for Kampo medicine. Yoshimasu Todo (1702–1732), one of the main scholars, revised the recipes of Shang Han Lun. Another Koho teacher, Toyo Yamawaki, disproved some parts of TCM theory in his book, *Zoshi.*[52]

Since western medicine and the progress of science strongly influenced Japan, Kampo lost popularity in the nineteenth century. However, it experienced a revival at the end of the twentieth century.

Similar to TCM, Kampo medicine also applies a holistic approach. The human being is considered an individual, but also part of society. Health is thought to be a state of a harmonic balance between the individual and his or her environment. Similar to the TCM theory, diseases are thought to be consequences of imbalance. Kampo physicians monitor early symptoms, which is a great advantage in the prevention of disease. Association of diagnosis, disease pattern (*Sho*) and corresponding treatment (*Ho*) represents an important principle of Kampo. The purpose is to reconstitute the balance of the organism.

Interestingly, Kampo plants also consist of many bioactive ingredients. The Japanese could cultivate only 20 percent of the medicinal plants found in China, and as a result a selection of effective drugs and an improvement of pharmaceutical regeneration took place. Overall, this process led to higher utility of established therapeutic agents and decreased dosage. As a result, Kampo pharmacies have only about 250 drugs, whereas in China there are approximately 500 drugs.[58]

Over 200 Kampo recipes have been reported that use between two and fifteen components.[59] In cancer treatment, Kampo medicine has been used to promote physical reconditioning and to reduce adverse effects of chemotherapy and radiation, enhancing the quality of life.[59] Due to the varied absorption, distribution, metabolism and excretion of Kampo drugs among patients, Kampo is best viewed as a system of individualized medicine.

Most medical plants used in Kampo medicine are also used in TCM, especially in cancer therapy, as there are strong overlaps between the two systems.

One interesting Kampo phytochemical is Shikonin from *Lithospermum erythrorhizon*.[59] Shikonin is a naphthoquinone that possesses considerable antibacterial, anti-inflammatory and antitumor activity. It has been shown to inhibit TNF-α and angiogenesis in mice. It blocks the expression of some integrins and is also able to inhibit endothelial cell proliferation and migration and induces apoptosis. Many studies of Kampo formulations have reported anti-metastatic effects.[60] The formulation Shichimotsu-koka-to (SKT) is used for the treatment of hypertension and artheriosclerosis. It consists of seven medicinal drugs: *Paeoniae radix, Angelicae radix, Astragali radix, Rehmanniae radix, Cnidii Rhizoma, Phellodendri cortex* and *Uncariae uncus et ramulus*. SKT shares some plants in common with other formulations that have been reported to inhibit metastasis. The fact that SKT quenches superoxide anions may explain its impact on metastasis, as radical oxygen species stimulate metastasis. Ohno *et al.*[60] found that long-term treatment with SKT reduced the number of pulmonary metastatic nodules and significantly extended the life span of lung cancer patients. Although the molecular mechanism of SKT's anti-metastatic effect is still not fully understood, it represents a promising new candidate in cancer therapy.

Kampo medicine significantly extended the survival time of patients with uterine cervical cancer and improved the patient's quality of life.[61]

8.3. Traditional Indian Medicine (Ayurveda)

The word 'Ayurveda' is composed of the two Sanskrit words *Ayur* meaning "life in" and *Veda* meaning "science."[62] Four thousand years ago, Sushruta defined the basic principle of Ayurveda: "Every individual is different from another and, hence, should be considered as a different entity. As many variations are there in the universe, all are seen in human being."[62]

Sushruta, who lived in the sixth century BC, was a famous surgeon in India. As "father of surgery," he described 120 surgical instruments and more than 300 operations in the *Sushruta Samhita*. Another classical ayurvedic textbook is the *Charak Samhita*. These two basic works describe over 700 botanicals between them, including classification and pharmacology. The Rasayanas, one part of the *Materia Medica* of Ayurveda, deals with the enhancement of the human body's resistance. In fact, Rasayana stands for 'nourishing and rejuvenating drugs' and investigates various factors impacting a drug's strength, e.g. immunmodulation.[62]

In Ayurvedic medicine, the five main elements are combined into three basic energies or ever-present functional principles. Space and air constitute Vata, fire and water form Pitta and water and earth are combined into Kapha. These three energetic principles are called *doshas*, which may be translated as "mood status." They are responsible for the psychobiological functions of the human body and can be found in every cell, tissue and organ. If there is a balance among these three principles, a healthy state is reached. Imbalances lead to disease. The three *doshas* represent a great number of individual features of human beings.

Ayurvedic medicine uses a six-step method to treat patients. First, detailed anamnesis and observations must be made. The second step is the Ayurvedic diagnosis, in which the nature, degree and extent of the imbalance is determined. Approximately 6,000 clinical signs and symptoms of diseases are described in Ayurvedic texts.[62] In the third step, the doctor estimates the influence of season, time and environment on the *doshas*. Fourth, advice on lifestyle and diet are given. Fifth, the doctor supports healing by a purification process referred to as *panchakarma*. If the patient's imbalance is too severe to be controlled by these five steps, a medicinal treatment is used, and the doctor may design an individual mixture of drugs.

Modern pharmaceutical agents are often hindered by the development of drug resistance or their own toxicity. Therefore, it might be useful to go back to traditional herbs, such as those of TCM or Ayurveda, as bases for new studies to discover novel drugs. The Ayurvedic Pharmacopoeia contains more than 1,200 plant species, nearly 100 minerals and over 100 animal products. Thousands of pure substances as well as many combinations and processed formulations have been described along with details of their drug interactions.[62]

If this time-won experience could be combined with new scientific methods, there might be a great potential to identify new lead compounds for drug development.

A novel development in Ayurveda is the combination of pharmacogenetics, pharmacogenomics and pharmacoproteomics, which has been termed AyuGenomics®.[62] AyuGenomics® is a tool for designing individualized therapies using different *dravyas* (Drugs) and *guna* (actions) based on the patient's individual genotype, phenotype and the influence of the environment.

Whitania somniferia (Solanaceae) provides an example of an ayurvedic plant with considerable therapeutic potential.[63] Also commonly known as *ashwagandha* or English-Indian ginseng/winter cherry, it is described in the Indian Herbal and Ayurvedic Pharmacopoeias. *Withania somnifera* exhibits varied effects, e.g. anti-inflammatory, antitumor, anti-stress, anti-oxidant, immunomodulatory, hemopoietic and rejuvenating impacts.[62] Its antitumor mechanism depends on many factors. The plant exerts both antioxidative and pro-oxidative activity and reduces unregulated cell growth through tumor suppressor p53. The active ingredients are with anolides. Furthermore, *Withania somnifera* inhibits angiogenesis and reduces NF-κB activity in tumor cells.

Another botanical product that has been suggested for cancer treatment is the dietary supplement PC-SPES. This is a complex matrix of eight plant materials developed by three Taiwanese scientists in 1990. It consists of *Isatis indigotica* (Brassicaceae; root), *Glycyrrhiza glabra* (Fabaceae; root), *Panax pseudoginseng* var. *notoginseng* (Araliaceae; root), *Ganoderma lucidum* (Ganodermataceae; spores), *Dendranthema morifolium* (Asteraceae; flower), *Scutellaria baicalensis* (Lamiaceae; root), *Rabdosia rubescens* (Lamiaceae; root), and *Serenoa repens* (Arecaceae; berry). However, the specific ratio of these plant materials is not disclosed. Unfortunately PC-SPES, which was manufactured in China, was adulterated leading to its withdrawal from the market in 2002. PC-SPES's successful reduction of prostate-specific antigen levels (PSA) in prostate cancer treatment gave it its name: PC stand for prostate cancer and *spes* is the Latin word for hope. PC-SPES has been reported to detoxify many other toxins and neutralise free radicals by antioxidant mechanisms.[6] It balances the red and white blood cell counts and exerts favourable effects on the central nervous system.[6] Its anti-cancer activity has been studied *in vitro, in vivo* and in clinical studies. The results summarized in the following section must be viewed with some scepticism, since PC-SPES batches were not standardized and a considerable variation in activity can be expected.

PC-SPES decreased proliferation and induced apoptosis in many different cancer cell lines.[6] It down-regulated levels of proliferating cell nuclear antigen (PCNA) and intracellular and secreted PSA.[6] Furthermore, it was shown by microarray analysis that hundreds of genes were differentially regulated by PC-SPES, including genes associated with the cytoskeleton, apoptosis, stress, cell cycle, proliferation and androgen-regulated genes.[64] Antitumor effects were not only revealed in rat and mouse models, but also in clinical studies with prostate cancer patients.[6,65,66] PC-SPES not only improved the patient's quality of life, e.g. more energy, less pain, increased appetite, but also decreased PSA and testosterone levels, showed estrogenic activity, reduced the size of the prostate, delayed disease progression for several months, prolonged survival time of patients and even led to a reduction of lung cancer metastasis.[6] PC-SPES was heralded an effective drug in decreasing tumor incidence and improving the effectiveness of conventional chemotherapy. It was used as an adjuvant or complementary treatment to hormonal therapy. However, side effects of PC-SPES have been reported, including gynecomastia, gynecodynia, breast and nipple tenderness, breast enlargement, hair loss, leg cramps, cardiovascular or thromboembolic events, loss of libido and potency, erectile dysfunction, hot flashes and diarrhea.[6] Although PC-SPES showed significant promise in cancer therapy, especially for prostate cancer, its clinical use is not recommended. The optimal formulation, dose, duration and schedule have yet to be determined and standardized. Further clinical trials are needed to prove its efficacy profile, contraindications and drug-herb interactions. First efforts in this direction have already been carried out.[66]

9. QUALITY ASSURANCE

Traditional healers often make use of growth conditions for plant collections that are difficult to reproduce in greenhouses or farms. Thus, the composition of herbal substances may differ from wild collection to large-scale production.[38,44]

Plants are a unique, renewable, cost-efficient and exceptionally varied resource. Natural health products usually consist of complex mixtures of chemicals, some of which may mitigate undesirable side effects, others which may act synergistically, leading to improved efficiencies at lower concentrations.[28] If ethnoparmacologists are able to demonstrate the efficiency of polyvalent of traditional extracts in clinical studies, a shift in thinking could be reached moving away from the 'single active compound' model of modern medicine to an approach based on using multiple phytochemical agents at low doses.[55,67]

Though ethnopharmacology and phytotherapy bear a great potential for modern medicine, the complexity of such an approach is challenging. Integration of these traditional techniques into modern medical practices is further complicated by quality issues regarding safety and efficacy, which are strongly modulated both by intrinsic and extrinsic factors.[68] Individual patient differences, organ specificity, or diurnal and seasonal variations may all lead to different qualitative and quantitative accumulation of active chemical constituents in medicinal plants. Environmental conditions, cultivation, field collection and storage methods, processing and manufacturing practice, age, contaminations, and adulteration, may further increase the variability of herbal activities. Therefore, appropriate legal regulations for standardization are required.[6,28,68]

Traditional herbal medicines come with strong evidence as to their safety due to their usage over millennia.[17] However, according to modern criteria, they must be monograph conformable and proven in clinical studies before they can be approved by regulatory agencies.[17] Therefore, they are subject to the same requirements in quality, efficacy and safety as chemically synthesized compounds.[17] According to the regulations of the US Food and Drug Administration identity, purity, quality, potency, and consistency of botanical drugs must be ensured.[69] Dietary Supplements are regulated separately from synthetic drugs and botanical drugs in the USA.

For drugs recorded in the pharmacopoeia, the requirements are clearly defined. Reliable documentation is required to avoid contaminations and adulterations.[70] As a negative example, cases of Viagra admixtures have been reported in traditionally used aphrodisiac plant drugs.[71] Quality assurance requires that certain criteria be met before a plant is considered safe and marketable. The main quality criteria include plant selection, origin and production (Wagner *et al.*, 2007).[17] Plant selection indicates whether a herb was obtained from wild collections or specific cultivations in greenhouses and fields. Wild harvests have lower production costs, but carry additional risks e.g. erroneous species identification as well as quantitative and qualitative variability of different plant batches from diverse sites.[44]

According to soil and climate conditions, different batches of plants may reveal considerable heterogeneity in their composition. A great part of the variability in quality of herbal drugs can be attributed to wild harvesting. In some cases, preservation becomes an issue in wild harvesting, as some plant species have almost been driven to extinction by herb pickers.[72] One example is the Devil's Claw (*Harpagophytum procumbens*), which is used against arthritis. While collecting the necessary parts, harvesters often end up inadvertently causing harm to the plant, leading to dehydration and death. For this reason, attempts were made to cultivate Devil's Claw to reduce the species' risk of extinction.[73]

The difficulties associated with wild harvests have led to guidelines and recommendations for quality assurance, e.g. the "good agricultural practice" (GAP).[74] WHO guidelines exist on "Good Agricultural and Collection Practices for Medical Plants 2004" (GACP), "Guidance for Industry Botanical Drug Products" (FDA, 2004) and "Points to Consider on Good Agricultural and Collection Practice for Starting Materials of Herbal Origin" (EMEA/HMPWP/31/99 Rev. 3, 2 May 2002).[75] These guidelines define and explain rules for the wild harvest and cultivation of medicinal plants with regard to climatic and geographical conditions, harvesting, storage and inspection of herbal drugs. Such policies protect consumers by ensuring consistent quality of medicinal plants, and also protect nature against overharvesting.

Quality control is one of the most important tasks for processing herbal drugs. Regardless of whether drugs are directly used (e.g. tea) or further processed e.g. as extracts, tinctures, syrups, etc. contamination by pesticides, heavy metals, radioactivity, solvents or toxins from plants pathogens must be avoided.[44,70] In a German study of Chinese herbal drugs, values considerably above the recommended maximum levels for lead and mercury were detected.[70] TLC (thin layer chromatography), HPLC (high performance liquid chromatography), GC (gas chromatography) and LC-MS (liquid chromatography-mass spectrometry) as well as microscopic studies are several of the most common laboratory methods used in quality assurance.

For biodiversity prospecting projects, databases such as NAPRALERT[76] or Natural Products Repository[77] can be helpful. Experienced plant collectors and taxonomists not only collect and identify plants, but also document their knowledge on medical concepts. Social, cultural, religious, ethnohistorical and linguistic aspects must be considered when assessing information on a given plant.[78] It is important to know the method of collection, processing and usage of a species as natural medicine. Therefore, ethnopharmacologists often live with ethnic tribes and talk to traditional healers to learn from their knowledge. Collaborations with indigenous tribes should be based on mutual information exchange and benefit-sharing. The rights of indigenous people and the importance of conservation of biological diversity are laid out in declarations such as the Code of Ethics of the International Society of Ethnobiology (2005) and the Convention of Biological Diversity.[78,79] Signing an ethical Memorandum of Agreement before starting a project can be mutually beneficial.

10. RISKS OF NEW DRUGS

Written guidelines or monographs for the use of herbal drugs from traditional medicine help to manage correct use and avoid toxicity.[44] In one documented example, *Stephania tetrandra* was used as an effective component in a slimming tea.[80] By mistake, birthwort, which contains highly toxic aristolochic acids, was added to the herbal mixture. This led to numerous cases of renal failure and bladder cancers. This and other examples illustrate the importance of competence and full understanding in dealing with traditionally applied plants. Nowadays, highly sensitive chromatographic procedures can detect even trace amounts of the *Aristolochia* species in herbal mixtures.[81]

Unknown or unrecognized interactions between herbs and conventional drugs represent another danger in using traditional herbal remedies. As shown in a recent study, a large portion of patients take herbal medications for supportive treatment and attenuation of side effects.[82] A well-known example of an adverse interaction is that of St. John's Wort (*Hypericum perforatum*) and cystostatic drugs in cancer therapy. Cancer patients often seek help from herbal medicines to manage their mental depression. However, many patients do not report their use of herbal medicines to their physicians.[83] St. John's Wort induces the activity of cytochrome P450 monooxygenases in the liver, and cancer patients who take it unconsciously hurt themselves, as the effect of anti-cancer drugs fails due to the increase in metabolism from cytochrome P450. Treatment failure causes further tumor progression with fatal consequences for patients.

11. CONCLUSIONS

The widespread assumption held by the public that natural medicines are gentle can be refuted with numerous examples. The fact that many established anticancer drugs are derived from plants, such as paclitaxel from *Taxus baccata*, convincingly shows that plants can contain highly toxic compounds. Whether chemically synthesized in a laboratory or isolated from a medicinal plant, the interaction between a chemical molecule and other macromolecules in the body determine the therapeutic effects.

The aforementioned use of yam roots by American Natives tribes to inhibit ovulation points to possibly serious effects of phytohormonal natural products.[38] Phytochemicals intervening with the hormonal balance of humans can carry great risk. Herbal products are popular among women to ease menopausal symptoms.[84] While phytoestrogens may regulate hormonal balance, it has been suggested that they may exert carcinogenic side effects as well. Taking phytohormone-containing herbal drugs as part of hormone replacement therapy may increase the risk of developing hormone-dependent gynecological tumors.[85]

Natural products frequently show multi-target effects. This may lead to improved therapeutic efficacy in complex diseases, but may also increase adverse targeting effects and toxicity.[17,83] Therefore, phytochemicals may serve as useful lead compounds, which can be chemically modified in the laboratory to improve their pharmacological properties.

In conclusion, ethnopharmacology and phytotherapy hold great potential in modern medicine, as is illustrated by the many examples for cancer treatment. In addition, adopting a more holistic view of the sort that has been common in Asian medicine over a period of thousands of years might supplement conventional therapy options of western medicine today. Traditional medicine can play an important function in industrialised countries and not only in non-industrialised societies where herbs and traditional practices are still the primary mode of therapy.[86] Studying traditional medicine can help to improve current methods of therapy in modern medicine and to reach a higher success rate in treating patients.

REFERENCES

1. World Health Organization (2000) General Guidelines for Methodologies on Research and Evaluation of Traditional Medicine; Geneva.
2. Karikas GA. (2010) Anticancer and chemopreventing natural products: some biochemical and therapeutic aspects; *Journal of B. U. ON.* (Official journal of the Balkan Union of Oncology) **15**: 627–638.
3. Reitz M. (2004) Krebs als ständiger Begleiter der Menschheit. Vom göttlichen Tumor zur molekularen Medizin. Neue Züricher Zeitung.
4. http://info.cancerresearchuk.org/cancerstats/causes/infectiousagents/epstein-barrvirus/
5. Schäfer B. (2011) Taxol — Hoffnung unserer Zeit. Wiley-Vch Verlag. *Chemie unserer Zeit* **1**: 32–46.
6. Cordell GA. (2002) PC-SPES: A brief overview. *Integr Cancer Ther* **1**: 271–286.
7. Graham JG, Quinn ML, Fabricant DS, Farnsworth NR. (2000) Plants used against cancer — an extension of the work of Jonathan Hartwell. *J Ethnopharmacol* **73**: 347–377.
8. Cragg GM, Boyd MR, Cardellina JH 2nd, *et al.* (1994) Ethnobotany and drug discovery: the experience of the US National Cancer Institute. *Ciba Found Symp* **185**: 178–190, discussion 190–196.
9. Heinrich M, Bremner P. (2006) Ethnobotany and ethnopharmacy — their role for anti-cancer drug development. *Curr Drug Targets* **7**: 239–245.
10. Ichikawa H, Nakamura Y, Kashiwada Y, Aggarwal BB. (2007) Anticancer drugs designed by mother nature, ancient drugs but modern targets. *Curr Pharm Des* **13**: 3400–3416.
11. Eckart WU. (2009) Geschichte der Medizin, 6. Auflage, Springer Verlag, pp. 28–46.
12. Leven K-H. (2008) Geschichte der Medizin: Von der Antike bis zur Gegenwart. CH Beck, Wissen.
13. Mayer JG, Goehl K, Englert K. (2009) Die Pflanzen der Klostermedizin in Darstellung und Anwendung. Deutscher Wissenschafts-Verlag.
14. Mayer JG. (2003) Kräuterbuch der Klostermedizin, Der 'Macer Floridus' Medizin des Mittelalters. Reprint Verlag Leipzig.

15. Efferth T. (2006) *Molekulare Pharmakologie und Toxikologie.* Springer-Verlag Berlin Heidelberg, p. 19
16. Marzell H. (1967) Geschichte und Volkskunde der Deutschen Heilpflanzen. 2. Auflage, Wissenschaftliche Buchgesellschaft, pp. 58–83.
17. Wagner H, Vollmar A, Bechthold A. (2007) Pharmazeutische Biologie 2 — Biogene Arzneistoffe und Grundlagen von Gentechnik und Immunologie. 7. Auflage, Wissenschaftliche Verlagsgesellschaft mbH Stuttgart.
18. Ostermann T, Raak C, Büssing A. (2009) Survival of cancer patients treated with mistletoe extract (Iscador): a systematic Literature review. *BMC Cancer* **9**: 451.
19. Melzer J, Iten F, Hostanska K, Saller R. (2009) Efficacy and safety of mistletoe preparations (Viscum album) for patients with cancer diseases. A systematic review. *Forsch Komplementmed* **16**: 217–226.
20. Kienle GS, Glockmann A, Schink M, Kiene H. (2009) *Viscum album* L. extracts in breast and gynaecological cancers: a systematic review of clinical and preclinical research. *J Exp Clin Cancer Res* **28**: 79.
21. Monteiro R, Becker H, Azevedo I, Calhau C. (2006) Effect of hop (Humulus lupulus L.) flavonoids on aromatase (estrogen synthase) activity. *J Agric Food Chem* **54**: 2938–2943.
22. Lust S, Vanhoecke B, Janssens A, *et al.* (2005) Xanthohumol kills B-chronic lymphocytic leukemia cells by an apoptotic mechanism. *Mol Nutr Food Res* **49**: 844–850.
23. http://www.heilpflanzen-suchmaschine.de/berberitze/berberitze_synonyme.shtml
24. Diogo AV, Machado NG, Barbosa IA, *et al.* (2011) Berberine as a promising safe anti-cancer agent — is there a role of mitochondria? *Curr Drug Targets* **12**: 850–859.
25. http://www.botanik.univie.ca.at/hbv/download/ib_taxus baccata.pdf
26. Fintelmann V, Menßen HG, Siegers K-P. (1989) Phytotherapie Manual. Hippokrates Verlag Stuttgart, p. 107.
27. Greenlee H, Abascal K, Yarnell E, Ladas E. (2007) Clinical applications of silybum marianum in oncology. *Integr Cancer Ther* **6**: 158–165.
28. Sagar SM. (2007) Future directions for research on *Silybum marianum* for cancer patients. *Integr Cancer Ther* **6**: 166–173.
29. Deep G, Agarwal R. (2007) Chemopreventive efficacy of silymarin in skin and prostate cancer. *Integr Cancer Ther* **6**: 130–145.
30. Tamayo C, Diamond S. (2007) Review of clinical trials evaluating safety and efficacy of milk thistle (Silybum marianum [L.] Gaertn.). *Integr Cancer Ther* **6**: 146–157.

31. Deep G, Agarwal R. (2010) Antimetastatic efficacy of silibinin: molecular mechanisms and therapeutic potential against cancer. *Cancer Metastasis Rev* **29**: 447–463.
32. Hanahan D, Weinberg RA. (2011) Hallmarks of cancer: the next generation. *Cell* **144**: 646–674.
33. Ramasamy K, Agarwal R. (2008) Multitargeted therapy of cancer by silymarin. *Cancer Lett* **269**: 352–362.
34. Zhan T, Digel M, Küch E-M, *et al.* (2011) Silybin and dehydrosilybin decrease glucose uptake by inhibiting GLUT proteins. *J Cellular Biochem* **112**: 849–859.
35. Trichopoulou A, Lagiou P, Kuper H, Trichopoulos D. (2000) Cancer and mediterranean dietary traditions. *Cancer Epidemiol Biomarkers Prev* **9**: 869–873.
36. Fernandez E, Gallus S, La Vecchia C. (2006) Nutrition and cancer risk. An overview. *J Br Menopause Soc* **12**: 139–142.
37. Bosetti C, Pelucchi C, La Vecchia C. (2009) Diet and cancer in Mediterranean countries: carbohydrates and fats. *Public Health Nutr* **12**: 1595–1600.
38. Frohn B, Uber H. (1996) Xokonoschtletl: Medizin der Mutter Erde. Die alten Heilweisen der Indianer, Mosaik Verlag GmbH, München, pp. 19–93.
39. Rätsch C. (1997) *Medizin aus dem Regenwald: Die Weisheit der Naturvölker*. Midena Verlag, Augsburg, pp. 40–83.
40. Kaptchuk TJ, Goldman P, Stone DA, Stason WB. (2000) Do medical devices have enhanced placebo effects? *J Clinical Epidemiol* **53**: 786–792.
41. Beecher HK. (1955) The powerful placebo. *JAMA* **159**: 1602–1606.
42. Haake M, Mueller HH, Schade-Brittinger C, *et al.* (2007) German Acupuncture Trials (GERAC) for chronic low back pain: randomized, multicenter, blinded, parallel-group trial with 3 groups. *Arch Intern Med* **167**: 1892–1898.
43. Pfeifer HJ, Greenblatt D, Koch-Weser J. (1976) Clinical toxicity of reserpine in hospitalized patients: a report from the Boston Collaborative Drug Surveillance Program. *Am J Med Sci* **271**: 269–276.
44. Hänsel R, Sticher O. (2008) Pharmakognosie-Phytopharmazie. Springer Medizin Verlag Heidelberg, **8**: 194–196.
45. Cancer facts & figures 2010: *American cancer society*. http://www.cancer.org/acs/groups/content/@epidemiologysurveilance/documents/document/acspc-026238.pdf
46. Behr HG. (1992) Von Hanf ist die Rede — Kultur und Politik einer Droge, Rowohlt Taschenbuch Verlag.
47. Walsh D, Nelson K, Mahmoud F. (2003) Established and potential therapeutic application of cannabinoids in oncology. Springer-Verlag. *Support Care Cancer* **11**: 137–143.

48. Russo EB. (2008) Cannabinoids in the management of difficult to treat pain. *Ther Clin Risk Management* **4**: 245–259.
49. Ware A, Doyle CR, Woods R, *et al.* (2002) Cannabis use for chronic non-cancer pain: results of a prospective survey. Elsevier Science BV.
50. Efferth T, Li PC, Konkimalla VS, Kaina B. (2007) From traditional Chinese medicine to rational cancer therapy. *Trends Mol Med* **13**: 353–361.
51. Schmincke C. (2004/2007) Chinesische Medizin für die westliche Welt. Springer Verlag, 3. Auflage.
52. Efferth T. (2008) Pharmaceutical biology of traditional Chinese medicine for cancer therapy. In: Eddouks M (ed.). *Handbook of Ethnopharmacology*, Research Signpost, Kerala, India, pp. 105–128.
53. Chang YS, Seo E-K, Gyllenhaal C, Block KI. (2003) *Panax ginseng*: A role in cancer therapy? *Integr Cancer Ther* **2**: 13–33.
54. Chadwick LR, Pauli GF, Farnsworth NR. (2006) The pharmacognosy of Humulus lupulus L. (hops) with an emphasis on estrogenic properties. *Phytomedicine* **13**: 119–131.
55. Wong R, Sagar CM, Sagar SM. (2001) Integration of Chinese medicine into supportive cancer care, a modern role for an ancient tradition. *Cancer Treat Rev* **27**: 235–246.
56. Auerbach L, Meng A, Berkemeier H. (2005) Ernährung bei Krebs nach den 5 Elementen der TCM: Traditionelle Chinesische Medizin. Springer Verlag, pp. 41–64.
57. Seely D, Mills EJ, Wu P, *et al.* (2005) The effects of green tea consumption on incidence of breast cancer and recurrence of breast cancer: a systematic review and meta-analysis. *Integr Cancer Ther* **4**: 144–155.
58. Eberhard U. (2003) *Leitfaden Kampo-Medizin: Japanische Phytotherapie.* Urban & Fischer Verlag/Elsevier GmbH.
59. Efferth T, Miyachi H, Bartsch H. (2007) Pharmacogenomics of a traditional Japanese herbal medicine (Kampo) for cancer therapy. *Cancer Genomics & Proteomics* **4**: 81–92.
60. Ohno T, Inoue M, Ogihara Y. (2002) Suppressive effect of Shichimotsu-koka-to (Kampo medicine) on pulmonary metastasis of B16 melanoma cells. Department of Pharmacognosy, Graduate School of Pharmaceutical Sciences, Nagoya City University, Nagoya, Japan. *Biol Pharm Bull* **25**: 880–884.
61. Takegawa Y, Ikushima H, Ozaki K, *et al.* (2008) Can Kampo therapy prolong the life of cancer patients? Department of Health Sciences, The University of Tokushima, Tokushima, Japan. *J Med Invest* **55**: 99–105.
62. Patwardhan B, Gautam M, Bani S, *et al.* (2008) Ayurveda, drug discovery and immunomodulation: review and case study of Withania somnifera. *Handbook of Ethnopharmacology*, pp. 393–420.

63. Winters M. (2006) Ancient medicine, modern use: *Withania somnifera* and its potential role in integrative oncology. *Altern Med Rev* **11**: 269–277.
64. Hudson J, Altamirano M. (2006) The application of DNA micro-arrays (gene arrays) to the study of herbal medicines. *J Ethnopharmacol* **108**: 2–15.
65. Huerta S, Arteaga JR, Irwin RW, *et al.* (2002) PC-SPES inhibits colon cancer growth *in vitro* and *in vivo. Cancer Res* **62**: 5204–5209.
66. Shabbir M, Love J, Montgomery B. (2008) Phase I trial of PC-Spes2 in advanced hormone refractory prostate cancer. *Oncology Reports* **19**: 831–835.
67. Houghton PJ, Howes M-J, Lee CC, Steventon G. (2007) Uses and abuses of *in vitro* tests in ethnopharmacology: visualizing an elephant. *J Ethnopharmacol* **110**: 391–400.
68. Fong HHS. (2002) Integration of herbal medicine into modern medical practices: issues and prospects. *Integr Cancer Ther* **1**: 287–293.
69. U.S. Department of Health and Human Services Food and Drug Administration Center for Drug Evaluation and Research (CDER). (2004) Chemistry; Guidance for Industry Botanical Drug Products.
70. Ihrig M, Baumann J, Orbig H, *et al.* (2004) Qualitätsmängel bei TCM- Drogen. Pharmazeutische Zeitung online, GOVI-Verlag.
71. http://www.impotenz-selbsthilfe.de/therapie/pflanzliche-mittel.html
72. http://www.focus.de/wissen/wissenschaft/artenschutz/tid-17524/artenvielfalt-die-heilkraft-der-teufelskralle-_aid_488715.html
73. http://www.teufelskrallenstiftung.de/
74. http://www.globalgap.org/cms/front_content.php?idcat=9
75. Franz Ch.M. (1989) Good agricultural practice (GAP) for medicinal and aromatic plant production. *Acta Horticulturae* (ISHS) **249**: 125–128.
76. http://www.napralert.org/
77. http://dtp.nci.nih.gov/branches/npb/repository.html
78. Heinrich M, Bremner P. (2006) Ethnobotany and ethnopharmacy — their role for anti-cancer drug development. *Curr Drug Targets* **7**: 239–245.
79. Soejarto DD, Fong HHS, Tan GT, *et al.* (2005) Ethnobotany/ethnopharmacology and mass bioprospecting: issues on intellectual property and benefit-sharing. *J Ethnopharmacol* **100**: 15–22.
80. Stiborová M, Martínek V, Frei E, Arlt VM, Schmeiser HH. (2013) Enzymes metabolizing aristolochic acid and their contribution to the development of aristolochic acid nephropathy and urothelial cancer. *Curr Drug Metab* **14**: 695–705.
81. Michl J, Jennings HM, Kite GC, Ingrouille MJ, Simmonds MS, Heinrich M. (2013) Is aristolochic acid nephropathy a widespread problem in developing countries?: A case study of Aristolochia indica L. in Bangladesh using an ethnobotanical-phytochemical approach. *J Ethnopharmacol* **149**: 235–244.

82. Gratus C, Wilson S, Greenfield SM, *et al.* (2009) The use of herbal medicines by people with cancer: a qualitative study. *BMC Complement Alternat Med* **9**: 14.
83. Efferth T, Koch E. (2011) Complex interaction between phytochemicals. The multi-target therapeutic concept of phytotherapy. *Current Drug Targets* **12**: 122–132.
84. http://www.krebshilfe.net/information/vorsorge/frauen/hormone.shtm
85. http://www.lamedica.at/krebs/Phytohormone%20und%20Krebs.pdf
86. Cohen I, Tagliaferri M, Tripathy D. (2002) Traditional Chinese medicine in the treatment of breast cancer. *Semin Oncol* **29**: 563–574.

CHAPTER 8

Contribution of African Flora in a Global Fight against Cancer

Victor Kuete and Thomas Efferth†*

ABSTRACT

Throughout Africa, the use of traditional herbal medicines for the treatment of many ailments is well established. Plant extracts and their constituents are being investigated extensively in African research institutes in order to tackle the increasing incidence of cancer in the continent. This chapter highlights some important results of the anti-proliferative potential of plants from Africa as collected in popular science websites such as PubMed, Scopus, Sciencedirect, Scirus, Web-of-knowledge, Scholar Google, etc. It discusses the state-of-art research in Phytotherapy in Africa, with emphasis on most prominent results.

1. INTRODUCTION

The mortality rate in Africa is directly correlated with the high incidence of infectious diseases, pregnancy and childbirth, in addition to cancers, cardiovascular diseases and chronic respiratory diseases, which are also the leading causes of death in the developed world.[1] It has been estimated, that more than 60,000 deaths that occur in the continent are cancer related.[2] It has also been estimated that by 2020, the incidence of cancer will touch 15 million, 70 percent of which will be in developing countries, where

*Department of Biochemistry, Faculty of Science, University of Dschang, P.O. Box 67, Dschang, Cameroon

†Institute of Pharmacy and Biochemistry, Johannes Gutenberg University, Staudinger Weg 5, 55128 Mainz, Germany

governments are least prepared to address the growing burden of cancer and where survival rates are often less than half those in more developed countries.[2]

Awareness of this impeding epidemic in Africa should be a research priority today, and all possible resources should be mobilized for both the prevention and management of such tragedies. The socio-economic burden associated with cancer reported in Africa, is mainly because of the high cost of treatments and the low income of the population. Traditional herbal preparations play an essential role in African culture as they provide the primary health care needs for a large majority (80 percent) of the population.[3] In sub-Saharan as well as in the Maghreb regions of Africa, there is a rich tradition in the use of herbal medicine for treating several ailments. However, the integration of traditional medicine in the public health system in general is not effective throughout the continent. In response to the popularity of herbal treatment in traditional and complementary medicine, a growing number of countries in Africa have established structures and budgets for herbal drugs research. There has been a corresponding increase in the number of scientific reports during the last two decades and several research institutions have made considerable progress in developing basic scientific understanding and the mechanism of active samples. However, limited funding is available currently to tackle this disease in African countries.

2. BIODIVERSITY AND FLORA OF AFRICA

Considering biodiversity as genetic variation within populations, the number, relative abundance and uniqueness of species and the varieties, and finally the extent and condition of ecosystems, it can be observed that Africa is well endowed both in variety and abundance of living things.[4] Ecosystems are broadly arranged in a latitudinal pattern, with increasing species richness towards the equator.[4] It has been reported that plant species richness is also high in the winter-rainfall Mediterranean climate regions of Northern Africa and the southern Cape. In between are the subtropical deserts, which are generally a zone of lower diversity: for example, a vast part of the Sahara, the Ténéré, is home to only 20 plant species in an area of about 200,000 km^2.[4] In latitude from Ethiopia to the Cape, mountains contain several centers of endemism for birds, mammals, and plants. One of the most globally important centers of endemism is the coastal mountain range in the eastern part of Madagascar.[4] The increasing richness of plants and vertebrates toward the equator is related primarily to climatic factors, such as water availability. However the diversity of land variations, such as topography, is also important. There are exceptions to this: some areas with harsh climates including the Namib Desert and the Karoo in the west of South Africa have an estimated 4,500 plant species, a third to one-half of which are endemic.[4] Overall plant richness at species, genus and family level is lower than that of other tropical areas. The African mainland has between 40,000 and 60,000 plant species, of which approximately 35,000 are endemic.[4] South America, by comparison, has about 90,000 plant species in an area 40 percent smaller endemic.[4] Parts of the Congo basin have moderate levels of plant species richness, comparable to many parts of Central Europe. Five of the 20 global centers of plant diversity are located in Africa. More than 3,000 plant species per 10,000 km^2 can be found in the Cameroon-Guinea centre, the Capensis centre, the Maputaland-Pondoland centre, the Albertine Rift centre and the endemic Madagascar centre.[4] At least a sixth of the world's plant species are in Africa. The Cape Floral Kingdom, a global centre of plant endemism has about 9,000 vascular plant species found over an area of 90,000 km^2 of which about 69 percent are endemic.[4] There are more than 12,000 plant species in Madagascar, at least 81 percent of which are endemic. This is an exceptionally high proportion by endemic global standards.[4]

3. CYTOTOXIC PLANTS IN AFRICA

Since ancient times, humans have derived many benefits from medicinal plants. A variety of medicinal plants have traditionally been used in Asian cultures as medicinal plants to treat cancers.[5] Screening for pharmacologically active substances in plants, is an important step in the discovery of newer, safer and more effective drugs. Over 50,000 plants would possess therapeutic virtues in the world and about 80 percent of humans use herbal medicines at least once in their life.[6,7] Medicinal plants through the multiplicities of their chemical constituents are so important in the discoveries of new substances that are active against cancers. Nevertheless, reports on plants traditionally used for the treatment of cancer are rare in several parts of the world as well as in Africa.[8–10] In the last two decades investigations on natural products have been particularly successful in the field of anti-proliferative drug research in other parts of the world. Early examples of cytotoxic agents developed from higher plants are the antileukemic alkaloids (vinblastine and vincristine), which were both obtained from African plant known as the Madagascar periwinkle (*Catharanthus roseus*), paclitaxel isolated from *Taxus baccata* and used in treatment of lung, ovarian, breast, neck cancers and advanced forms of Kaposi's sarcoma as well as camptothecin from *Camptotheca acuminata* with synthetic derivatives such as topotecan and irinotecan are used in cancer.[11–13] In the US National Cancer Institute (NCI) plant screening program, a crude extract is generally considered to have *in vitro* cytotoxic activity if the IC_{50} value following incubation between 48 and 72 h, is less than 20 μg/mL.[14] Despite the exceptional biodiversity of Africa, insufficient scientific studies have been carried out in most of the countries regarding the anti-proliferative properties of medicinal plants. Nevertheless, efforts currently are being made, and some important results are continuously reported on both medicinal plant extracts and compounds. Some of the most active plant extracts include *Albizia gummifera*;[15] *Gomphocarpus physocarpus, Cussonia paniculata, Gymnosporia tenuispina, Kalanchoe thyrsiflora, Physalis peruviana*;[16] *Elaeodendron alluaudianum*;[17] *Acanthospermum hispidum*;[18] *Croton barorum and C. goudotii*;[19] *Xylopia aethiopica, Echinops giganteus, Imperata cylindrica, Dorstenia psilirus, Piper capense Zinziber officinalis* (Table 1).[20] The active fraction from *Xylopia aethiopica* was shown to induce DNA damage, cell cycle arrest in G1 phase, and apoptotic cell death in U2OS osteosarcoma cell line.[21] Some of these extracts were found to be good inhibitors of the angiogenic process.[22] It is well known that excessive angiogenesis is an important factor for pathogenesis in many industrialized western countries.[23] Therefore, compounds or extracts with anti-angiogenic properties are of importance in the treatment and prevention of malignancies as well as other chronic diseases.[24,25] African scientists are investigating the anti-angiogenic properties of natural products and this is creditable as such work has not been reported in the past. Many compounds have also been isolated as bioactive principles of some African medicinal plants against cancer cells.

Table 1 Hit Anticancer Plants of Africa

Species/Family	Area of Plant Collection	Potentially Active Compounds	Screened Activity#
Acanthospermum hispidum DC./ Asteraceae	Nigeria	Hispidunolides A and B;[26] acanthospermal B[27]	IC_{50} of 8.87 μg/mL (COR-L23) for roots MeOH extract[18]
Albizia gummifera (J. F. Gmel.) C. A. Sm. var. *gummifera*/Fabaceae	Madagascar	Gummiferaosides A (**1**); B and C;[15] budmunchiamine G; budmunchiamine K; 6′ξ-hydroxybudmunchiamine K; 9-normethylbudmunchiamine K[28]	IC_{50} of 7.2 μg/mL (A2780) for roots EtOH extract[15]
Bersama engleriana Engl./Melianthaceae	Cameroon	—	IC_{50} of 8.6 μg/mL (MCF-7), 15.7 μg/mL (DU-145) for leaves MeOH extract; 8.6 μg/mL (MCF-7), 10.9 μg/mL (HeLa) and 19.5 μg/mL (HepG2) for roots MeOH extract; 18.7 μg/mL (MCF-7) for bark MeOH extract[29]
Croton barorum Leandri/ Euphorbiaceae	Madagascar	Crotobarin[19] (**8**)	100% inhibition concentration at 10 μg/mL against the P388 murine lymphocytic leukemia cell line[19]
Croton goudotii Baill./ Euphorbiaceae	Madagascar	Crotogoudin[19] (**9**)	100% inhibition concentration at 10 μg/mL against the P388 murine lymphocytic leukemia cell line[19]
Cupressus lusitanica Mill./ Cupressaceae	Cameroon	—	IC_{50} of 13.1 μg/mL (MCF-7) for bark[29] MeOH extract[29]

(Continued)

Table 1 (Continued)

Species/Family	Area of Plant Collection	Potentially Active Compounds	Screened Activity#
Cussonia paniculata E.&Z./ Araliaceae	South Africa	—	100% inhibition concentration of 1.00 (RPMI-8226); 1.45 (HCT-116); 2.69 (KM12) for the leaves DCM:MeOH extract[16]
Dorstenia psilirus Welwitch/ Moraceae	Cameroon	Psoralen; 2-sitosterol glucoside[30,31]	IC_{50} of 9.17 (MiaPaCa-2); 7.18 (CCRF-CEM); 7.79 μg/mL (CEM/ADR5000) for roots MeOH extract[20]
Echinops giganteus var. lelyi (C. D. Adams) A. Rich./ Compositae	Cameroon	Lupeol sitosteryl; *β-D*-glucopyranoside[32,33]	IC_{50} of 9.84 (MiaPaCa-2); 6.68 (CCRF-CEM); 7.96 μg/mL (CEM/ADR5000) for seeds MeOH extract[20]
Elaeodendron alluaudianum H. Perrier/Celastraceae	Madagascar	elaeodendroside V (**6**) and W (**7**); sarmentosigenin-3 *β-O-β*-6-deoxyguloside[17]	IC_{50} of 3.3 μg/mL (A2780) for EtOH extract[17]
Gomphocarpus physocarpus Schltr./Apocynaceae	SouthAfrica	—	100% inhibition concentration of 2.04 (A549/ATCC); 2.40 (DU-145); 2.69 μg/mL (RXF-393) for the roots DCM:MeOH extract[16]
Guibourtia tessmannii Harms Leonard/Leguminosae	Cameroon	—	IC_{50} of 13.1 μg/mL (MCF-7), 8.8 μg/mL (HeLa) for bark and leaves MeOH extracts, respectively[29]

(*Continued*)

Table 1 (*Continued*)

Species/Family	Area of Plant Collection	Potentially Active Compounds	Screened Activity#
Gymnosporia tenuispina (Sond.) Szyszyl/	South Africa	—	100% inhibition concentration of 0.06 (CCRF-CEM); 0.10 (MOLT-4); 0.28 μg/mL (SR) for the flowers and leaves MeOH extract[16]
Kalanchoe thyrsiflora Harv./ Crassulaceae	South Africa	—	100% inhibition concentration of 1.00 (SF-539); 1.86 (A549); 2.09 (HOP-62) for the roots and leaves MeOH extract[16]
Imperata cylindrica Beauv. var. koenigii Durand et Schinz/ Gramineae (Poaceae)	Cameroon	Jaceidin; quercetagetin-3,5,6,3′-tetramethyl ether; *β*-sitosterol-3-*O*-*β*-*D*-glucopyranosy1-6″-tetradecanoate[34]; imperanene[35]	IC_{50} of 12.11 (MiaPaCa-2); 8.4 (CCRF-CEM); 7.18 μg/mL (CEM/ADR5000) for roots MeOH extract[20]
Physalis peruviana L./ Solanaceae	South Africa	—	100% inhibition concentration of 0.27 (PC-3); 1.07 (RPMI-8226); 4.37 μg/mL (SR)[16]
Piper capense L.f./Piperaceae	Cameroon	Kaousine; Z-antiepilepsirine[36]	IC_{50} of 8.92 (MiaPaCa-2); 7.03 (CCRF-CEM); 6.56 μg/mL (CEM/ADR5000) for seeds MeOH extract[20]
Salvia africana L./*Lamiaceae*	South Africa	—	8.72 μg/mL (SF-268) for MeOH-chloroform 1:1 extract of the aerial part[37]

(*Continued*)

Table 1 (Continued)

Species/Family	Area of Plant Collection	Potentially Active Compounds	Screened Activity[#]
Salvia stenophylla Burch. ex Benth./Lamiaceae	South Africa	—	17.41 μg/mL (HT-29) for MeOH-chloroform 1:1 extract of the aerial part[37]
Salvia radula Benth/Lamiaceae	South Africa	—	9.69 μg/mL (MCF-7) for MeOH-chloroform1:1 extract of the aerial part[37]
Xylopia aethiopica (Dunal) A. Rich./Annonaceae	Cameroon	Volatile oil[38]; ent-15-oxokaur-16-en-19-oic acid[21]	IC_{50} of 6.68 (MiaPaCa-2); 3.91 (CCRF-CEM); 7.4 μg/mL (CEM/ADR5000) for seeds MeOH extract[20]; 12 μg/mL (HCT116), 7.5 μg/mL (U937) for 70% EtOH extract[21]
Zinziber officinalis Roscoe/ Zinziberaceae	Cameroon	3-hydroxy-1-methoxy-10-methyl-9-acridone; 1-hydroxy-3-methoxy-10-methyl-9-acridone (4), 1-hydroxy-2,3-dimethoxy-10-methyl-9-acridone (5),1,3-dihydroxy-2-methoxy-10-methyl-9-acridone[39]	IC_{50} of 16.33 (MiaPaCa-2); 8.82 (CCRF-CEM); 6.83 μg/mL (CEM/ADR5000) for rhizome MeOH extract[20]

[#]Cancer cell lines: Brain (SF-539, SF-268); Breast (MCF-7); Colon (HT29; HCT-116; KM12); Cervix (HeLa); Leukaemia (HL60; CCRF-CEM; CEM/ADR5000; MOLT-4; RPMI-8226; SR); Liver (HepG2) Lung (A549; COR-L23; HOP-62); Ovary (A2780); Oral (KB); Pancreatic (MiaPaCa-2); Prostate (DU145; PC-3); renal (RXF-393); dichloromethane (DCM); methanol (MeOH), Ethanol (EtOH).

4. CYTOTOXIC COMPOUNDS FROM AFRICAN PLANTS

In the NCI plant screening program, a pure compound is generally considered to have *in vitro* cytotoxic activity if the IC_{50} value following incubation between 48 and 72 h, is less than 4 μg/mL.[14] The cut-off point for good cytotoxic compound has also been established to be 10 μM.[40] These criteria will be used below to discuss the activity of compounds isolated from African plants.

4.1. Terpenoids

Some terpenoids (Fig. 1) isolated from African plants have shown good activities against various cancer cells. This includes oleanane-type triterpenoid saponins, gummiferaosides A (**1**), B and C (IC_{50} of 0.8, 1.5, and 0.6 μg/mL, respectively on A2780 human ovarian cancer cell line) obtained from the roots of the Madagascan plant *Albizia gummifera*;[15] caseanigrescen A (**2**), B (**3**), C (**4**) and D (**5**) (IC_{50} of 1.4, 0.83, 1.0, and 1.0 μM, respectively against A2780 cancer cells) isolated from *Casearia nigrescens*;[41] cardenolide glycosides, elaeodendroside V (**6**) and W (**7**) (IC_{50} of 0.12 and 0.07 μM respectively against A2780 cancer cells; 0.15 and 0.08 μM against the U937 human histiocytic lymphoma cell line, respectively) isolated from *Elaeodendron alluaudianum*;[17] crotobarin (**8**) [IC_{50} of 2.5; 2.1; 0.79 and 0.56 μM respectively against KB (human oral epidermoid carcinoma), HT29 (human colon adenocarcinoma), A549 (human lung adenocarcinoma), and HL60 (human promyelocytic leukemia) cell lines] and crotogoudin (**9**) (IC_{50} of 2.5; 2.1; 0.79 and 0.56 μM respectively against KB, HT29, A549, and HL60 cells) both isolated from *Croton barorum* and *Croton goudotii*.[19] Mechanistically, compounds **8** and **9** induced growth arrest at the G2/M phase in the cell cycle of the K562 human leukemia cell line at 4 μM.[19]

4.2. Phenolic Compounds

Different classes (Fig. 2) of phenolic compounds have been reported with high anti-proliferative effects against cancer cell lines. Flavonoids from the genus *Dorstenia*, gancaonin Q (**10**), 6-prenylapigenin (**11**), 6,8-diprenyleriodictyol (**12**), and 4-hydroxylonchocarpin (**13**), inhibited the

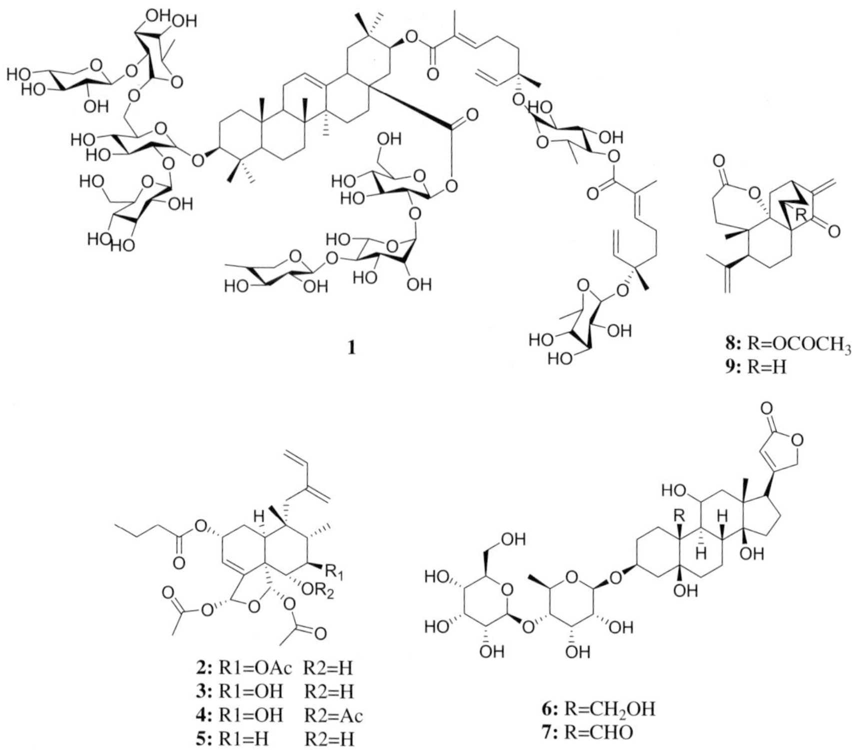

1

8: R=$OCOCH_3$
9: R=H

2: R1=OAc R2=H
3: R1=OH R2=H
4: R1=OH R2=Ac
5: R1=H R2=H

6: R=CH_2OH
7: R=CHO

Figure 1 Cytotoxic terpenoids from African medicinal plants.

proliferation of a panel of 14 cancer cell lines including human CCRF-CEM leukemia cells and their multidrug-resistant subline, CEM/ADR5000, PF-382 leukemia T-cells, and HL-60 promyelocytic leukemia (moderately differentiated), MiaPaCa-2 and Capan-1pancreatic adenocarcinoma, MCF-7 breast adenocarcinoma, SW-680 colon carcinoma cells, 786-0 renal carcinoma cells, U87MG glioblastoma-astrocytoma cells, A549 lung adenocarcinoma, Caski and HeLa cervical carcinoma cells, Colo-38 skin melanoma cells.[42] IC_{50} below or around 40 μg/mL was reported for compound **10** on PF-382 and HL-60 (4.8 μg/mL), MiaPaCa-2 (1.1 μg/mL), and MCF-7 (0.8 μg/mL), compound **11**, on PF-382 (3.8 μg/mL) and MCF-7 (0.6 μg/mL), compound **12**, on CCRF-CEM (4.9 μg/mL), MiaPaCa-2 (4.4 μg/mL) and MCF-7 (0.6 μg/mL) and compound **13**, on CCRF-CEM (1.6 μg/mL), CEM/ADR5000 (3.7 μg/mL), MiaPaCa-2 (3.8 μg/mL) and

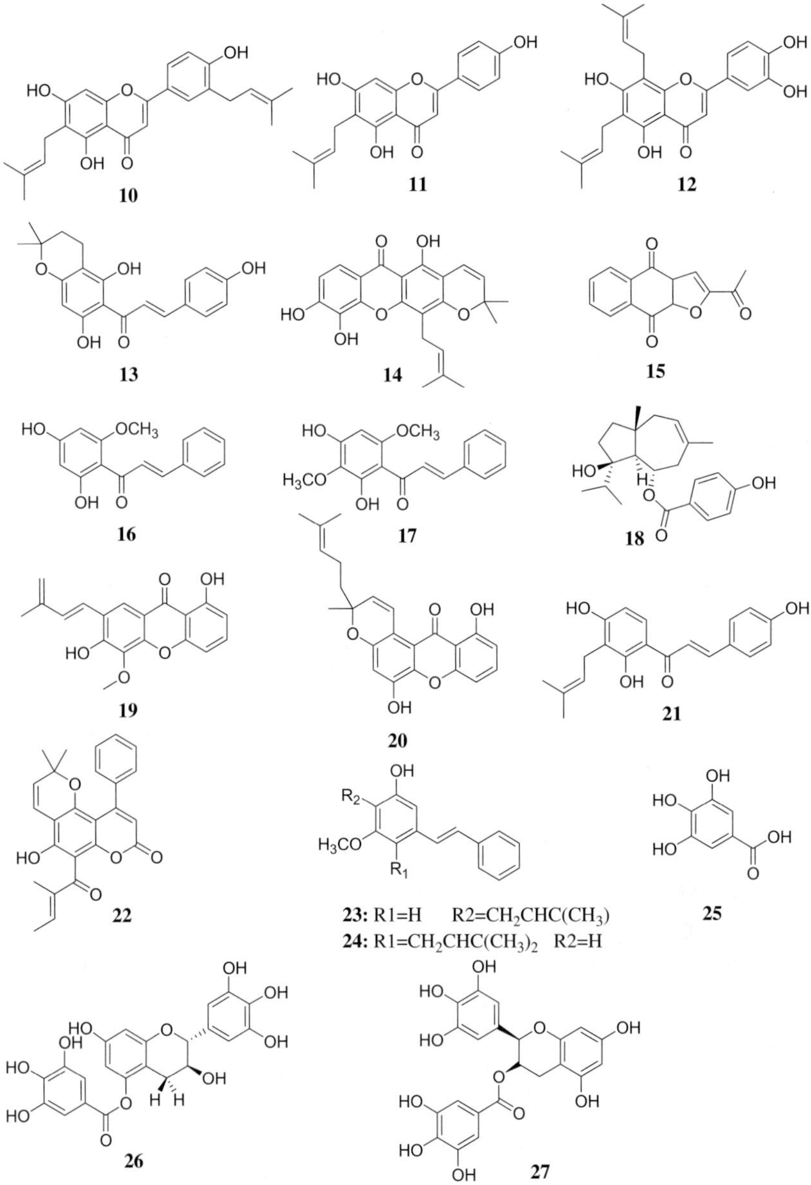

Figure 2 Cytotoxic phenolics from African medicinal plants.

MCF-7 (1.4 μg/mL).[42] Interestingly, compounds **10–13** were of low toxicity on the normal AML12 normal hepatocytes cells.[42] Caspase 3/7 activation was reported as one of the modes of induction of apoptosis for compounds **10**, **12**, and **13** in CCRF-CEM cells and they also showed anti-angiogenic properties.[42] Xanthone V_1 (**14**) isolated from the Cameroonian plant *Vismia laurentii* and 2-acetylfuro-1,4-naphthoquinone (**15**) from *Newbouldia laevis* were also tested on the panel of fourteen cell lines enumerated above.[22] IC_{50} below or around 4 µg/mL was reported for compound **14** on CCRF-CEM (4.9 μg/mL), HL60 (4.56 μg/mL), 786-0 (3.79 μg/mL), U87 MG (3.80 μg/mL), A549 (3.99 μg/mL), Colo-38 (1.19 μg/mL), and Caski (0.24 μg/mL).[22] Such values were also reported with compound **15** on PF-382 (0.57 μg/mL), Colo-38 (0.67 μg/mL), HeLa (0.40 μg/mL), and Caski (0.17 μg/mL).[22] Compounds **14** and **15** showed a respective, 65.8 percent and 59.6 percent inhibition of blood capillaries growth on the chorioallantoic membrane of quail eggs in the anti-angiogenic assay and induced cell cycle arrest in S-phase, and also significant apoptosis in CCRF-CEM leukemia cells.[22] Caspase 3/7 was activated by compound **14**.[22] Two flavonoids from the Cameroonian plant *Polygonum limbatum*, cardamomin (**16**) and 2′,4′-dihydroxy-3′,6′-dimethoxychalcone (**17**) induced significant inhibition of the proliferation of the leukemia THP-1 cells ($IC_{50} < 4\ \mu$g/mL).[43]

Jaeshkeanadiol *p*-hydroxybenzoate (**18**) isolated from the Egyptian plant Ferula *hermonis* showed good anti-proliferative effects on MCF-7 breast cancer cells (IC_{50}: 2.47 μg/mL);[44] Globulixanthones A (**19**) and B (**20**) (IC_{50} of 2.15 and 1.78 μg/mL against human KB cells respectively) isolated from *Symphonia globulifera*;[45] isobavachalcone (**21**) isolated from *Dorstenia barteri*[46] and *D. turbinata*[47] exhibited cytotoxic effects against many tumor cell lines, including ovarian carcinoma OVCAR-8 cells, prostate carcinoma PC-3 cells, breast carcinoma MCF-7 cells and lung carcinoma A549 cells.[48] Compound **21** significantly ablated Akt phosphorylation at Ser-473 and Akt kinase activity in cells, which subsequently led to inhibition of Akt downstream substrates and caused significant level of the mitochondrial pathway of apoptosis.[48] Nishimura *et al.*[49] demonstrated that compound **21** induced apoptotic cell death with caspase-3 and

-9 activation and Bax upregulation in neuroblastoma cell lines. Compound **21** also inhibited MMP-2 secretion from U87 glioblastoma cells.[46] A pyranocoumarin calophyllolide (**22**) isolated from a Cameroonian plants *Calophyllum inophyllum* exhibited a good cytotoxicity against KB nasopharynx cancer cell line with and IC50 value of 3.5 μg/mL.[50]

Two stilbenes isolated from a Nigerian plant *Cajanus cajan*, longistylins A (**23**) and C (**24**), showed high anti-proliferative activities with IC_{50} values of 0.7–14.7 μM against many different cancer cell lines including MCF-7 human breast adenocarcinoma, COR-L23 human large cell lung carcinoma and C32 human amelanotic melanoma.[18]

Gallic acid (**25**) and its derivatives were identified as the cytotoxic constituents of the Egyptian plant *Acacia nilotica*;[51] the IC_{50} values of 1.6 and 3.3 μg/mL against human uveal melanoma 92.1 and OCM3 cell lines, respectively were obtained with gallic acid, 4.8 and 5.1 μg/mL towards 92.1 cells, and 11 and 8.2 μg/mL towards OCM3 cells, respectively for gallocatechin 5-*O*-gallate (**26**) and epigallocatechin 3-*O*-gallate (**27**).[51]

4.3. Alkaloids

Compared to terpenoids and phenolics, a limited number of alkaloids isolated from African medicinal plants have been reported for their cytotoxic effects on cancer cells. However, data available from the screening of some plant compounds are rather moderate. It is thus, as, acridone alkaloids isolated from the fruits of *Zanthoxylum leprieurii* helebelicine A, 3-hydroxy-1-methoxy-10-methyl-9-acridone, 1-hydroxy-3-methoxy-10-methyl-9-acridone, 1-hydroxy-2,3-dimethoxy-10-methyl-9-acridone showed moderate activity against human lung carcinoma cells A549 (IC_{50} values of 31 to 52 μM) and colorectal adenocarcinoma cells DLD-1 (IC_{50} of 27 to 74 μM).[39] However, it should be noted that the well-known cytotoxic compounds, vinblastine and vincristine are present in the Madagascar periwinkle (*Catharanthus roseus*)[11,12] indicating that plants of Africa are not devoid of highly active alkaloids. In addition some well-known cytotoxic alkaloids such as veprisine and jatrorrhizine are also reported in Africa medicinal plants.[52]

5. CONCLUSIONS

In this chapter we have documented those plants from Africa's rich flora that are a potential source of cytotoxic drug. Here we have brought out the major hit research concerning plant extracts and naturally occurring compounds that have good anti-proliferative activities. However, this paper only provides an outline of anticancer research from African plants as there are also several of them with moderate or poor activities, published in low impact scientific journals that have not been reported herein. Finally, the findings reported in this chapter provide a useful tool for pre-clinical studies of the active samples as well as other pharmacognostical, pharmacological and toxicological studies.

REFERENCES

1. WHO. (2009) World health statistics 2009: Cause-specific mortality and morbidity. http://www.who.int/whosis/whostat/EN_WHS09_Table2.pdf. Accessed on May 03; 2010.
2. Vorobiof DA, Abratt R. (2007) The cancer burden in Africa. *South African Med J* **97**: 937–939.
3. WHO. (2002) WHO traditional medicine strategy 2002–2005. Geneva. http://whqlibdoc.who.int/hq/2002/WHO_EDM_TRM_2002.1.pdf. Accessed on Jan. 13; 2008.
4. United Nations Environment Programme. (2009) Biodiversity in Africa. McGinley M, Hogan CM, In: Cleveland CJ (ed.) *Encyclopedia of Earth.* Washington, D.C.; Accessed on June 11; 2012.
5. Lee SH, Lillehoj HS, Chun HK, *et al.* (2007) *In vitro* treatment of chicken peripheral blood lymphocytes, macrophages, and tumor cells with extracts of Korean medicinal plants. *Nutr Res* **27**: 362–366.
6. Bhakuni DS, Bittner M, Marticorena C, *et al.* (1974) Screening of Chilean plants for antimicrobial activity. *Lloydia* **37**: 621–632.
7. Farnsworth NR, Akerele O, Bingel AS. (1985) Medicinal plants in therapy. *Bulletin of the World Health Organization* **63**: 965–981.
8. Mongelli E, Pampuro S, Coussio J, *et al.* (2000) Cytotoxic and DNA interaction activities of extracts from medicinal plants used in Argentina. *J Ethnopharmacol* **71**: 145–151.

9. Steenkamp V, Gouws MC. (2006) Cytotoxicity of six South African medicinal plant extracts used in the treatment of cancer. *South African J Botany* 72: 630–633.
10. Kuete V, Efferth T. (2010) Cameroonian medicinal plants: pharmacology and derived natural products. *Front Pharmacol* **1**: 123.
11. Samuelsson G. (2004) *Drugs of Natural Origin: A Textbook of Pharmacognosy.* 5th edn. Stockholm: Swedish Pharmaceutical Press.
12. Voss C, Eyol E, Berger MR. (2005) Identification of potent anticancer activity in Ximenia Americana aqueous extracts used by African traditional medicine. *Toxicol Appl Pharmacol* **211**: 177–187.
13. Takimoto CH, Calvo E. (2008) Principles of Oncologic Pharmacotherapy. In: Pazdur R, Wagman LD, Camphausen KA, Hoskins WJ (eds), *Cancer Management: A Multidisciplinary Approach* 11th edn.
14. Boik J. (2001) *Natural Compounds in Cancer Therapy.* Minnesota, USA: Oregon Medical Press.
15. Cao S, Norris A, Miller JS, *et al.* (2007) Cytotoxic triterpenoid saponins of *Albizia gummifera* from the Madagascar rain forest. *J Nat Prod* **70**: 361–366.
16. Fouche G, Cragg GM, Pillay P, *et al.* (2008) *In vitro* anticancer screening of South African plants. *J Ethnopharmacol* **119**: 455–461.
17. Hou Y, Cao S, Brodie P, *et al.* (2009) Antiproliferative cardenolide glycosides of *Elaeodendron alluaudianum* from the Madagascar Rainforest. *Bioorg Med Chem* **17**: 2215–2218.
18. Ashidi JS, Houghton PJ, Hylands PJ, Efferth T. (2010) Ethnobotanical survey and cytotoxicity testing of plants of South-western Nigeria used to treat cancer, with isolation of cytotoxic constituents from Cajanus cajan Millsp. leaves. *J Ethnopharmacol* **128**: 501–512.
19. Rakotonandrasana OL, Raharinjato FH, Rajaonarivelo M, *et al.* (2010) Cytotoxic 3,4-seco-atisane diterpenoids from *Croton barorum* and *Croton goudotii. J Nat Prod* **73**: 1730–1733.
20. Kuete V, Youns M, Krusche B, *et al.* (2011) Cytotoxicity of some Cameroonian spices and selected medicinal plant extracts. *J Ethnopharmacol* **134**: 803–812.
21. Choumessi AT, Danel M, Chassaing S, *et al.* (2012) Characterization of the antiproliferative activity of *Xylopia aethiopica. Cell Div* **7**: 8.
22. Kuete V, Wabo HK, Eyong KO, *et al.* (2011) Anticancer activities of six selected natural compounds of some Cameroonian medicinal plants. *PLoS One* **6**: e21762.
23. Krenn L, Paper DH. (2009) Inhibition of angiogenesis and inflammation by an extract of red clover (*Trifolium pratense* L.). *Phytomedicine* **16**: 1083–1088.

24. Paper DH. (1998) Natural products as angiogenesis inhibitors. *Planta Med* **64**: 686–695.
25. Carmeliet P. (2003) Angiogenesis in health and disease. *Nat Med* **9**: 653–660.
26. Cartagena E, Bardon A, Catalan CAN, *et al.* (2000) Germacranolides and a new type of guaianolide from *Acanthospermum hispidum. J Nat Prod* **63**: 1323–1328.
27. Arena ME, Cartagena E, Gobbato N, *et al.* (2011) *In vivo* and *in vitro* antibacterial activity of acanthospermal B, a sesquiterpene lactone isolated from *Acanthospermum hispidum. Phytother Res* **25**: 597–602.
28. Rukunga GM, Waterman PG. (1996) New macrocyclic spermine (budmunchiamine) alkaloids from Albizia gummifera: with some observations on the structure–activity relationships of the budmunchiamines. *J Nat Prod* **59**: 850–853.
29. Mbaveng AT, Kuete V, Mapunya BM, *et al.* (2011) Evaluation of four Cameroonian medicinal plants for anticancer, antigonorrheal and antireverse transcriptase activities. *Environ Toxicol Pharmacol* **32**: 162–167.
30. Ngadjui TB, Dongo E, Happi NE, *et al.* (1998) Prenylated flavones and phenylpropanoid derivatives from roots of *Dorstenia psilurus. Phytochemistry* **48**: 733–737.
31. Kuete V, Metuno R, Ngameni B, *et al.* (2007) Antimicrobial activity of the methanolic extracts and compounds from *Treculia obovoidea* (Moraceae). *J Ethnopharmacol* **112**: 531–536.
32. Kojima H, Sato N, Hatano A, Ogura H. (1990) Sterol glucosides from *Prunella vulgaris. Phytochemistry* **29**: 2351–2355.
33. Tane P, Bergquist KE, Tene M, *et al.* (1995) Cyclodione, an unsymmetrical dimeric diterpene from *Cylicodiscus gabunensis. Tetrahedron* **51**: 11595–11600.
34. Mohamed GA, Abdel-Lateff A, Fouad MA, *et al.* (2009) Chemical composition and hepato-protective activity of *Imperata cylindrica Beauv. Pharmacogn Mag* **5**: 28–36.
35. Matsunaga K, Shibuya M, Ohizumi Y. (1995) Imperanene, a novel phenolic compound with platelet aggregation inhibitory activity from Imperata cylindrica. *J Nat Prod* **58**: 138–139.
36. Kaou AM, Mahiou-Leddet V, Canlet C, *et al.* (2010) New amide alkaloid from the aerial part of *Piper capense* L.f. (Piperaceae). *Fitoterapia* **81**: 632–635.
37. Kamatou GPP, Van Zyl RL, Davids H, *et al.* (2008) Antimalarial and anticancer activities of selected South African *Salvia* species and isolated compounds from *S. radula. South African J Botany* **74**: 238–243.
38. Kadou J, Etou Ossibi AW, Aklikoku K, *et al.* (2007) Chemical composition and hypotensive effects of essential oil of *Monodora myristica. J Biol Sci* **7**: 937–942.

39. Ngoumfo RM, Jouda JB, Mouafo FT, *et al.* (2010) *In vitro* cytotoxic activity of isolated acridones alkaloids from *Zanthoxylum leprieurii* Guill. et Perr. *Bioorg Med Chem* **18**: 3601–3605.
40. Brahemi G, Kona FR, Fiasella A, *et al.* (2010) Exploring the structural requirements for inhibition of the ubiquitin E3 ligase breast cancer associated protein 2 (BCA2) as a treatment for breast cancer. *J Med Chem* **53**: 2757–2765.
41. Williams RB, Norris A, Miller JS, *et al.* (2007) Cytotoxic clerodane diterpenoids and their hydrolysis products from *Casearia nigrescens* from the rainforest of Madagascar. *J Nat Prod* **70**: 206–209.
42. Kuete V, Ngameni B, Wiench B, *et al.* (2011) Cytotoxicity and mode of action of four naturally occuring flavonoids from the genus Dorstenia: gancaonin Q, 4-hydroxylonchocarpin, 6-prenylapigenin, and 6,8-diprenyleriodictyol. *Planta Med* **77**: 1984–1989.
43. Dzoyem JP, Nkuete AH, Kuete V, *et al.* (2012) Cytotoxicity and antimicrobial activity of the methanol extract and compounds from *Polygonum limbatum. Planta Med* **78**: 787–792.
44. Kuete V, Wiench B, Hegazy ME, *et al.* (2012) Antibacterial activity and cytotoxicity of selected Egyptian medicinal plants. *Planta Med* **78**: 193–199.
45. Nkengfack AE, Mkounga P, Fomum ZT, *et al.* (2002) Globulixanthones A and B, two new cytotoxic xanthones with isoprenoid groups from the root bark of *Symphonia globulifera. J Nat Prod* **65**: 734–736.
46. Ngameni B, Touaibia M, Belkaid A, *et al.* (2007) Inhibition of matrix metalloproteinase-2 secretion by chalcones from the twigs of *Dorstenia barteri Bureau. Arkivoc* **9**: 91–103.
47. Ngameni B, Kuete V, Simo IK, *et al.* (2009) Antibacterial and antifungal activities of the crude extract and compounds from *Dorstenia turbinata* (Moraceae). *South African J Botany* **75**: 256–261.
48. Jing H, Zhou X, Dong X, *et al.* (2010) Abrogation of Akt signaling by Isobavachalcone contributes to its anti-proliferative effects towards human cancer cells. *Cancer Lett* **294**: 167–177.
49. Nishimura R, Tabata K, Arakawa M, *et al.* (2007) Isobavachalcone, a chalcone constituent of *Angelica keiskei*, induces apoptosis in neuroblastoma. *Biol Pharm Bull* **30**: 1878–1883.
50. Yimdjo MC, Azebaze AG, Nkengfack AE, *et al.* (2004) Antimicrobial and cytotoxic agents from *Calophyllum inophyllum. Phytochemistry* **65**: 2789–2795.
51. Salem MM, Davidorf FH, Abdel-Rahman MH. (2011) *In vitro* anti-uveal melanoma activity of phenolic compounds from the Egyptian medicinal plant *Acacia nilotica. Fitoterapia* **82**: 1279–1284.
52. Kuete V, Efferth T. (2011) Pharmacogenomics of Cameroonian traditional herbal medicine for cancer therapy. *J Ethnopharmacol* **137**: 752–766.

CHAPTER 9

Natural Products Derived from Terrestrial Plants with Activity Towards Cancer Cells

Sonia Falenska[*], *Ina Kirmes*[*], *Stephanie Kletting*[*], *Irini Karagianni*[*] *and Karen Duffy*[†]

ABSTRACT

Because of the severe side effects associated with classical anticancer drugs, mainly due to inadequate differentiation between healthy and tumor tissues and the development of drug resistance, there is a need for development of new anticancer drugs and treatment strategies. In this chapter we have introduced several novel plant-based potential anticancer agents that are under investigation or in clinical trials. The modes of action of plant-derived compounds are manifold and include inhibition of DNA topoisomerase I/II, of nucleotide biosynthesis, of signal transduction pathways, of cyclooxygenase, of prostaglandin to inhibition of cell growth and induction of apoptosis. Furthermore, some natural compounds prevent the development of cancer.

1. INTRODUCTION

Cancer is one of the most deadly diseases facing the world today. According to the World Health Organization, cancer is responsible for roughly 13 percent of the total worldwide deaths in recent years. In 2008 alone, cancer claimed 7.6 million lives. The most common types of cancer include lung, stomach, liver, colorectal and breast cancer (WHO — Cancer fact sheet N°297 February 2011).

[*]Institute of Pharmacy and Biochemistry, Johannes Gutenberg University, Staudinger Weg 5, 55128 Mainz, Germany.

[†]Cornell University, Ithaca, NY, USA.

Many targets for cancer therapy have been suggested over the past decades. Some of the most important targets for therapeutic intervention are:

- Antimitotic agents: inhibition of (re)polymerization of microtubules
- Inhibitors of DNA topoisomerases I and II
- Inhibitors of nucleotide biosynthesis
- Inhibitors of growth factors and/or protein tyrosine kinase activity
- Inhibitors of signal transduction pathways (e.g. p21, RAS, protein kinases)
- Cyclooxygenase inhibitors
- Prostaglandin inhibitors

Because of the severe side effects associated with classical anticancer drugs, mainly due to inadequate differentiation between healthy and tumor tissues and the development of drug resistance, there is a continuing need for development of new anticancer drugs and treatment strategies. Combinatorial chemistry is now a major focus of chemotherapy research, and novel synthetic compounds are identified by scientific exploration of an enormous pool of synthetic, biological and natural products.

Natural products can be categorized as either primary or secondary metabolites. Whereas primary metabolites serve an organism's nutritional requirements, secondary metabolites are produced by the organism for defense. Secondary metabolites include polyketides, essential oils, terpenoids, alkaloids, lignans, steroids, flavonoids and others. Some secondary metabolites exert pharmacological activities, including cytotoxic and anticancer effects. Natural compounds are used for medical treatments all over the world. In the past, the activity of these natural compounds was orally reported by medicine men and shamans, or passed on as stories or anecdotal information between local tribes. Certainly, the use of plant extracts and herbs has its origin in ancient times. The earliest records of the use of such herbal remedies are from ancient China and Egypt.

Since 1955, the National Cancer Institute (NCI, USA) has screened more than 114,000 plant-derived extracts for anticancer activity. Three major types of chemo-preventive agents have been identified, namely inhibitors of carcinogen formation, agents suppressing cancer promotion and anti-progression agents. Natural substances may also be useful in structurally modified forms derived by chemical syntheses to achieve more desirable pharmacological properties such as water solubility or thermostability. Despite these advances, there are still some difficulties with using natural products that have to be considered. For example, botanicals have sensitizing potential, which can cause allergic reactions or photosensitization. Organ toxicity, such as hepato- or nephrotoxicity or natural compounds represents another common problem.

In this chapter, several novel plant-based potential anticancer agents that are under investigation or in clinical trials are introduced.

2. INDUCTION OF APOPTOSIS

Apoptosis, a form of programmed cell death, is an important process in regulating the balance of cell growth and cell death. Cell death following inflammation is provoked by injury and characterized by changes in mitochondrial structures and bursting of the cell. Apoptosis is a physiologically expedient process controlled by gene expression. Morphological indications of apoptosis include shrinkage of the cell due to caspase-catalyzed breakdown of cell proteins, condensation of chromatin, fragmentation of the DNA and blebbing of the cell membrane. The resulting apoptotic bodies are phagocytosed by surrounding cells. Apoptosis plays a major role in cell death when irreparable DNA damage. Mutations suppressing apoptosis are often oncogenic, leading to tumor initiation and progression. Therefore, proteins involved in the regulation of apoptosis such as p53, Bcl-2, NFκB and MAPKs are promising targets for cancer therapy. Proteins of the Bcl-2 family include both the pro-apoptotic proteins Bax and Bak, which trigger the release of cytochrome-c from the mitochondria into the cytoplasm, and the anti-apoptotic proteins Bcl-2 and Bcl-xL, which inhibit Bax translocation and thus cytochrome-c release.[1] Some important natural products from medicinal plants are described in the following paragraphs.

2.1. *Artemisia annua*

Artemisinin derived from *Artemisia annua* (Asteraceae) and its derivatives dihydroartemisinin, artesunate and artemether are widely used as

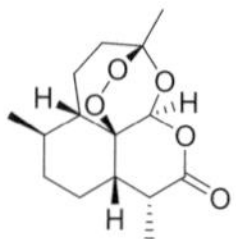

Artemisinin

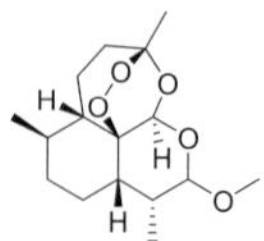

Artemether

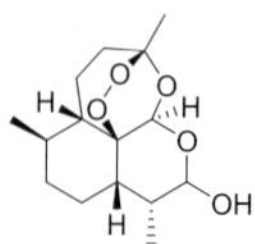

Dihydrortemisinin

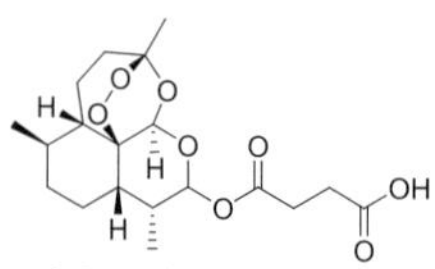

Artesunate

antimalarial agents and have been shown to have structure- and dose-dependent anticancer effects. Their activity is primarily due to the induction of apoptosis. The main incentive for developing these compounds as anticancer agents is their low host toxicity. Hou *et al.*[2] observed that artemisinin and dihydroartemisinin induce G1-phase cell cycle arrest in human hepatoma cells primarily through down-regulation of cyclins and Cdks, molecules that play an essential role in regulating cell cycle progression. Furthermore, it has also been shown that both compounds induce apoptosis of hepatoma cells by increasing the Bax/Bcl-2 ratio and by activating caspase-3. Consistent with the *in vitro* results, dihydroartemisinin administered alone and in combination (chemosensitization) with gemcitabine, a known inducer of apoptosis in human cancers produced strong *in vivo* antitumor effects on nude mice with human hepatocellular carcinoma xenografts.

2.2. *Vallaris solanaceae* and *Sica acuta*

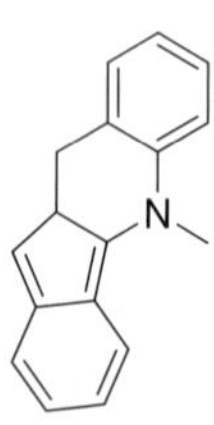

Cryptolepine

Vallaris solanaceae (Apocynaceae), a vine with fragrant white flowers, grows in Bangladesh and other Southeast Asian countries and is traditionally used against ringworms and skin infections. *Sica acuta* (Malvaceae), a shrub that grows in tropical countries such as Bangladesh, is reported to have diuretic, antipyretic, analgesic and free-radical-scavenging effects. Vallarisoside, a cardenolide glycoside from *Vallaris solanaceae* and cryptolepine, a *Sida acuta* alkaloid were isolated and evaluated for their activity in overcoming TRAIL resistance in human gastric adenocarcinoma cells. TRAIL (tumor-necrosis-factor-related apoptosis-inducing ligand), a member of the TNF superfamily, is a ligand with the ability to selectively kill tumor cells through induction of apoptosis. It binds to death receptors and provokes the formation of a death-inducing signaling complex (DISC), which leads to proteolytic activation of caspase-8 and caspase-3. However, especially highly malignant tumor cells are resistant to apoptosis induction by TRAIL. However, TRAIL combined with vallarisoside and cryptolepine, respectively, decreased the viability of human gastric adenocarcinoma cells compared to treatment with TRAIL alone, suggesting a possible synergism between TRAIL and the isolated compounds. Cryptolepine, a candidate antitumor agent, has been shown to have potent cytotoxic activity against a variety of cancer cells by means of induction of cell cycle arrest and activation of mitochondrial release of cytochrome c into the cytoplasm to induce apoptosis.[3]

2.3. *Pancratium littorale*

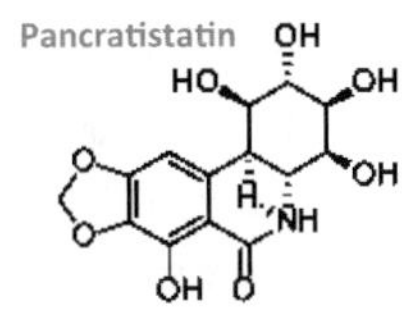

A perennial herb native to coastal regions of Central America, *Pancratium littorale* (Amaryllidaceae) exerts selective anticancer effects with potential for advancement to clinical trials. The active agent pancratistatin, a phenanthridone alkaloid, causes apoptosis in T-cell leukemia cells *ex vivo.* Interestingly, it does not enhance apoptosis in non-cancerous peripheral blood mononuclear cells, suggesting a basis for the development of non-toxic cancer therapy. It causes phosphatidyl serine to flip to the outer leaflet of the plasma membrane, and instigates permeabilization and collapse of the mitochondrial membrane potential, leading to apoptosis via generation of reactive oxygen species (ROS), cytochrome *c* leakage and caspase-3 activation.[4]

2.4. *Uncaria tomentosa*

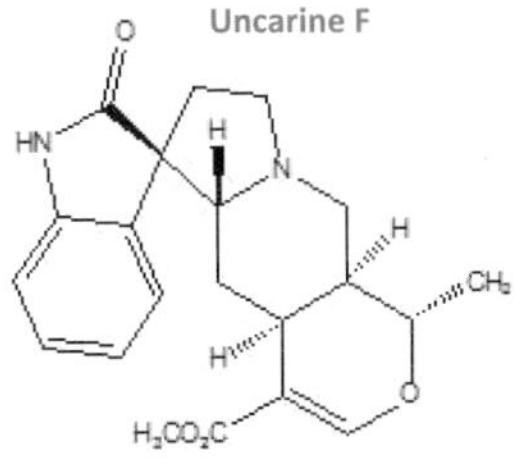

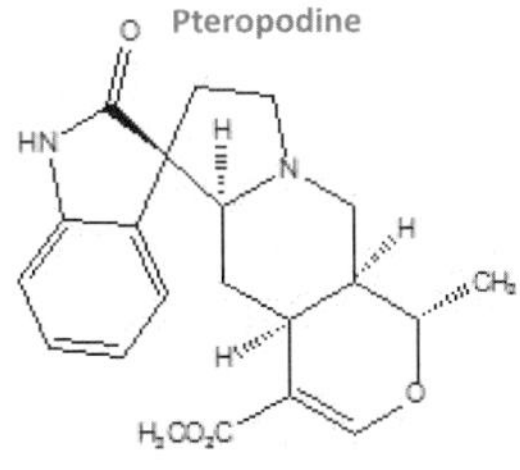

The South American Rubiaceae, also known as *Cat's claw,* has been used in Peruvian traditional medicine for treatment of infections, inflammation and cancer. Anti-oxidant and anti-inflammatory effects have been observed using aqueous extracts of *Uncaria tomentosa,* which are nearly devoid of alkaloid content. Ethanolic extracts are reported to have stronger anti-inflammatory and antioxidant activity than aqueous extracts, probably due to their greater amounts of phenolic compounds, proanthocyanidins and triterpenes.[5] Bacher *et al.*[6] examined highly purified *Uncaria tomentosa* oxindole alkaloids, namely isopteropodine, pteropodine, isomitraphylline and uncarine F, for antiproliferative and apoptotic effects on human acute lymphoblastic T-cells with the classical features of apoptosis. All tested alkaloids inhibited the growth of the T-cells and exerted cytostatic effects on both proliferating and non-proliferating G0/G1-arrested cells, with pteropodine and uncarine F being the most potent. Overexpression of anti-apoptotic Bcl-2 probably occurs in more than half of all hematological malignancies, rendering cells resistant to most cytotoxic anticancer drugs. Interestingly, Bcl-2 overexpression did not prevent the cells from programmed cell death induced by pteropodine or uncarine F, suggesting that the classic apoptotic pathway may not be essential for oxindole alkaloid-induced apoptosis. Since the exact mechanisms of apoptosis induced by pteropodine and uncarine F remain unclear, further investigations of the oxindole alkaloids *in vitro* and *in vivo* are warranted.

2.5. *Betula pubescens*

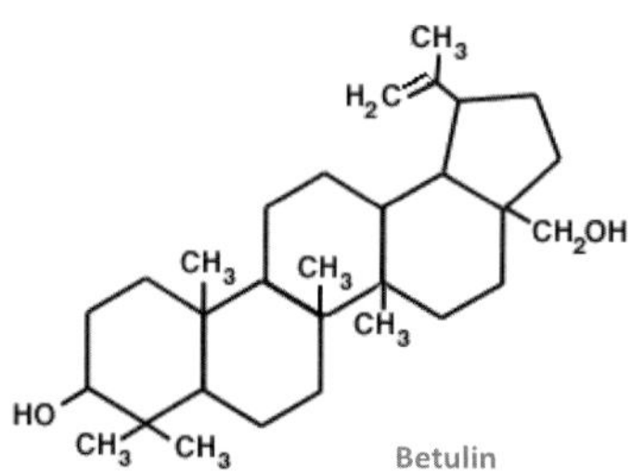

Betulin

Betulin and betulinic acid are pentacyclic triterpenes found in the bark of several plants such as the white birch, a deciduous tree that grows mainly in northern Europe and Asia. Both compounds have been shown to exhibit cytotoxicity toward several human cancer cell lines without cytotoxic effect in normal cells. Betulin-based oleogel, prepared from a standardized dry extract from birch bark, appears to be an effective new compound for treating actinic keratoses (squamous cell carcinomas *in situ*). Rzeski *et al.*[7] examined the anticancer effects of betulin in human tumor cell lines derived from cancers of the nervous system (medulloblastoma, neuroblastoma, glioma) and in primary tumor cultures isolated from patients (ovarian carcinoma, cervical carcinoma, glioblastoma multiforme). Betulin decreased proliferation

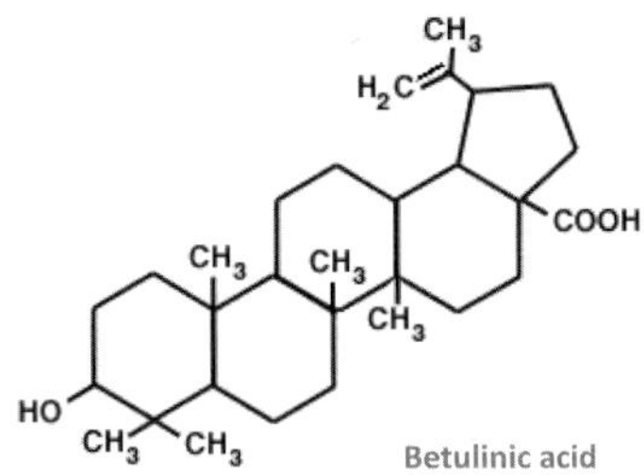

Betulinic acid

in all tested tumor cell cultures in a concentration-dependant manner. In primary cultures, betulin treatment revealed a more pronounced effect than in the stable cancer cell lines. It displayed a relatively modest cytotoxicity in healthy human skin fibroblasts. Betulin has been shown to induce apoptotic cell death and to inhibit migration in cancer cells, an important marker of tumor metastatic potential. While both compounds are very similar in their chemical structure (betulinic acid has a C-28 carboxyl group instead of a C-28 hydroxymethyl group), they differ in their biological activity. In contrast to betulinic acid, betulin does not alter the expression of the anti-apoptotic protein Bcl-2, pro-apoptotic Bax or cell cycle regulator cyclin D1. Its exact intracellular mechanism remains unknown.

3. INHIBITORS OF DNA TOPOISOMERASES I AND II

3.1. *Phyllanthus amarus*

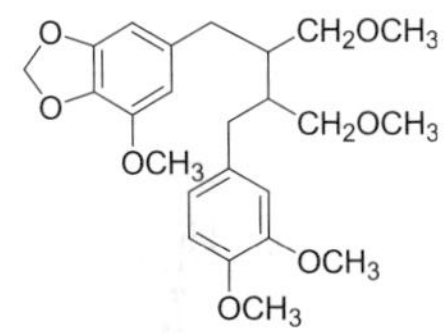

Niranthrin

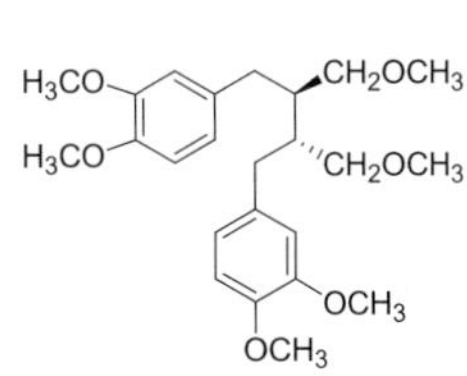

Phyllanthin

Phyllanthus amarus is a tropical plant from India containing numerous bioactive compounds. These compounds fall into three main groups: lignans, flavonoids and tannins. The most important lignans are nirtetralin, niranthrin, phyllanthin and phyltetralin.[8]

P. amarus extracts inhibit the function of P-glycoprotein (P-gp) and reduce the activity of DNA topoisomerases I and II in *Saccaromyces cerviacae* mutant cell cultures.[9,10] Tang *et al.*[11] showed that the extracts also inhibit cell proliferation in melanoma and prostate cancer cells *in vitro*. Furthermore, extracts induce cell cycle arrest and interfere with DNA repair,[12] and have an anti-angiogenetic effect. This was shown in mice with Lewis lung carcinoma.[13] Certain compounds in the plant also interfere with the migration of vascular endothelial cells.[13]

3.2. *Boswellia serrata*

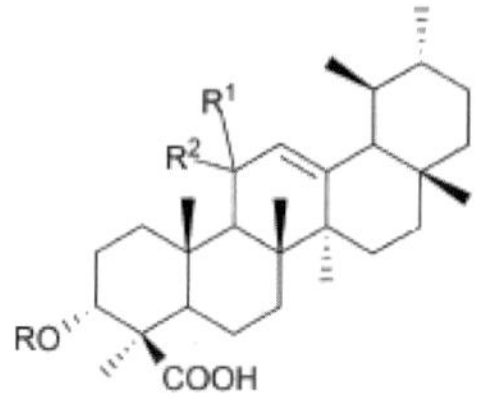

1. R1=R2=H, R=H (BA)
2. R1+R2=O, R=H (KBA)
3. R1=R2=H, R= $COCH_3$(ABA)
4. R1+R2=O, R=$COCH_3$(AKBA)

Native in tropical parts of Asia and Africa, *Boswellia serrata* contains a complex mixture of terpenoids, polysaccharides, essential oils, proteins, and inorganic compounds,[14,15] among which the pentacyclic triterpenoids are most pharmacologically relevant.[16] The four main triterpene compounds are β-boswellic acid (BA), 11-keto-β-boswellic acid (KBA), and the corresponding acetates 3-*O*-acetyl-β-boswellic acid (ABA) and 3-*O*-acetyl-11-keto-β-boswellic acid (AKBA).[14] These boswellic acids inhibit the synthesis of DNA, RNA and proteins in human leukemia HL-60 cells.[17,18] In addition, they exhibit antitumor activity in animal experiments.[19] Boswellic acids are reported to non-competitively inhibit lipoxygenase (5-LO)[20–22] through a novel non-redox mechanism. 3-*O*-acetyl-11-keto-β-boswellic acid has been described as a catalytic inhibitor of DNA topoisomerases I and IIα[23] that acts by directly binding to these enzymes (in contrast to other inhibitors that bind to DNA or form complexes with the enzyme and DNA). Besides, both ABA and AKBA inhibit NF-κB signaling,[24–26] which represents an important mechanism in halting the proliferation of cancer cells.[27]

4. INHIBITORS OF NUCLEOTIDE BIOSYNTHESIS

Synthesis of individual nucleotide monomers is necessary for RNA and DNA biosynthesis and is under tight regulatory control. Rapidly growing cells such as cancer cells require more nucleotides than many normal cells. Thus, the inhibition of nucleotide biosynthesis can be a selective target for cancer therapy.

4.1. *Aglaia foveolata*

Silvestrol has been isolated from the fruits and twigs of the Indonesian plant *Aglaia foveolata* of the family Meliaceae. It is a cyclopenta[b]benzofuran with an unusually bulky pendant dioxanyl ring.[28,29] It is unclear whether the dioxanyl side chain is responsible for its cytotoxicity. It inhibits nucleotide and/or protein biosynthesis and causes cell cycle arrest in the G2/M phase.[30–33,34] Furthermore, it inhibits NF-κB activity by blocking NF-κB DNA binding activity and I-κB degradation.[35] The inhibition mechanisms of Silvestrol consist of blocking ribosomal recruitment by binding to e1F4A, one of three subunits of the eukaryotic translation initiation factor 4F complex. By targeting elF4A, which is a DEAD-box RNA helicase that unwinds the 5′ mRNA structure, Silvestrol blocks ribosome recruitment. Silvestrol activates an unusual pathway of apoptosis in LNCaP cells, involving caspases-2, -9 and -10 but not caspases-3 and -7.[36]

Silvestrol showed inhibitory activity *in vivo*[73] in several human cancer cell lines that were cultivated in hollow fibers and implanted intraperitoneally in mice. It was also active in a P388 murine leukemia model in recent studies.[28,36,37] Silvestrol has also demonstrated activity against breast, prostate and lung cancer cell lines *in vitro*.[28,36,38]

5. INHIBITORS OF SIGNAL TRANSDUCTION PATHWAYS

5.1. *Veratrum californicum*

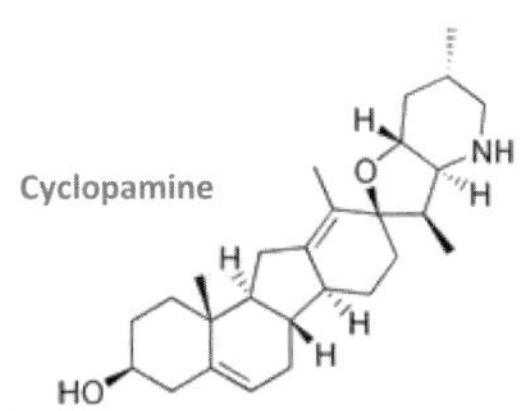

Cyclopamine, a steroidal alkaloid isolated from the corn lily (*Veratrum californicum*, Melanthiaceae), was the first phytochemical shown to target the *Hedgehog* (Hh) signaling pathway. This signaling pathway is involved in vertebrate embryonic developmental processes such as organogenesis and determining left-right asymmetry, as well as in self-renewal of adult stem cells. Genes encoding proteins involved in the *Hedgehog* pathway are frequently mutated or aberrantly activated in cancers, making them potential targets for anti-cancer therapy. Hedgehog signaling occurs when the Hh ligand binds to the transmembrane protein, patched (PTCH). In a resting state, PTCH inhibits the activity of another transmembrane protein, smoothened (Smoh). However, upon Hh binding, Smoh is released from inhibition by PTCH, thus allowing Smoh-mediated transcriptional activation of several cancer-related genes. Cyclopamine targets the self-renewal properties of cancer stem cells by inhibiting the activation of Smoh. It has been shown to reduce tumor burden in a mouse tumor allograft and to be cytotoxic towards cultured human medulloblastoma cells. Upon treatment of murine medulloblastoma with cyclopamine, proliferation was blocked and neuronal differentiation was induced, effectively decreasing the population of cancer stem cells.[39]

5.2. *Resveratrol*

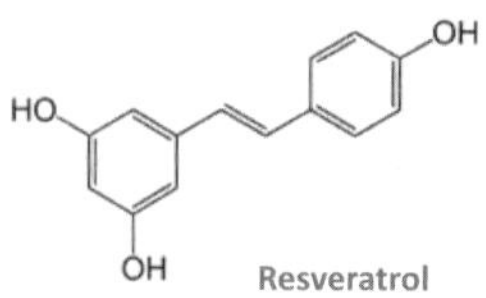

Resveratrol

The polyphenolic phytoalexin resveratrol, found in grapes, berries and peanuts, is well known for its cardioprotective and chemopreventive properties. Cecchinato *et al.*[40] demonstrated the ability of resveratrol to suppress cell proliferation and to induce apoptosis in acute lymphoblastic leukemia cells via the modulation of several pathways that regulate cell survival. Resveratrol inhibits the Notch signaling pathway by decreasing expression of the Notch protein. The Notch signaling pathway is involved in a variety of developmental processes including angiogenesis, organogenesis, central nervous system development and adult-type hematopoetic stem cell generation. Notch expression was found to be upregulated in brain tumor stem cells, and inhibition of the Notch receptor resulted in apoptosis and prevention of xenograft tumor formation.[39] Ligand binding to the extracellular domain of Notch, normally triggered by cell-to-cell contact, leads to the proteolytic cleavage of Notch receptor's intracellular domain. The cleavage product is then translocated to the nucleus, where it regulates gene expression. Resveratrol has been shown to increase the apoptotic proteins Bax and the tumor suppressor protein p53. Resveratrol also induces apoptosis via inhibition of the pro-survival PI3k/Akt pathway.[40]

6. ANTIOXIDANTS FOR CHEMOPREVENTION

In contrast to the direct anti-cancer activities described above, some medicinal plants contain ingredients that indirectly protect normal cells from becoming cancerous. For example, antioxidants can protect a cell's DNA from cancerous mutations. Oxygen radicals and their reaction products are generally toxic to cells and can cause DNA damage, leading to cancer.

Most endogenous oxygen radicals arise in oxygen-dependent species as byproducts of the mitochondria. Without enzymatic inactivation, these radicals start a chain reaction in the cytosol generating progressively more radicals. In daily life, we are confronted with many environmental poisons that generate mutagenic oxygen radicals. Many drugs and toxins become metabolically activated by normal cellular processes such as the cytochrom-P_{450} system and induce cytotoxic reactive oxygen species (ROS). Even the sunlight in the UV-A and UV-B range can generate radicals in skin cells exposed to the sun.

More than 20 different types of oxidative DNA damages are known today, with 8-oxoguanin being the most important oxidative pre-mutagenic lesion. Oxygen radicals attack cell lipids, chromatin and DNA. Damage to DNA is the most carcinogenic, because oxygen radicals induce double-strand breaks and thus changes in the regulation of specific genes. The human body has several protection mechanisms against oxidative damage such as the enzymes catalase and superoxide-dismutase, which accelerate reactions consuming oxygen radicals, and glutathione-associated enzymes. Antioxidant agents operate as free-radical scavengers; they arrest the chain reaction of radical formation without themselves becoming radicals. Antioxidants have been associated with decreased incidence of cancer and subsequently lower mortality rates in several human studies.[41]

One of the best documented herbal antioxidants is the green tea extract of *Camellia sinensis.* Green tea contains high concentration of oligomeric proanthocyanidins, precursors of anthocyanidins that give fruits such as cranberries their characteristic red-blue color. Anthocyanidins absorb light at 270–290 nm (UV) and 465–560 nm (blue/green). Furthermore, these molecules bind free radicals. Besides proanthocyanidins, green tea contains derivatives of catechin and epicatechin, the most important of which is epigallocatechin. Catechins are polyphenolic plant metabolites, and are thought to give green tea its bitter flavor. In nature, polyphenols filter UV-B radiation to protect the photosynthesis apparatus of the plant. In humans, oral consumption as well as topical application protects against chemical- and UV-induced carcinogenesis. Polyphenols also prevent inflammation, the first step in the formation of skin tumors. In black tea, which is produced from fermented green tea, it was shown that the aflavins, the fermentation product of catechins, exhibit the same antioxidant properties as catechins.

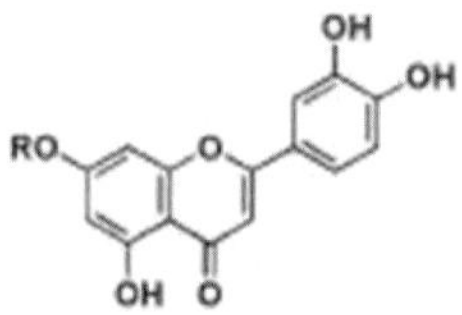

26 R = H
27 R = β-Glucose

Another famous natural stimulant with antioxidant activity is the extract of *Coffea arabica.* The fruits contain polyphenols, condensed proanthocyanidins, quinic acid and ferulic acid.

Sonneratia caseolaris is a small tree with large red flowers that grows in the Sundarbans mangrove wetlands of Bangladesh. Traditionally, *S. caseolaris* is used as an astringent and antiseptic, as well as against hemorrhage. The plant exhibits antioxidant activity. The ingredients responsible for its antioxidative quality are the flavones luteolin[14] and its 7-O-β-glucoside cynaroside.[15]

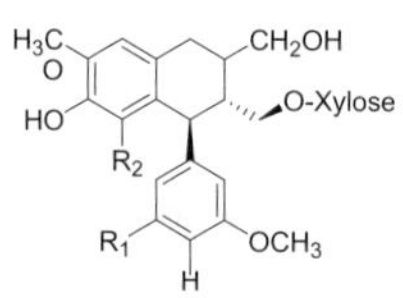

32 R1=OCH3 R2=OCH3
36 R1=H R2=H

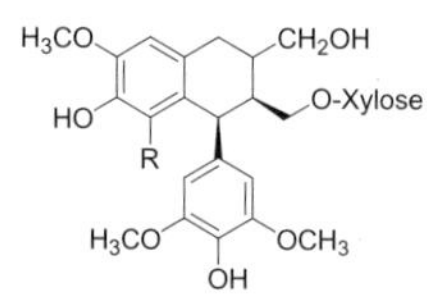

33 R=OCH3
34 R=H

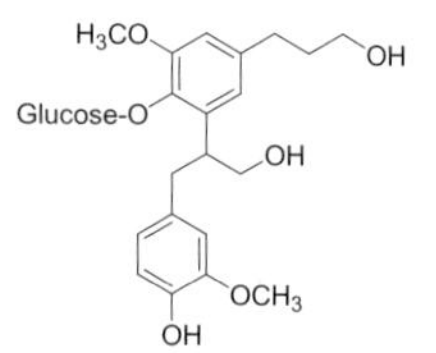

35

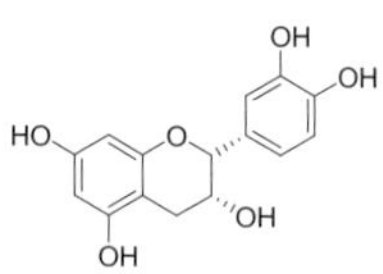

37

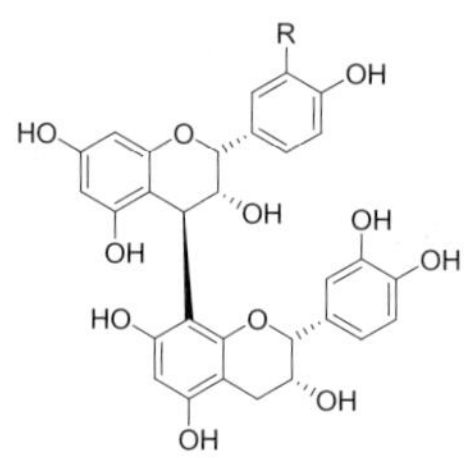

38 R=H
39 R=OH

Another promising plant with antioxidant properties is *Saraca asoca* (synonym: *Saraca indica*), a medicinal plant from Bangladesh. Its bark is an astringent, and is used against menorrhagia, hemorrhoids and hemorrhagic dysentery. Besides tannins, flavonoids, proanthocyanidins and leucoanthocyanidins, the main effective substances in the extract are the lignan glycosides lyoniside, nudiposide, 5-methoxy-9-β-xylopyranosyl(−)-isoariciresinol, icariside E3 and schizandriside. The potent flavonoids are (−)-epicatechin, epiafzelechin-(4β->8)-epicatechin and procyanidin B2.[25,42–48]

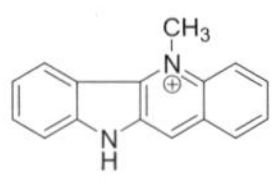

74

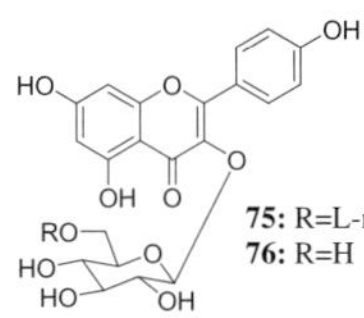

75: R=L-rhamnopyranose
76: R=H

Phlebodium aureum, also called *Polypodium leucotomos*, is a tropical cabbage palm fern that delivers an extract displaying photoprotective properties *in vitro* and *in vivo*. The presumed mechanism for its activity is via the inhibition of UV-induced photoisomerisation of *trans*-urocanic acid. *Trans*-urocanic acid is a photoreceptor located in the *Stratum corneum* of the skin and an intermediate in the catabolism of *L*-histidine. When exposed to UV-B irradiation, *trans*-urocanic acid is converted to its *cis* isomer. Extract of the palm fern blocks the photodecomposition of the receptor in the presence of oxidizing agents.

Sida acuta, called *Berela* by the natives of Bangladesh, is a shrub that grows throughout the tropical biome. Traditionally, the roots are used as a diuretic and an antipyretic; it is also thought to possess antiplasmodial, antimicrobial and analgesic properties. The isolated extract contains alkaloids, tocoferols, sterols, polyphenols and glycosides. The alkaloid cryptolepine and two kaempferol glycosides have been shown to be radical-scavenging ingredients.[29,30,35]

A flavonoid complex isolated from the milk thistle, *Silybum marianum*, and called Silymarin showed antioxidative, anticarcinogenic and anti-inflammatory properties in animal models. It is presumed to be a safe agent, and thus may be applicable for chemoprevention, sun protection and antiphotocarcinogenic protection. In primary rat hepatocytes, protocatechuic acid (PCA) isolated from *Hibiscus sabdariffa* reduced oxidative damage from *t*-butyl hydroperoxide. For this reason, it was classified as a cancer preventive agent.[49] Other dietary botanicals are under discussion as antioxidants and UV protection agents, e.g. apigenin, curcumin, resveratrol, rosemary, propolis produced by honeybees, red ginseng, genistein of soybeans, pomegranate extracts and many more. Caffeic acid phenethyl ester (CAPE), the active component in honeybee propolis, can impair the growth of human cancer cell lines *in vitro*. It has been especially effective in oral cancer cells, and is not cytotoxic to normal cells. CAPE can induce cell cycle arrest and apoptosis as well as inhibit NF-κB. Concentrations of 10 μM result in significant antiproliferative effects in C6 glioma cells. In addition, CAPE's antioxidant effect interferes with the oxidative activation of cells. However, *in vivo* studies are not yet available.[49]

Esculetin is a coumarin isolated from the *Plumbaginacea* family and is thought to have antioxidant and anti-tumor properties.[50–52] It can lead to cell cycle arrest, and inhibit the growth of H3B hepatocytes. Active agents of the coumarin family are attractive for therapeutic use, because they are components in many plants that are already widely distributed. Furthermore, esculetin acts as an anti-inflammatory agent by inhibiting the synthesis of prostaglandins and leucotrienes in the same way that acetylsalicylic acid does. Synthethic coumarin derivatives are being considered for chemoprevention and therapy.[49]

Glycyrrhiza glabra is very common medicinal herb in Asia. It is an antioxidant against reactive oxygen species and an anti-tumor agent.[53] The flavonols quercetin, kaempherol and myricetin can be isolated from *G. glabra*. Quercetin inhibits lipid oxidation caused by free radicals *in vitro*. Another active component of *Glycyrrhiza glabra* is glycyrrhizin, which possesses antioxidant and anti-tumor properties as well. It has been shown to reduce damage to the plasma membrane, but the mechanism is not yet understood. The ability of these two components, quercetin and glycyrrhizin, to scavenge free-radicals and to inhibit lipid peroxidation is promising in their development as chemopreventive agents.[49] Components of the Asian tree *Zanthoxylum ailanthoides* can also scavenge free radicals and protect cell membranes. The specific activity of ailanthoidol is probably due to its antagonism of the adenosine receptor.[49]

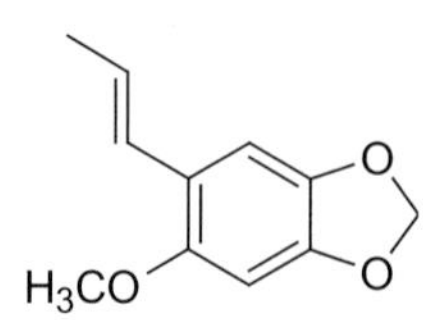

Carpacin

Carpacin is another herbal agent that has been shown to prevent cancer. It is isolated from *Piperacea* species. In Asia, it is used as an antidepressant. Its cancer-preventing activity is based on the induction of glutathione-S-transferase (GST) and the resulting inhibition of unscheduled DNA synthesis. GST plays a major role in the Phase II cellular detoxification processes. Glutathione reacts readily with electrophilic molecules resulting in less toxic products that are water soluble and can be easily eliminated. Most carcinogens are electrophiles, and are detoxified via this pathway. Stimulating GST-activity therefore seems to be cancer-protective mechanism. Guided by carpacin as a lead compound, synthethic phenols 2,4,5-trihydroxybenzaldehyde (2,4,5-THBA) and dihydroxybenzaldehyde (DHBA) have been synthesized, which have potent antioxidant effects *in vitro*, as shown by DPPH assay and *t*-butyl hydroperoxide test.[49]

7. CYCLOOXYGENASE INHIBITORS

Cyclooxygenase (COX), also known as prostaglandin-H-synthase, catalyzes the conversion of arachidonic acid to prostaglandin in two steps. In the first step, cyclooxygenase removes hydrogen from C13 of arachidonic acid and adds two molecules of oxygen, forming prostaglandin G_2 (PGG_2). In the following step, PGG_2 is reduced to prostaglandin H_2 (PGH_2) by peroxidase activity. Tissue-specific isomerases then convert PGH_2 into prostaglandins (PGD_2, PGE_2), prostacyclin (PGI_2) or thromboxane A_2.[54]

Two isoforms of COX can be distinguished. While COX-1 is constitutively expressed, COX-2 is induced by cytokines, growth factors and tumor promoters.[55] Recent data indicate that both isoforms are also potent anti-angiogenic agents. Angiogenesis, or the formation of new blood vessels, is critical for tumor growth and metastasis; the expansion of a tumor is limited in the absence of angiogenesis.[56] However, COX also has the ability to activate carcinogens that damage genetic material.[57]

7.1. *Cassia quinquangulata* and *Aiphanes aculeata*

Cassia quinquangulata (Leguminosae), native to the tropics, contains resveratrol and serves as a potent inhibitor of cyclooxygenase, specifically COX-1.[57]

Aiphanes aculeate (Arecaceae) is a palm, which grows in tropical rain forests in the northwest of South America. Pezzuto[58] has established the cancer chemopreventive potential of *trans*-resveratrol (3,5,4′-trihydroxy-*trans*-stilbene in this plant. The stilbenolignan, aiphanol, represents a novel carbon skeleton with a stilbene-phenylpropane unit with a dioxane moiety. Aiphanol was identified together with other constituents in this

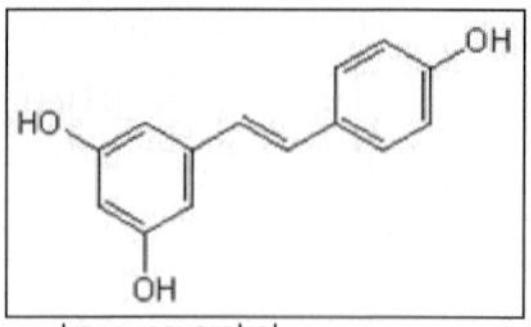

trans-resveratrol

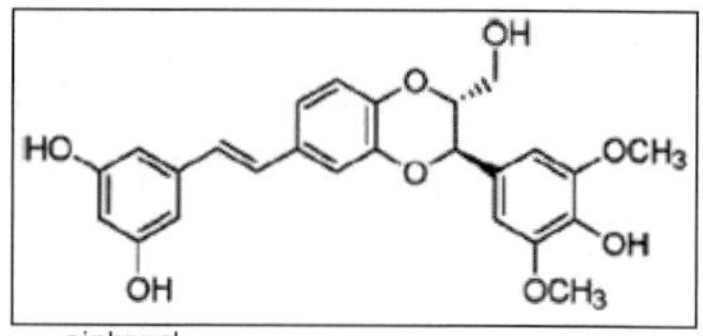

aiphanol

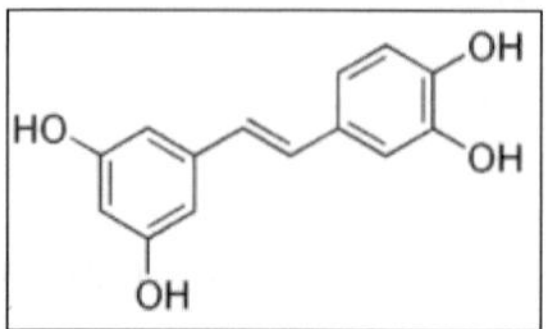

piceatannol

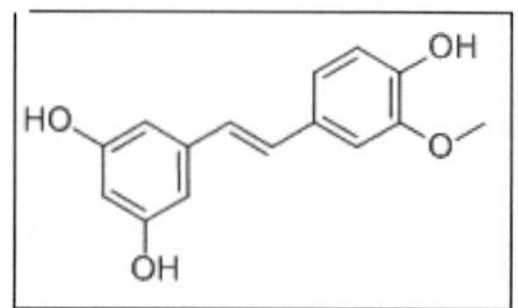

isorhapontigenin

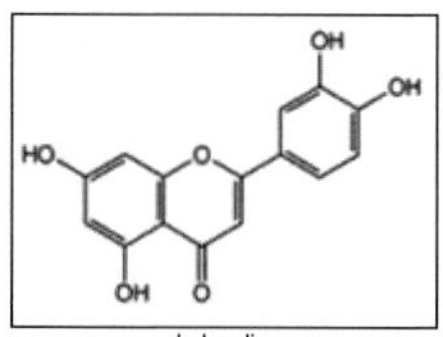

luteolin

1) http://img.hisupplier.com/var/userImages/2008-12/02/dndfan$173952477%28s%29.jpg;
2) Lee *et al.*, 2001;
3) http://upload.wikimedia.org/wikipedia/commons/thumb/8/8e/Piceatannol.png/200px-Piceatannol.png;
4) http://www.chemicalbook.com/CAS%5CGIF%5C32507-66-7.gif;
5) http://upload.wikimedia.org/wikipedia/commons/thumb/f/f9/Luteolin.png/275px-Luteolin.png

plant such as isorhapontigenin and piceatannol and with flavones such as luteolin.[59]

7.2. *Plantago major*

Plantago major (Plantaginaceae) is an herbaceous perennial that originated in Europe but is now widespread throughout the world. Traditionally, *Plantago major* was known for its wound-healing qualities. The plant

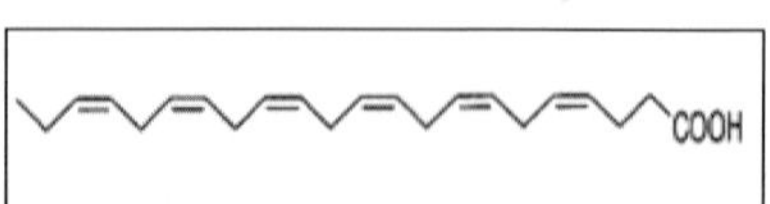

docosahexaeonic acid

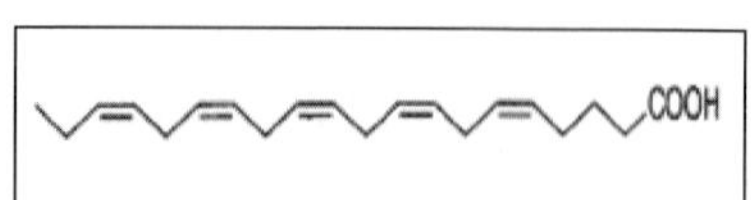

eicosapentaeonic acid

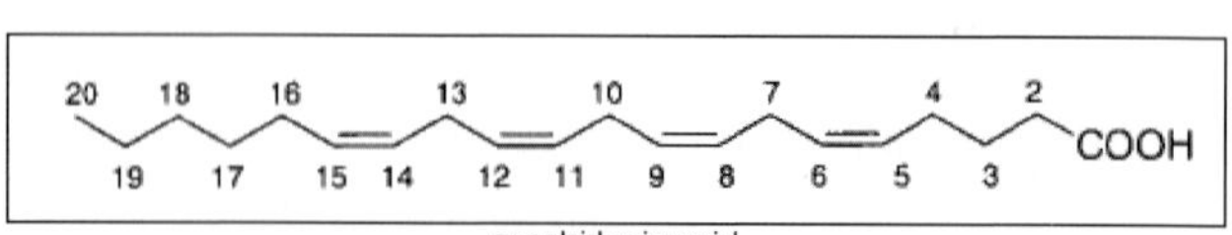

arachidonic acid

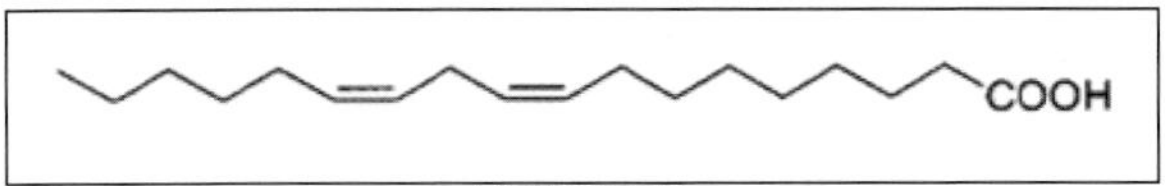

Linolenic acid

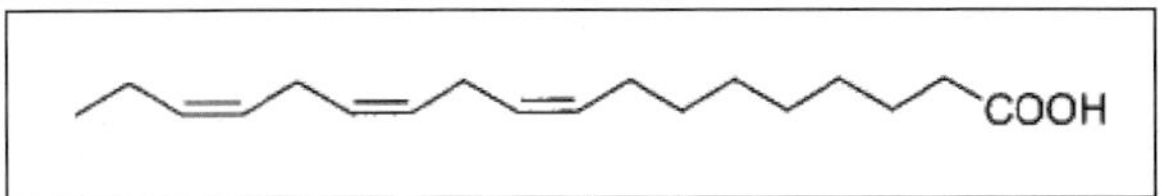

α-linolic acid

contains compounds from different classes of substances including carbohydrates, fatty acids, alkaloids, caffeic acid derivatives, flavonoids, iridoid glycosides and triterpenoids. It contains the structural triterpenoid isomer oleandric acid as well as glycyrrethic acid, both of which inhibit COX-2 and COX-1-catalyzed prostaglandin biosynthesis. Docosahexaenoic acid and eicosapentaeonic acid compete with arachidonic acid as substrates for COX. It has been proposed that docosahexaenoic acid (DHA) is resistant to COX-1 enzymatic oxidation and so DHA reacts as an inhibitor and not as a substrate.[60]

It has also been shown that fatty acids with more than 20 carbons can act as COX inhibitors. Compounds consisting of 18 carbon atoms such as linolenic acid, a metabolic precursor of arachidonic acid, were partially active. Caffeic acid phenethyl ester as well as several flavonoids isolated from *Plantago major* suppressed COX-2 enzymatic activity and PGE_2 production.[54] Thioethers compounds were also found active. Hence, a thioether moiety seems to improve COX-2 inhibition.

In conclusion, the fatty acids linolenic acid and α-linolenic acid isolated from *Plantago major* are probably responsible for the COX-2 and COX-1 inhibitory effects of the plant extract [54]. The triterpenoids may also play a role in the inhibition.

7.3. *Allium sativum*

Allium sativum (Allioideae), a plant that is native to grassy regions of central and East Asia, also shows inhibition of COX-2. The responsible compound is thioester ajoene, which inhibits the lipopolysaccaride-induced prostaglandin E2 releasing process in RAW 264.7 macrophages by a mechanism similar to that of non-steroidal anti-inflammatory drugs.[61]

Because of the enhanced effect of COX-2 inhibition by thioether-containing fatty acids, exploration of sulphur-containing compounds such as glycolipids as selective inhibitors of COX-2 warrants further research.[54]

8. PROSTAGLANDIN INHIBITORS

Prostaglandins are tissue hormones derived from arachidonic acid in a cellular process involving cyclooxygenases. Prostaglandins contribute to inflammation, constrict blood vessels, enhance blood coagulation and increase the perception of pain. They also enhance the gastric secretion of mucus and hydrogen carbonate and inhibit the production of gastric acid.

Several investigations have reported higher levels of prostaglandins in human tumors than in surrounding normal tissues.[62]

8.1. *Leucas aspera*

Leucas aspera (Labiatae) a common aromatic herb grows in abundance in Bangladesh. Traditionally, a decoction produced from the plant is taken as analgesic or antipyretic, as well as an antirheumatic or anti-inflammatory treatment. Its paste is applied topically on inflamed areas. It is also used as an insecticide. Because of its anti-inflammatory and analgesic properties, *Leucas aspera* was tested for prostaglandin inhibitory activity[63] and four new diterpenes, leucasperones A and B and leucasperols A and B;

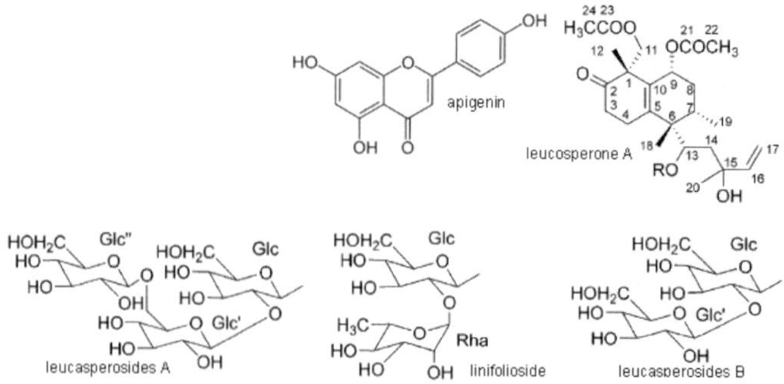

three new isopimarane glycosides, leucasperosides A, B and C; and several other previously known compounds asperphenamate, maslinic acid, (−)-isololiolide and linifolioside[64] were identified. Leucasperone A, leucasperosides A and B, and linifolioside inhibited prostaglandin activity.[65] Subsequent reports have suggested that apigenin, a flavone found in chamomile or celeriac, also has an inhibitory effect on prostaglandin E_2 release.[64]

9. CELL GROWTH INHIBITORS

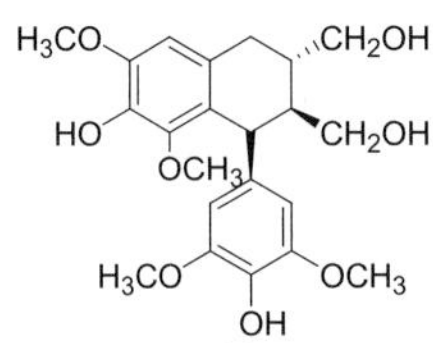

28

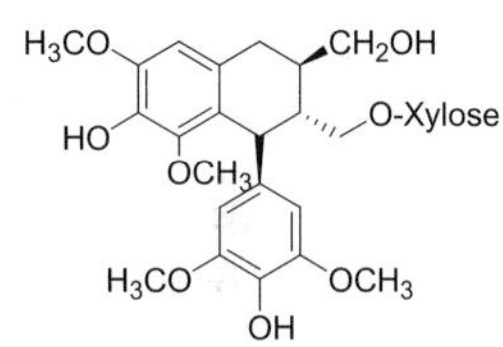

29

Tumor progression is triggered by a myriad of factors, each of which provides a potential target for therapy. Many plant extracts possess the capacity to inhibit cell growth. An important parameter to quantify the growth-inhibitory potential of a compound is the 50 percent inhibition concentration (IC50), the concentration of an inhibitor at which half of the maximum inhibition is achieved.

9.1. *Aphanamixis polystachya*

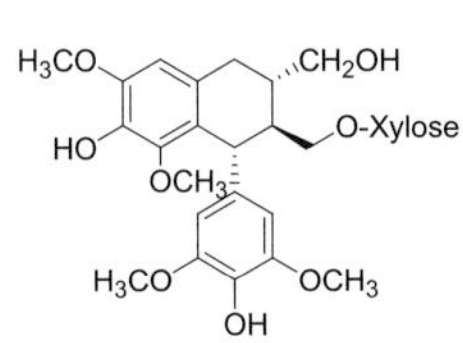

30

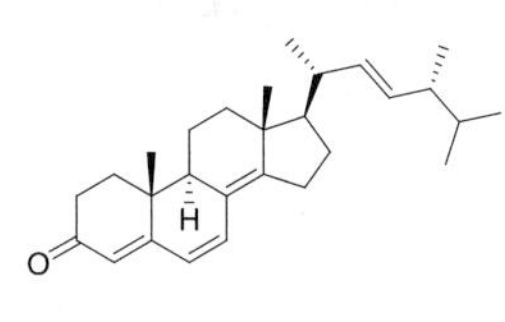

31

Aphanamixis polystachya is a medicinal plant found in Bangladesh. Its bark is traditionally used to treat spleen and liver diseases, abdominal complaints and tumors. Methanol extract of this plant has an IC_{50} value of 50 μg/mL in HeLa cervical carcinoma cells. The butanol and *n*-hexane extracts are even more potent (IC_{50} of 22 and 33 μg/mL). The isolated compounds include limonoids, terpenoids, glycosides and a saponin alkaloid. The methanolic, butanolic and *n*-hexane fractions contain several active compounds, including sigmasterol, linoleic and oleic acids.[63]

9.2. *Zingiber spectabile*

Zingiber spectabile, known as *beehive* ginger, is another traditional medicinal herb that grows in Bangladesh. The rhizome is used to treat cough and asthma. Its extract is germicidal and acts as a tonic. Even though ginger is predominantly used as culinary spice, it is believed that *Zingiber spectabile* also has anticancer properties. Terpinen-4-ol, labda-8, 12-diene-15,16-dial, α-terpineol, α-pinene, limonene and related chemical compounds were first isolated from the rhizome of *Zingiber spectabile*.[63]

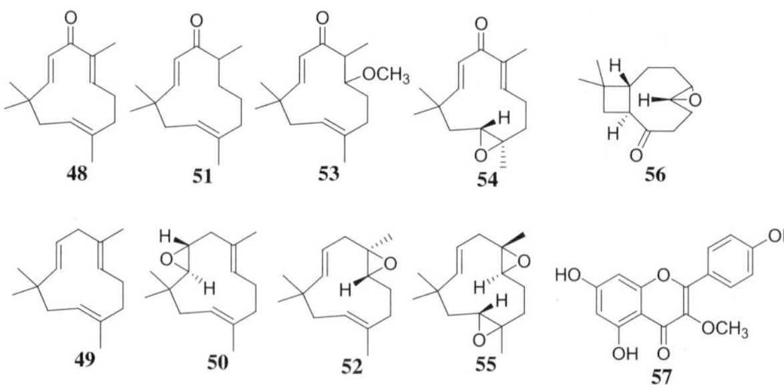

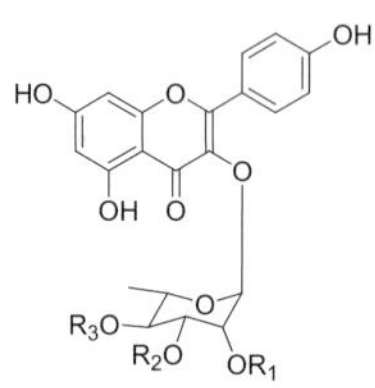

58	R_1=H	R_2=$COCH_3$	R_3=$COCH_3$
59	R_1=$COCH_3$	R_2=$COCH_3$	R_3=H
60	R_1=$COCH_3$	R_2=H	R_3=$COCH_3$
61	R_1=H	R_2=H	R_3=$COCH_3$
62	R_1=H	R_2=$COCH_3$	R_3=H
63	R_1=$COCH_3$	R_2=H	R_3=H
64	R_1=H	R_2=H	R_3=H

10. CLINICAL TRIALS

Extensive preclinical investigations of all potential drugs must be conducted before clinical trials can be initiated. Preclinical experiments include *in vitro* and *in vivo* experiments concerning efficacy, toxicity and pharmacokinetics of the compound. Clinical studies further determine the efficacy and safety of plant-derived compounds in patients. It is generally a long road from isolation of a phytochemical to its clinical application. Even if an active natural product has not yet reached clinical use, investigation of the proposed product can provide critical clues for the development of new targeted cancer drugs.

Clinical trials are commonly separated into four phases each with different purposes. In Phase I trials, the candidate drug is given to between 20 and 40 healthy participants to evaluate its safety, determine a proper dosage range, decide on the best form of application and identify side effects. In Phase II trials, the compound is administered to a larger group of patients (50–100) to assess the effectiveness of the drug. In Phase III trials, the drug is given to a large group of participants (1,000–3,000) to further confirm its effectiveness, monitor side effects and collect information on the results of the treatment on a broad patient collective. This is necessary to evaluate the safety of the compound for the general population. After successful completion of phase III trials, drugs are approved for routine market and clinical use. However, Phase IV post-approval studies or post-marketing surveillance trials continue to be performed. Phase IV trials are aimed at gaining additional information on rare side effects or reactions to continuing and long-term use of the drug. Thousands of people can be involved in Phase IV trials over long periods of time.

The following drugs are in the process of clinical trials.

10.1. *Curcuma longa*

Curcumin is an active ingredient of *Curcuma longa*. It is an East Indian plant also known as "turmeric" and a member of the ginger family. Curcumin belongs to diarylheptanoids and is responsible for the intensely yellow color of turmeric rhizome. It has been shown to inhibit cell proliferation of a wide variety of cell lines.[66,67]

In a currently running Phase III trial, its effectiveness is being examined in patients with advanced or inoperable pancreatic cancer and in patients with metastatic colon cancer.[68,69] Curcumin's activity in the prevention of colon and gastric cancers has been demonstrated in rodents.[70] It acts by reducing the expression of human epidermal growth factor receptor 2 (HER2) and epidermal growth factor receptor (EGFR).[65] In addition, curcumin downregulates the expression of tumor necrosis factor (TNF), NFκB, epidermal growth receptor 1 (EGR-1), activator-protein 1 (AP-1), cycloxygenase 2 (COX2), lysyl oxidase (LOX), nitric oxide synthase (NOS) and matrix metallopeptidase 9 (MMP-9).[66,67,71]

Its mechanism includes the downregulation of the anti-apoptotic protein Bcl-XL and integrin-associated protein (IAP).[72–74] Furthermore, the compound and its derivatives inhibit VEGF and bFGF-mediated corneal neovascularization as well as angiogenesis *in vivo* and *in vitro.*[75–82] Overall, there is a strong base of well-documented evidence that curcumin inhibits proliferation and interferes with cell cycle progression.[72–74,83–87]

10.2. *Indigofera tinctoria, Polygonum tinctorium* and *Isatis tinctoria*

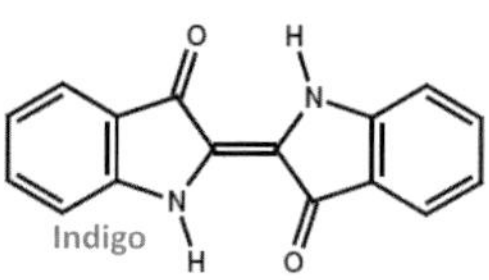
Indigo

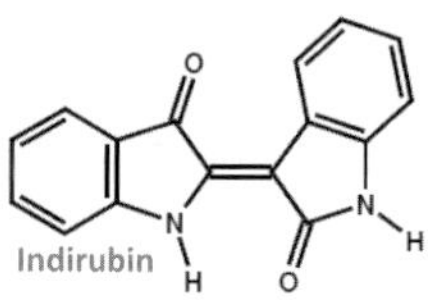
Indirubin

Another herbal compound with anticancer activity is indirubin, a red colored component of indigo dye. It is a 2,3′-bisindole and an isomer of the blue coloring agent, indigo. It is a minor constituent of various plants including *Indigofera tinctoria, Polygonum tinctorium* and *Isatis tinctoria.* Indirubin's flat heterogenic ring system fits into the ATP binding site of kinases. It competes effectively with ATP for the binding site and at sufficient concentrations is able to block cell cycle progression. It is an especially potent inhibitor of cyclin-dependent kinases. The compound also inhibits glycogen-synthase kinase-3b (GSK-3), blocks the c-Src-kinase and induces cell cycle arrest in G2 and/or G2/M phase leading to apoptosis.[42,43] Another mechanism by which indirubin inhibits tumor growth and prevents angiogenesis by blocking VEGFR2 mediated JAK/STAT3 signaling in endothelial cells was shown by Zhang *et al.*[44]

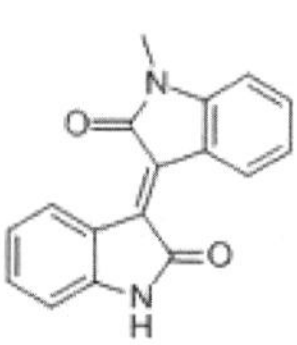
Meisoindigo

N-methylisoindigotin, abbreviated meisoindigo, is a semisynthetic derivative of indirubin. Meisoindigo inhibits the growth of cancer cells by inducing marked apoptosis and moderate cell-cycle arrest in the G0/G1 phase.[88–91] Furthermore, the compound upregulated p21, p27, pro-apoptotic Bak and Bax and downregulated anti-apoptotic Bcl-2 in primary AML cells.[91] In a Phase III clinical trial, it was shown that the efficacy of the compound was comparable to that of hydroxyurea and busulfan. Interestingly, meisindigo was well tolerated without side effects such as myelosuppression, which frequently occur with established anticancer drugs.[92–94]

10.3. *Cephalotaxus*

The alkaloid homoharringtonine (HHT) is a major constituent of the genus *Cephalotaxus* (Cephalotaxaceae). HHT inhibits protein biosynthesis by inhibiting the elongation of amino acid chains, and thus leads to apoptosis. Its effect on ribosomes is both dose- and time-dependent. HHT exerts synergistic effects with Ara-C. Its cytotoxicity is cell-cycle specific, primarily blocking the progression of cells from G1- into S-phase and from G2- into M-phase.[48,95]

One Phase II clinical trial is currently being conducted focusing on HHT treatment of chronic myelogenous leukemia and another is being conducted investigating its activity against advanced solid and hematologic tumors.[96,97] Semisynthetic homoharringtonine (ssHHT) is now being evaluated in Phase II clinical trials for the treatment of chronic myelogenous leukemia and acute myelogenous leukemia patients. Furthermore, ssHHT induced apoptosis independently of the expression of Bax in both HL60 and multidrug-resistant HL60/MRP cell lines in a time- and dose-dependent manner.[11]

10.4. *Maytenus serrata* and *Actinosynnema pretiosum*

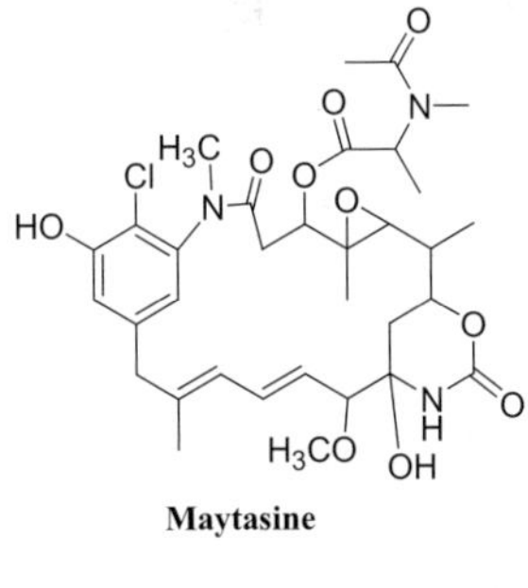

Maytasine

Maytansinoids have been isolated from various plants, mosses and the actinomycete, *Actinosynnema pretiosum*. Its parent substance, maytansine, is an ansa macrolide that was isolated from the Ethopian shrub *Maytenus serrata*. Maytansine is a potent microtubule-targeting compound that induces mitotic arrest.[98,99,100] Maytansine binds to tubulin at the rhizoxin binding site, inhibiting microtubule assembly and inducing microtubule disassembly, thereby disrupting mitosis.[101,102] A Phase I study was completed treating patients with solid tumors with a maytansinoid DM4-conjugated humanized monoclonal antibody to assess the safety and pharmacokinetics of huC242-DM4.[103]

10.5. *Dysoxylum binectariferum*

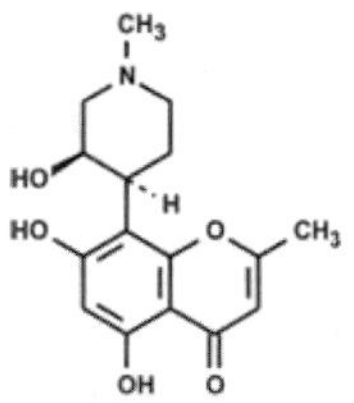

Flavopiridol

Flavopiridol, also known as Alvocidib, is a semi-synthetic flavone based on the alkaloid rohitukine. It is commonly extracted from *Dysoxylum binectariferum.*[104–106] Flavopiridol is used to treat chronic lymphocytic leukemia, and its mechanism may be related to the inhibition of CDKs (pan-cyclin-dependent kinases), main regulators of cell cycle progression. It may also release p53 by inhibiting Mdm-2, a molecule that targets p53 for degradation.[106–108] As a consequence, the cell cycle is arrested in the G1/S and G2/M phases.[110,111] Flavopiridol causes apoptosis in non-small cell lung cancer cells and inhibits tumor cell growth *in vivo.* Several phase II studies investigated the effectiveness of flavopiridol in patients with metastatic malignant melanoma; with previously untreated or relapsed mantle cell lymphoma; and with recurrent, locally advanced or metastatic soft tissue sarcoma.[103,112–114] In addition, combination therapy with flavopiridol and other standard chemotherapeutic agents has been assessed.[115–117]

10.6. *Papaver somniferum*

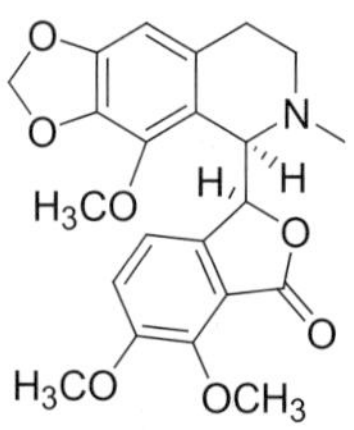

Noscapine

Noscapine is a phthalideisoquinoline alkaloid isolated from the capsules of opium poppy (*Papaver somniferum*). The compound is not a sedative and has been widely used as an oral antitussive medication.[118] Noscapine also demonstrated potent antitumor activity through a tubulin-binding mechanism with minimal toxicity towards normal cells.[106,119] When noscapine binds to tubulin, it prevents normal microtubule assembly reactions so that the cell is arrested at metaphase.[120] *In vivo* studies using human non-small cell lung cancer patients and gastric cancer xenograft models showed that noscapine suppresses tumor growth and up-regulates PARP, Bax and cytochrome C.[121,122] Furthermore, it activates caspase-3 and caspase-9 and represses Bcl-2 expression.[122] Noscapine was evaluated in a Phase I clinical trial in patients with advanced or relapsed multiple myeloma.[122]

10.7. *Combretum caffrum*

Combretum caffrum

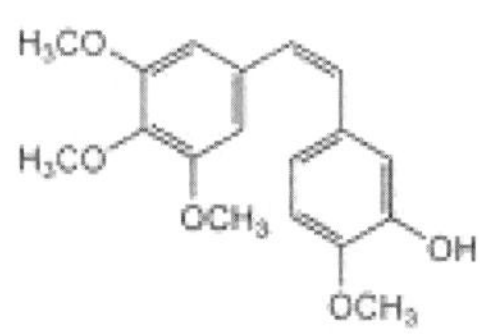

Combretastatin A-4

Combretastatins are well known stilbenoids from the South African tree *Combretum caffrum*, commonly known as "Bush Willow." They bind to tubulin at colchicin binding sites, which leads to inhibition of microtubule polymerization.[124–127] Combretastatin A-4 and its phosphate prodrug (CA-4-P) are cis-stilbenes, which inhibit leukemic P388 and L1210 cell lines.[128] The two compounds target vascular antigens and induce rapid and selective vascular dysfunction in tumors. CA-4-P was shown to reduce blood flow and subsequently induce tumor cell death in a variety of preclinical models. However, the isomerisation of cis-stilbene derivatives to the thermodynamically more stable trans-isomers led to a rather low *in vivo* efficacy.[129] A number of Phase I and II studies have been initiated to determine the effect of using combretastasin A-4 in combination with established anti-cancer drugs for the treatment of anaplastic thyroid cancer and advanced solid tumors.[130–132]

11. CONCLUSIONS AND PERSPECTIVES

As outlined in this chapter, many phytochemicals have the ability to inhibit tumors because of their natural anti-oxidant, anti-proliferative, or pro-apoptotic properties. Recently, increasing evidence has suggested that a tiny, biologically unique population of stem-like cells in most neoplasms may be responsible for tumor progression, metastasis and relapse. Characterization of these cancer stem cells has led to the identification of key cellular activities that may provide vulnerable targets for therapeutic intervention. The self-renewal of stem cells is an essential element of tumor survival and propagation.

Genes expressed in embryonic stem cells encoding proteins involved in the Hedgehog, Wnt/β-catenin, and Notch signaling pathways, are key factors in regulating self-renewal. These genes are expressed in normal stem cells, but they mutate frequently and are thus a potential therapeutic target for treatment with phytochemicals.[39]

In addition to targeting signaling pathways, many natural products interact with ATP-binding cassette (ABC) transport proteins to reverse the multidrug-resistant phenotype. For example, combination treatment of curcumin with established chemotherapeutic drugs such as vincristine or doxorubicin enhanced the cellular accumulation of vincristine and doxorubicin. Thus, plant-derived substances have the ability to sensitize drug-resistant cells.

Targeting stem cell-like cancer cells and modulating multidrug resistance are just two examples of how phytochemicals may help to improve treatment options, especially when used in combination with established anticancer drugs. The integration of phytochemical approaches into standard treatment procedures represents a prominent task in the years to come.

REFERENCES

1. Liu JJ, Lin M, Yu JY, *et al.* (2011) Targeting apoptotic and autophagic pathways for cancer therapeutics. *Cancer Lett* **300**: 105–114.
2. Hou J, Wang D, Zhang R, Wang H. (2008) Experimental therapy of hepatoma with artemisinin and its derivatives: *in vitro* and *in vivo* activity, chemosensitization, and mechanisms of action. *Clin Cancer Res* **14**: 5519–5530.
3. Ahmed F, Sadhu SK, Ishibashi M. (2010) Search for bioactive natural products from medicinal plants of Bangladesh. *J Nat Med* **64**: 393–401.
4. Griffin C, Hamm C, McNulty J, Pandey S. (2010) Pancratistatin induces apoptosis in clinical leukemia samples with minimal effect on non-cancerous peripheral blood mononuclear cells. *Cancer Cell Int* **10**: 6.
5. Gonzales GF, Valerio LG. Jr. (2006) Medicinal plants from Peru: a review of plants as potential agents against cancer. *Anticancer Agents Med Chem* **6**: 429–444.
6. Bacher N, Tiefenthaler M, Sturm S, *et al.* (2006) Oxindole alkaloids from Uncaria tomentosa induce apoptosis in proliferating, G0/G1-arrested and

bcl-2-expressing acute lymphoblastic leukaemia cells. *Br J Haematol* **132**: 615–622.

7. Rzeski W, Stepulak A, Szymański M, *et al.* (2009) Betulin elicits anti-cancer effects in tumour primary cultures and cell lines *in vitro. Basic Clin Pharmacol Toxicol* **105**: 425–432.
8. Desai AG, Qazi GN, Ganju RK, *et al.* (2008) Medicinal plants and cancer chemoprevention. *Curr Drug Metab* **9**: 581–591.
9. Leite DF, Kassuya CA, Mazzuco TL, *et al.* (2006) The cytotoxic effect and the multidrug resistance reversing action of lignans from Phyllanthus amarus. *Planta Med* **72**: 1353–1358.
10. Rajeshkumar NV, Kuttan R. (2000) Phyllanthus amarus extract administration increases the life span of rats with hepatocellular carcinoma. *J Ethnopharmacol* **73**: 215–219.
11. Tang R, Faussat AM, Majdak P, *et al.* (2006) Semisynthetic homoharringtonine induces apoptosis via inhibition of protein synthesis and triggers rapid myeloid cell leukemia-1 down-regulation in myeloid leukemia cells. *Mol Cancer Ther* **5**: 723–731.
12. Rajeshkumar NV, Joy KL, Kuttan G, *et al.* (2002) Antitumour and anticarcinogenic activity of Phyllanthus amarus extract. *J Ethnopharmacol* **81**: 17–22.
13. Huang ST, Yang RC, Lee PN, *et al.* (2006) Anti-tumor and anti-angiogenic effects of Phyllanthus urinaria in mice bearing Lewis lung carcinoma. *Int Immunopharmacol* **6**: 870–879.
14. Bhahwal AS, Ghulam NQ, Subhash CT. (2009) Boswellic acids: a group of medicinally important compounds. *Nat Prod Rep* **26**: 72–89.
15. Ammon HP. (2002) Boswellic acids (components of frankincense) as the active principle in treatment of chronic inflammatory diseases. *Wien Med Wochenschr* **152**: 373–378. [Article in German]
16. Ammon HP. (2006) Boswellic acids in chronic inflammatory diseases. *Planta Med* **72**: 1100–1116.
17. Shao Y, Ho CT, Chin CK, *et al.* (1998) Inhibitory activity of boswellic acids from Boswellia serrata against human leukemia HL-60 cells in culture. *Planta Med* **64**: 328–331.
18. Tsukada T, Nakashima K, Shirakawa S. (1986) Arachidonate 5-lipoxygenase inhibitors show potent antiproliferative effects on human leukemia cell lines. *Biochem Biophys Res Commun* **140**: 832–836.
19. Wang LG, Liu XM, Ji XJ. (1991) Determination of DNA topoisomerase II activity from L1210 cells-a target for screening antitumor agents. *Zhongguo Yao Li Xue Bao* **12**: 108–114.

20. Schneider I, Bucar F. (2005) Lipoxygenase inhibitors from natural plant sources. Part 1: medicinal plants with inhibitory activity on arachidonate 5-lipoxygenase and 5-lipoxygenase [sol]cyclooxygenase. *Phytother Res* **19**: 81–102.
21. Ammon HP, Mack T, Singh GB, Safayhi H. (1991) Inhibition of leukotriene B4 formation in rat peritoneal neutrophils by an ethanolic extract of the gum resin exudate of Boswellia serrata. *Planta Med* **57**: 203–207.
22. Shrivastava SK, Singh GK, Basniwal PK. (2003) Boswellic acid: an alternative anti-inflammatory therapy. *Indian J Nat Prod* **19**: 14–20.
23. Syrovets T, Büchele B, Gedig E, *et al.* (2000) Acetyl-boswellic acids are novel catalytic inhibitors of human topoisomerases I and II alpha. *Mol Pharmacol* **58**: 71–81.
24. Tak PP, Firestein GS. (2001) NF-κB: a key role in inflammatory diseases. *J Clin Invest* **107**: 7–11.
25. Makarov SS. (2000) NF-κB as a therapeutic target in chronic inflammation: recent advances. *Mol Med Today* **6**: 441–448.
26. Yamamoto Y, Gaynor RB. (2001) Therapeutic potential of inhibition of the NF-κB pathway in the treatment of inflammation and cancer. *J Clin Invest* **107**: 135–142.
27. Syrovets T, Gschwend JE, Büchele B, *et al.* (2005) Inhibition of IκB kinase activity by acetyl-boswellic acids promotes apoptosis in androgen-independent PC-3 prostate cancer cells *in vitro* and *in vivo. J Biol Chem* **280**: 6170–6180.
28. Hwang BY, Su BN, Chai H, *et al.* (2004) Silvestrol and episilvestrol, potential anticancer rocaglate derivates from Aglaia silvestris. *J Org Chem* **69**: 3350–3358.
29. Kinghorn AD, Chin YW, Swanson SM. (2009) Discovery of natural product anticancer agents from biodiverse organisms. *Curr Opin Drug Discov Devel* **12**: 189–196.
30. Bohnenstengel FI, Steube KG, Meyer C, *et al.* (1999) 1H-cyclopenta[b]-benzofuran lignans from Aglaia species inhibit cell proliferation and alter cell cycle distribution in human monocytic leukemia cell lines. *Z Naturforsch C* **54**: 1075–1083.
31. Lee SK, Cui B, Mehta RR, *et al.* (1998) Cytostatic mechanism and antitumor potential of novel 1H-cyclopenta[b]benzofuran lignans isolated from *Aglaia elliptica. Chem Biol Interact* **115**: 215–228.
32. Ohse T, Ohba S, Yamamoto T, *et al.* (1996) Cyclopentabenzofuran lignin protein synthesis inhibitors from Aglaia odorata. *J Nat Prod* **59**: 650–652.

33. Edwards RB, Lucas DM, West DA, *et al.* (2007) The plant-derived adent Silvestrol has B-cell selective activity in vitro in chronic lymphocytic leukemia patient cells and in vivo in the Tcl-1 mouse model of CLL. *49th Annual Meeting of the American Society of Hematology*, pp. 8–11.
34. Bohnenstengel FI, Steube KG, Meyer C, *et al.* (1999) Structure activity relationships of antiproliferative rocaglamide derivates from Aglaia species (Meliaceae). *Z Naturforsch* C **54**: 55–60.
35. Baumann B, Bohnenstengel F, Siegmund D, *et al.* (2002) Rocaglamide derivatives are potent inhibitors of NF-κB activation in T-cells. *J Biol Chem* **277**: 44791–44800.
36. Kim S, Hwang BY, Su BN, *et al.* (2007) Silvestrol, a potential anticancer rocaglate derivate from Aglaia foveolata, induces apoptosis in LNCaP cells through the mitochondrial/apoptosome pathway without activation of executioner caspase-3 or -7. *Anticancer Res* **27**: 2175–2183.
37. Rose WC, Shurig JE, Meeker JB. (1998) Correlation of *in vitro* cytotoxicity with preclinical *in vivo* antitumor activity. *Anticancer Res* **8**: 355–367.
38. Su BN, Chai H, Mi Q, *et al.* (2006) Activity-guided isolation of cytotoxic constituents from the bark of Aglaia crassinervia collected in Indonesia. *Bioorg Med Chem* **14**: 960–972.
39. Kawasaki BT, Hurt EM, Mistree T, Farrar WL. (2008) Targeting cancer stem cells with phytochemicals. *Mol Interv* **8**: 174–184.
40. Cecchinato V, Chiaramonte R, Nizzardo M, *et al.* (2007) Resveratrol-induced apoptosis in human T-cell acute lymphoblastic leukaemia MOLT-4 cells. *Biochem Pharmacol* **74**: 1568–1574.
41. Szarka CE, Grana G, Engstrom PF. (1994) Chemoprevention of cancer. *Curr Probl Cancer* **18**: 6–79.
42. Hoessel R, Leclerc S, Endicott JA, *et al.* (1999) Indirubin, the active constituent of a Chinese antileukaemia medicine, inhibits cyclin-dependent kinases. *Nat Cell Biol* **1**: 60–67.
43. Merz KH, Schwahn S, Hippe F, *et al.* (2004) Novel indirubin derivatives, promising anti-tumor agents inhibiting cyclin-dependent kinases. *Int J Clin Pharm Th* **42**: 656–658.
44. Zhang X, Song Y, Wu Y, *et al.* (2011) Indirubin inhibits tumor growth by anti-tumor angiogenesis through blocking VEGFR2 mediated JAK/STAT3 signaling in endothelial cell. *Int J Cancer* **129**: 2502–2511.
45. Tujebajeva RM, Graifer DM, Karpova GG, Ajtkhozhina NA. (1989) Alkaloid homoharringtonine inhibits polypeptide chain elongation on human ribosomes on the step of peptide bond formation. *FEBS Lett* **257**: 254–256.

46. Zhou JY, Chen DL, Shen ZS, Koeffler HP. (1990) Effect of homoharringtonine on proliferation and differentiation of human leukemic cells *in vitro. Cancer Res* **50**: 2031–2035.
47. Quintás-Cardama A, Kantarjian H, Garcia-Manero G, *et al.* (2007) Phase I/II study of subcutaneous homoharringtonine in patients with chronic myeloid leukemia who have failed prior therapy. Department of Leukemia, University of Texas MD Anderson Cancer Center, Houston, Texas 77030, USA. Erratum in: *Cancer* **109**: 2625.
48. Zhou DC, Zittoun R, Marie JP. (1995) Homoharringtonine: an effective new natural product in cancer chemotherapy. *Bull Cancer* **82**: 987–995.
49. Tseng TH, Lee YJ. (2006) Evaluation of natural and synthetic compounds from East Asiatic folk medicinal plants on the mediation of cancer. *Anticancer Agents Med Chem* **6**: 347–365.
50. Sardari S, Nishibe S, Horita K, *et al.* (1999) Cyclic AMP phosphodiesterase inhibition by coumarins and furanocoumarins. *Pharmazie* **54**: 554–556.
51. Rios JL, Mañez S, Paya M, Alcaraz MJ. (1992) Antioxidant activity of flavonoids from *Sideritis javalambrensis. Phytochemistry* **31**: 1947–1950.
52. Hofmanová J, Kozubík A, Dusek L, Pacherník J. (1998) Inhibitors of lipoxygenase metabolism exert synergistic effects with retinoic acid on differentiation of human leukemia HL-60 cells. *Eur J Pharmacol* **350**: 273–284.
53. Akamatsu H, Komura J, Asada Y, Niwa Y. (1991) Mechanism of anti-inflammatory action of glycyrrhizin: effect on neutrophil functions including reactive oxygen species generation. *Planta Med* **57**: 119–121.
54. Jachak SM. (2006) Cyclooxygenase inhibitory natural products: current status. *Curr Med Chem* **13**: 659–678.
55. Ristimäki A. (2004) Cyclooxygenase 2: from inflammation to carcinogenesis. *Novartis Found Symp* **256**: 215–221; discussion 221–226, 259–269.
56. Folkman J. (2006) Angiogenesis. *Annu Rev Med* **57**: 1–18.
57. Jang M, Cai L, Udeani GO, *et al.* (1997) Cancer chemopreventive activity of resveratrol, a natural product derived from grapes. *Science* **275**: 218–220.
58. Pezzuto JM. (2011) The phenomenon of resveratrol: redefining the virtues of promiscuity. *Ann N Y Acad Sci* **1215**: 123–130.
59. Lee D, Cuendet M, Vigo JS, *et al.* (2001) A novel cyclooxygenase-inhibitory stilbenolignan from the seeds of *Aiphanes aculeata.* Org Lett **3**: 2169–2171.
60. Corey EJ, Shih C, Cashman JR. (1983) Docosahexaenoic acid is a strong inhibitor of prostaglandin but not leukotriene biosynthesis. *Proc Natl Acad Sci USA* **80**: 3581–3584.

61. Dirsch VM, Vollmar AM. (2001) Ajoene, a natural product with non-steroidal anti-inflammatory drug (NSAID)-like properties? *Biochem Pharmacol* **61**: 587–593.
62. Oshima H, Oshima M. (2012) The inflammatory network in the gastrointestinal tumor microenvironment: lessons from mouse models. *J Gastroenterol* **47**: 97–106.
63. Prajapati MS, Patel JB, Modi K, Shah MB. (2010) *Leucas aspera*: a review. *Pharmacogn Rev* **4**: 85–87.
64. Sadhu SK, Okuyama E, Fujimoto H, Ishibashi M. (2003) Separation of Leucas aspera, a medicinal plant of Bangladesh, guided by prostaglandin inhibitory and antioxidant activities. *Chem Pharm Bull (Tokyo)* **51**: 595–598.
65. Sadhu SK, Okuyama E, Fujimoto H, Ishibashi M. (2006) Diterpenes from *Leucas aspera* inhibiting prostaglandin-induced contractions. *J Nat Prod* **69**: 988–994.
66. Aggarwal BB, Kumar A, Bharti AC. (2003) Anticancer potential of curcumin: preclinical and clinical studies. *Anticancer Res* **23**: 363–398.
67. Shao ZM, Shen ZZ, Liu CH, *et al.* (2002) Curcumin exerts multiple suppressive effects on human breast carcinoma cells. *Int J Cancer* **98**: 234–240.
68. Prof. Nadir Arber. (2006) Phase III Trial of Gemcitabine, Curcumin and Celebrex in Patients With Metastatic Colon Cancer. ClinicalTrials.gov Identifier: NCT00295035.
69. Prof. Nadir Arber. (2007) Phase III Trial of Gemcitabine, Curcumin and Celebrex in Patients With Advance or Inoperable Pancreatic Cancer. ClinicalTrials.gov Identifier: NCT00486460.
70. Ikezaki S, Nishikawa A, Furukawa F, *et al.* (2001) Chemopreventive effects of curcumin on glandular stomach carcinogenesis induced by N-methyl-N′-nitro-N-nitrosoguanidine and sodium chloride in rats. *Anticancer Res* **21**: 3407–3411.
71. Surh YJ, Chun KS, Cha HH, *et al.* (2001) Molecular mechanisms underlying chemopreventive activities of anti-inflammatory phytochemicals: down-regulation of COX-2 and iNOS through suppression of NF-κB activation. *Mutat Res* **480–481**: 243–68.
72. Jana NR, Dikshit P, Goswami A, Nukina N. (2004) Inhibition of proteasomal function by curcumin induces apoptosis through mitochondrial pathway. *J Biol Chem* **279**: 11680–11685.
73. Rashmi R, Kumar S, Karunagaran D. (2004) Ectopic expression of Bcl-XL or Ku70 protects human colon cancer cells (SW480) against curcumin-induced apoptosis while their down-regulation potentiates it. *Carcinogenesis* **25**: 1867–1877.

74. Woo JH, Kim YH, Choi YJ, *et al.* (2003) Molecular mechanisms of curcumin-induced cytotoxicity: induction of apoptosis through generation of reactive oxygen species, down-regulation of Bcl-XL and IAP, the release of cytochrome c and inhibition of Akt. *Carcinogenesis* **24**: 1199–1208.
75. Toi M, Matsumoto T, Bando H. (2001) Vascular endothelial growth factor: its prognostic, predictive, and therapeutic implications. *Lancet Oncol* **2**: 667–673.
76. Kerbel R, Folkman J. (2002) Clinical translation of angiogenesis inhibitors. *Nat Rev Cancer* **2**: 727–739.
77. Arbiser JL, Klauber N, Rohan R, *et al.* (1998) Curcumin is an *in vivo* inhibitor of angiogenesis. *Mol Med* **4**: 376–383.
78. Houck KA, Ferrara N, Winer J, *et al.* (1991) The vascular endothelial growth factor family: identification of a fourth molecular species and characterization of alternative splicing of RNA. *Mol Endocrinol* **5**: 1806–1814.
79. Houck KA, Leung DW, Rowland AM, *et al.* (1992) Dual regulation of vascular endothelial growth factor bioavailability by genetic and proteolytic mechanisms. *J Biol Chem* **267**: 26031–26037.
80. Niklińska W, Burzykowski T, Chyczewski L, Nikliński J. (2001) Expression of vascular endothelial growth factor (VEGF) in non-small cell lung cancer (NSCLC): association with p53 gene mutation and prognosis. *Lung Cancer* **34** Suppl **2**: S59–S64.
81. Shima DT, Deutsch U, D'Amore PA. (1995) Hypoxic induction of vascular endothelial growth factor (VEGF) in human epithelial cells is mediated by increases in mRNA stability. *FEBS Lett* **370**: 203–208.
82. Shweiki D, Itin A, Soffer D, Keshet E. (1992) Vascular endothelial growth factor induced by hypoxia may mediate hypoxia-initiated angiogenesis. *Nature* **359**: 843–845.
83. Chen HW, Huang HC. (1998) Effect of curcumin on cell cycle progression and apoptosis in vascular smooth muscle cells. *Br J Pharmacol* **124**: 1029–1040.
84. Rashmi R, Santhosh Kumar TR, Karunagaran D. (2003) Human colon cancer cells differ in their sensitivity to curcumin-induced apoptosis and heat shock protects them by inhibiting the release of apoptosis-inducing factor and caspases. *FEBS Lett* **538**: 19–24.
85. Rashmi R, Kumar S, Karunagaran D. (2004) Ectopic expression of Hsp70 confers resistance and silencing its expression sensitizes human colon cancer cells to curcumin-induced apoptosis. *Carcinogenesis* **25**: 179–187.
86. Duvoix A, Morceau F, Delhalle S, *et al.* (2003) Schmitz M, Schnekenburger M, Galteau MM, Dicato M, Diederich M. Induction of apoptosis by curcumin:

mediation by glutathione S-transferase P1-1 inhibition. *Biochem Pharmacol* **66**: 1475–1483.

87. Deeb D, Xu YX, Jiang H, *et al.* (2003) Curcumin (diferuloyl-methane) enhances tumor necrosis factor-related apoptosis-inducing ligand-induced apoptosis in LNCaP prostate cancer cells. *Mol Cancer Ther* **2**: 95–103.
88. Wang Y, Lu L, Li Z, *et al.* (2006) Anti-angiogenesis effects of meisoindigo on chronic myelogenous leukemia *in vitro*. *Leuk Res* **30**: 54–59.
89. Wang Y, Zhu XF, Xiao ZJ, *et al.* (2005) Inducement effect of meisoindigo on apoptosis of leukemia cell line HL-60 and its mechanism. *Ai Zheng* **24**: 1464–1468. [Article in Chinese]
90. Weng Y, Liu B, Peng Z, Xiao Z. (2005) Effects of meisoindigo on acute promyelocytic leukemia cell line NB4 cells *in vitro*. *Baixuebing Linbaliu* **14**: 136–139.
91. Lee CC, Lin CP, Lee YL, *et al.* (2010) Meisoindigo is a promising agent with in vitro and in vivo activity against human acute myeloid leukemia. *Leuk Lymphoma* **51**: 897–905.
92. Xiao Z, Hao Y, Liu B, Qian L. (2002) Indirubin and meisoindigo in the treatment of chronic myelogenous leukemia in China. *Leuk Lymphoma* **43**: 1763–1768.
93. Xiao Z, Qian L, Liu B, Hao Y. (2000) Meisoindigo for the treatment of chronic myelogenous leukemia. *Brit J Haematol* **111**: 711–712.
94. Cooperative Study of Phase III Clinical Trial on Meisoindigo. Tianjin; People's Republic of China. *Zhonghua Xueyexue Zazhi* **18**: 69–72.
95. Jin J, Jiang DZ, Mai WY, *et al.* (2006) Homoharringtonine in combination with cytarabine and aclarubicin resulted in high complete remission rate after the first induction therapy in patients with de novo acute myeloid leukemia. *Leukemia* **20**: 1361–1367.
96. Chemotherapy and Biological Therapy in Treating Patients with Chronic Phase Chronic Myelogenous Leukemia. UT MD Anderson Cancer Center (Kantarjian, Hagop, MD) (2010) History of Changes. ClinicalTrials.gov Identifier: NCT00003239.
97. Pharmacokinetic (PK) Study of Homoharringtonine (Omacetaxine Mepesuccinate) Administered Subcutaneously to Patients with Advanced Solid and Hematologic Tumors. *ChemGenex Pharmaceuticals* (2009) History of Changes. ClinicalTrials.gov Identifier: NCT00675350.
98. Kupchan SM, Komoda Y, Court WA, *et al.* (1972) Maytansine, a novel antileukemic ansa macrolide from Maytenus ovatus. *J Am Chem Soc* **94**: 1354–1356.

99. Remillard S, Rebhun LI, Howie GA, Kupehan SM. (1975) Antimitotic activity of the potent tumor inhibitor maytansine. *Science* **189**: 1002–1005.
100. Wolpert-Defilippes MK, Adamson RH, Cysyk RI, Johns DG. (1975) Initial studies on the cytotoxic action of maytansine, a novel ansa macrolide. *Biochem Pharmacol* **24**: 751–754.
101. Lopus M, Oroudjev E, Wilson L, *et al.* (2010) Maytansine and cellular metabolites of antibody-maytansinoid conjugates strongly suppress microtubule dynamics by binding to microtubules. *Mol Cancer Ther* **9**: 2689–2699.
102. Oroudjev E, Lopus M, Wilson L, *et al.* (2010) Maytansinoid-antibody conjugates induce mitotic arrest by suppressing microtubule dynamic instability. *Mol Cancer Ther* **9**: 2700–2713.
103. Sankhala KK, Mita AC, Ricart AD, *et al.* (2010) Maytansinoid DM4-conjugated humanized monoclonal antibody huC242 in treating patients with solid tumors. History of Changes. ClinicalTrials.gov Identifier: NCT00352131.
104. Harmon AD, Weiss U, Silverton JV. (1979) The structure of rohitukine, the main alkaloid of Amoora rohituka (syn. Aphanamixus polystachya) (Meliaceae). *Tetrahedron Lett* **20**: 721–724.
105. Cragg GM, Kingston DGI, Newman DJ. (2005) Anticancer agents from natural products; Boca Raton, FL, CRC Taylor&Francis. 553–571.
106. Lucas DM, Still PC, Pérez LB, *et al.* (2010) Potential of plant-derived natural products in the treatment of leukemia and lymphoma. *Curr Drug Targets* **11**: 812–822.
107. Byrd JC, Lin TS, Dalton JT, *et al.* (2007) Flavopiridol administered using a pharmacologically derived schedule is associated with marked clinical efficacy in refractory, genetically high-risk chronic lymphocytic leukemia. **109**: 399–404.
108. Wang LM, Ren DM. (2010) Flavopiridol, the first cyclin-dependent kinase inhibitor: recent advances in combination chemotherapy. *Mini Rev Med Chem* **10**: 1058–1070.
109. Demidenko ZN, Blagosklonny MV. Flavopiridol induces p53 via initial inhibition of Mdm-2 and p21 and, independently of p53, sensitizes apoptosis-reluctant cells to tumor necrosis factor. *Cancer Res* **64**: 3653.
110. Senderowicz AM. (1999) Flavopiridol: the first cyclin-dependent kinase inhibitor in human clinical trials. *Invest new Drugs* **17**: 313–320.
111. Raju U, Nakata E, Mason KA, *et al.* (2003) Flavopiridol, a cyclin-dependent kinase inhibitor, enhances radiosensitivity of ovarian carcinoma cells. *Cancer Res* **63**: 3263.

112. Burdette-Radoux S, Tozer RG, Lohmann R, *et al.* (2010) A Phase II study of Flavopiridol in treating patients with metastatic malignant melanoma. History of Changes. ClinicalTrials.gov Identifier: NCT00005971.
113. Kouroukis CT, Belch A, Crump M, *et al.* (2010) A Phase II study of Flavopiridol in treating patients with previously untreated or relapsed mantle cell Lymphoma. History of Changes. ClinicalTrials.gov Identifier: NCT00005074.
114. PhD Donald G Morris. (2010) A Phase II study of Flavopiridol in treating patients with recurrent, locally advanced, or metastatic soft tissue sarcoma. History of Changes. ClinicalTrials.gov Identifier: NCT00005974.
115. National Cancer Institute. (2010) Flavopiridol and Imatinib mesylate in treating patients with hematologic cancer. History of Changes. ClinicalTrials.gov Identifier: NCT00064285.
116. Carvajal RD, Tse A, Shah MA, *et al.* (2009) A Phase II study of Flavopiridol(Alvocidib) in combination with Docetaxel in refractory, metastatic pancreatic cancer. History of Changes; *Pancreatology* **9**: 404–409; Clinical Trials.gov Identifier: NCT00331683.
117. Rathkopf D, Dickson MA, Feldman DR, *et al.* (2009). Phase I study of flavopiridol with oxaliplatin and fluorouracil/leucovorin in advanced solid tumors. *Clin Cancer Res* **15**: 7405–7411; History of Changes. ClinicalTrials.gov Identifier: NCT00080990.
118. Robbers JE, Spcedie MK, Tyler VE. (1996) *Pharmacognosy and Pharmacobiotechnology*. Baltimore MD: Williams & Wilkins.
119. Ke Y, Ye K, Grossniklaus HE, *et al.* (2000) Noscapine inhibits tumor growth with little toxicity to normal tissues or inhibition of immune responses. *Cancer Immunol Immunother* **49**: 217–225.
120. Ye K, Ke Y, Keshava N, *et al.* (1998) Opium alkaloid noscapine is an antitumor agent that arrests metaphase and induces apoptosis in dividing cells. *The National Academy of Sciences.* **95**: 1601–1606.
121. Jackson T, Chougule MB, Ichite N, *et al.* (2008) Antitumor activity of noscapine in human non-small cell lung cancer xenograft model. *Cancer Chemother Pharmacol* **63**: 117–126.
122. Liu M, Luo XJ, Liao F, *et al.* (2011) Noscapine induces mitochondria-mediated apoptosis in gastric cancer cells *in vitro* and *in vivo. Cancer Chemother Pharmacol* **67**: 605–612.
123. PhD Haqq C. (2011) A Study of Noscapine HCL(CB 3304) in patients with released or refractory multiple myeloma. Cougar Biotechnology, Inc.; History of Changes. Clinical Trials.gov Identifier: NCT00912899.

124. McGown AT, Fox BW. (1989) Structural and biochemical-comparison of the antimitotic agents colchicine, combretastatin-A4 and amphethinile. *Anti-Cancer Drug Des* **3**: 249–254.
125. Lin CM, Singh SB, Chu PS, *et al.* (1988) Interactions of tubulin with potent natural and synthetic analogs of the antimitotic agent combretastatin: a structure-activity study. *Mol Pharmacol* **34**: 200–208.
126. Pettit GR, Singh SB, Hamel E, *et al.* (1989) Antineoplastic agents 145. Isolation and structure of the strong cell-growth and tubulin inhibitor combretastatin-A4. *Experientia* **45**: 209–211.
127. Petit I, Karajannis MA, Vincent L, *et al.* (2008) The microtubule-targeting agent CA4P regresses leukemic xenografts by disrupting interaction with vascular cells and mitochondrial-dependent cell death. *Blood* **111**: 1951–1961.
128. Pinney KG, Jelinek C, Edvardsen K, *et al.* (2005) *Anticancer Agents from Natural Products.* Taylor&Francis, pp. 23–46.
129. Woods JA, Hadfield JA, Pettit GR, *et al.* (1995). The interaction with tubulin of a series of stilbenes based on Combretastatin A-4. *Br J Cancer* **71**: 705–711.
130. Nathan P, *et al.* (2008) Safety study of increasing doses of Combretastatin in combination with Bevacizumab (Avastin) in patients with advanced solid tumors. OXiGENE. History of Changes. Clinical Trials.gov Identifier: NCT00395434.
131. Huntsman Cancer Institute. (2008) Safety and effectiveness of Combretastatin A-4 phosphate combined with chemotherapy in advanced solid tumors. OXiGENE. History of Changes. Clinical Trials.gov Identifier: NCT00113438.
132. Mooney CJ, Nagaiah G, Fu P, *et al.* (2010) Combretastatin A-4 phosphate in treating patiens with advances anaplastic thyrois cancer. History of Changes. ClinicalTrials.gov Identifier: NCT00060242.

CHAPTER 10

Established Anticancer Drugs from Natural Origin

Clara Becker, Kerstin Hoffmann, Laura Hoffmann, Tanya King, Franziska Faulstich, Katrin Viertel, Victor Kuete and Thomas Efferth

ABSTRACT

Plants have been used to treat diseases for thousands of years and possess an arsenal of chemotherapeutic agents. Microorganisms and especially some fungi species are also a source of natural compounds used to fight several ailments. Some of these products have already been exploited and are the basis of many anti-cancer drugs. There is continuous search all over the world for naturally occurring antiproliferative compounds. Their mechanisms of action remain a key factor to understand and explain their cytotoxic potential. In this chapter we will discuss the state of discovery of anti-cancer drugs of natural origin, while emphasizing their therapeutic properties. We will also highlight those substances that are currently in the late-stage of discovery and hit compounds acting as tubulin, topoisomerase, heat shock proteins, and cell cycle inhibitors.

1. INTRODUCTION

The main goal in cancer treatment is to inhibit or kill cancer cells without harming healthy tissue by acting on pathways that are unique for cancer cells. This specificity is achieved by targeting cancer cells on a molecular level. The advantage of this therapy is undoubtedly the fewer side effects and the improved acceptation of the drug by both medical practitioners

Institute of Pharmacy and Biochemistry, Johannes Gutenberg University, Staudinger Weg 5, 55128 Mainz, Germany.

and cancer patients. Plants have been used to treat diseases for thousands of years. Numerous natural products have been found to be active against cancer cells and till now, many agents of plant origin and their derivatives form a part of the standard repertory of anti-cancer chemotherapy. It is noticeable that many of these plants were used in traditional Chinese medicine long before the Western world recognized and paid attention to the enormous scientific potential.[1] In this chapter, we discuss tubulin targeting agents, important topoisomerase inhibitors, heat shock proteins (Hsps) and natural products affecting the cell cycle.

2. MICROTUBULE INHIBITORS

Many tubulin targeting agents are currently in clinical development, including microtubule stabilizing and destabilizing compounds.[2] Here we will focus on naturally occurring anticancer drugs. Taxanes, docetaxel and paclitaxel, are representatives of clinically successful chemotherapeutics that act through stabilization of GDP-bound tubulin. Epothilones are a novel class of microtubule-stabilizing drugs that share some of the binding characteristics of the taxanes. Another prominent group that blocks microtubule assembly is that of the vinca alkaloids.

Tubulin, as a potential antitumor agent, continues to attract attention both for drug discovery and development as disruption of the tubulin-microtubule dynamic equilibrium leads to cell cycle arrest at G2/M phase and apoptotic cell death.[2] To understand the mode of action of microtubule inhibitors it is necessary to comment on the mechanism of microtubule formation. The microtubule is a dynamic system that shortens or lengthens depending on the local concentration of GTP and microtubulin associated proteins (MAPs).[3] Tubulin is a globular, heterodimeric protein that consists of two subunits: α-tubulin and β-tubulin, which are tightly linked and polymerize to form the microtubules. On the β-tubulin subunit there are specific binding sites, hence the target for taxane, vinca and colchicine.[4] Vinca alkaloids and taxane bind to the β-tubulin subunit at the interdimer interface.[5] The dynamic equilibrium between polymerization and depolymerization plays an important role in different cellular processes such as cell division, locomotion and intracellular transport.[6] The key role of microtubules during mitosis makes them one of the best known targets in cancer therapy.[7] For this reason, agents that interfere with tubulin function have a broad antitumor spectrum and represent one of the most significant classes of agents in widespread use.[3] The development and approval of new drugs is necessary because the established drugs pose several important limitations. Especially exposure of cancer cells to a single antineoplastic agent can lead to resistance to multiple agents, also known as "multi-drug-resistance" (MDR). This mechanism includes the drug-efflux P-glycoprotein pump, which belongs to a family of ATP-binding cassette transporters and removes chemotherapeutic drugs by pumping them out of the cell before an effect occurs.[8,9] Changes in β-tubulin isotype expression, mutations, and therefore an increase in microtubule dynamics results in compromised binding of drug. Intense production of microtubule associated protein 4 (MAP4), which stabilizes tubulin, is an additional mechanism of drug resistance.[3]

2.1. Microtubule Stabilizers

Tubulin targeting agents are classified into microtubule stabilizing (taxanes and epothilones) and destabilizing compounds (vinca- and colchicine-site binders). As disruption of the dynamic equilibrium between tubulin polymerization and depolymerization leads to cell cycle arrest and apoptotic cell death, tubulin continues to attract attention for both drug discovery and development.

2.1.1. *Taxane site-binding agents*

Taxanes are widely used cytotoxic agents that exhibit good antitumor activities against a range of solid tumors, especially of the breast, lung, and ovary, and are routinely used in a variety of settings.[6–8] Paclitaxel (**1**) (Taxol®) and docetaxel (**2**) (Taxotere®) are diterpenes derived from plants of the genus *Taxus* (yews), e.g. the European yew tree *Taxus baccata* and the Pacific yew, *Taxus brevifolia*. By harvesting and extracting the needles, Baccatin III can be provided as precursor in large amounts. Furthermore, taxanes can be obtained from cell cultures under optimized conditions. The successful introduction of paclitaxel to the market in 1993 has spurred the isolation of many additional taxoids and so far, we have knowledge of more than 400 naturally occurring analogs.[6]

Taxanes block cell cycle progression as they prevent assembly of tubulin monomers by binding to microtubules.[6] Their binding effects mimic that of the GTP nucleotide.[7] The subsequent stabilization of microtubule dynamics interferes with the function of the microtubule leading to apoptosis.[3] The taxane ring system and the side chain at position C-13 have been considered as essential determinants for the activity of paclitaxel and docetaxel.[10,12] Because of the enhancement of tubulin polymerization they are also known as mitotic spindle poisons. The mechanism of action stands in contrast to other microtubule poisons such as vinca alkaloids and colchicines.[7] Still part of ongoing efforts,[12] the combination of taxanes with novel biological and cytotoxic drugs is conducted to further improve the structures' properties and effectiveness. The taxanes are widely used for treatment of various malignancies, but primary and acquired resistance to chemotherapy remains a significant clinical concern.[14] Also, problems concerning water solubility, bioavailability, and toxicity profile have not yet been adequately resolved.[3] The poor solubility necessitates formulation in cremophore (polyoxyethylated castor oil), an agent that can cause hypersensitive reactions.[14] Abraxane® is an albumin-bound formulation of paclitaxel with increased solubility and is approved for breast cancer.[14,15] Another paclitaxel derivate that does not require cremophore use is ANG1005. It allows treatment of brain cancer as it is transported across the blood-brain barrier.[14,16] The semisynthetic taxoids XRP9881 (Larotaxel®) and TPI287 (Tapestry) circumvent the common mechanism of taxane resistance and are currently undergoing clinical trials.[8]

2.1.2. *Epothilones*

Epothilones represent a novel group of cytotoxic agents of microbial origin that have been extensively studied in recent and ongoing clinical trials.[8] The group of macrolide antibiotics[8] contains a unique pharmacophore that binds at or near the taxane-binding site on microtubules.[14,17] The mode of action is similar to that of the taxanes as they cause microtubular stabilization and cell cycle arrest.[8,14] Structure-activity relationships suggest that the C3-C8 of the 16-membered macrolide ring is responsible for cytotoxic activity and microtubule polymerization potential.[17]

An attractive feature of epothilones is their poor substrate suitability for P-gp-mediated transport. Consequently, they are applicable for taxane-resistant tumors expressing the MDR phenotype.[8,14] They also show significant brain availability as MDR plays a crucial role in preventing compounds from crossing the blood-brain barrier.[2] Additionally, some analogs have reduced toxicity and increased water solubility.[17]

The natural epothilones A (**3**) and B (**4**) (Fig. 1) are produced by the myxobacterium *Sorangium cellulosum*, but their use was limited by pharmacokinetic difficulties and metabolic instability. To overcome these problems, several structure modifications have been undertaken[8,18] and presently more than 100 trials are in progress.[14]

Ixabepilone (Ixempra®) is a semisynthetic derivative of natural epothilone B and was approved for locally advanced and metastatic breast cancer in 2007.[19] It is more metabolically stable and induces apoptosis even in tumor cells that overexpress P-gp and/or contain tubulin mutations.[17] Thus, it has the potential to treat further metastatic malignancies.[17]

Other epothilones in late stage of clinical development include epothilone B (Patupilone®), ZK-EPO (Sagopilone®) and KOS-1584 (an

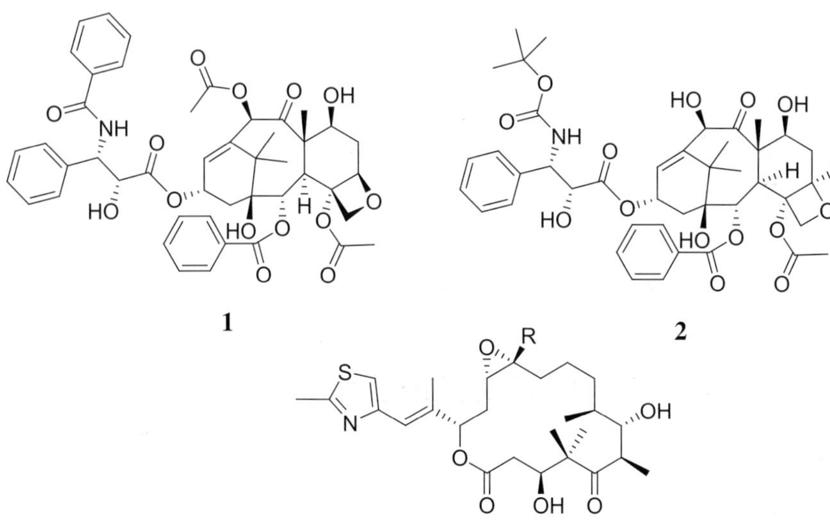

1 **2**

3 (R=H) **4** ($R=CH_3$)

Figure 1 Chemical structures of four microtubule inhibitors.

epothilone D analog).[14,20] These analogs show potent tumor inhibitory activity, tolerability *in vivo* (Ixabepilone, ZK-EPO, KOS-1584), brain and bone penetration (KOS-1584, ZK-EPO), and water solubility that permits oral administration (KOS-1584).[21] Unfortunately, peripheral neuropathy has been observed with epothilones as well, and thus remains a significant dose-limiting problem.[14,18]

2.2. Microtubule Destabilizers

2.2.1. *Vinca site-binding agents*

The *Vinca* alkaloids derive from *Catharanthus roseus* (synonyms: *Vinca rosea*, Madagaskar evergreen) and were discovered in a systematic search for the activity of the plant.[6] Vinblastine (**5**) (Velban®) and vincristine (**6**) (Oncovin®) (Fig. 2) were introduced into the clinic in the late 1950s.[14] Second generation semi-synthetic derivates are vinorelbine (Navelbine®) and vindesine (Eldisine®) that have been developed as commercial drugs and are already in clinical use. Table 1 shows indications and adverse effects for each of these compounds.

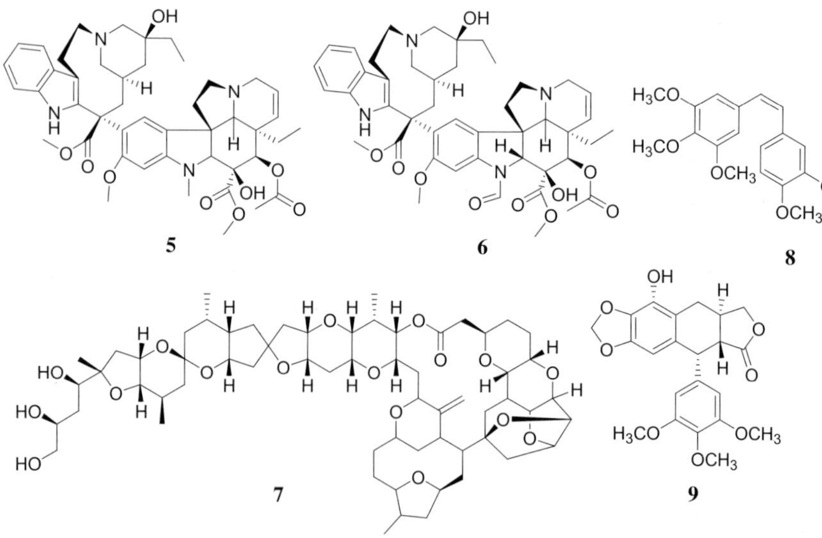

Figure 2 Chemical structures of selected microtubule destabilizers.

Table 1 Indications and Adverse Effects of the *Vinca* Alkaloids

	Vinblastine	Vincristine	Vinorelbine	Vindesine
Indication	Hodgkin's disease non-Hodkin's lymphoma other lymphomas testicular cancer breast cancer lung cancer head and neck cancer	Hodgkin's disease non-Hodkin's lymphoma acute leukemia (AML) Acute lymphoblastic leukemia (ALL) breast cancer lung cancer cervical cancer rapidly proliferating neoplasms nephroblastoma	breast cancer lung cancer	lymphomas acute leukemia (AML) Acute lymphoblastic leukemia (ALL) breast cancer lung cancer
Adverse effects	nausea and vomiting bone marrow depression loss of reflexes alopecia...	nausea and vomiting bone marrow depression neurotoxicity alopecia...	nausea and vomiting bone marrow depression granulocytopenia diarrhea...	nausea and vomiting bone marrow depression weakness diarrhea...

The success of first-generation *Vinca* alkaloids in the treatment of hematologic malignancies and different solid tumors[2,5] has prompted further research for novel analogs with improved clinical efficacy and safety. Such efforts have led to the development of vinflunine, the newest member of the *Vinca* alkaloid family, which is under active investigation for its unique antitumor activity.[5] Vinflunine has a relatively low affinity for tubulin, but the tubulin-binding affinity of *Vinca* alkaloids appears to be inversely correlated with their *in vivo* antitumor activity.[5,22] The reduced affinity for tubulin may underlie the low incidence of peripheral neuropathy observed.[5] The antitumor activity of Vinflunine is at least in part mediated via an antivascular mechanism.[22] *Vinca* alkaloids have been shown to induce vascular disruption that leads to tumor necrosis.[2] In addition, Vinflunine is a less potent inducer of drug resistance.[5] *Vinca* alkaloid side-binding prevents microtubule assembly as they bind to free tubulin subunits[6] and β-tubulin near the GTP-binding site.[14] This disrupts the dynamic equilibrium between tubulin and its polymeric form, the microtubule. Thus blocked mitotic spindle formation prevents cell division and therefore causes tumor inhibition.[6]

Unfavorable suppression of microtubule dynamics leads to many side effects of this class of compounds since cells typically depend on a functioning cytoskeleton. Neurotoxicity is the dose limiting fall-out associated with vincristine treatment, while neutropenia is often observed with other *Vinca* alkaloids.[14]

Other naturally occurring microtubule destabilizers including Halichondrin B (**7**) (Eribulin), a polyether isolated from marine sponges *Halichondria okadai*, and a hemiasterlin derivate E7974 are currently undergoing clinical trials. Dolastatins and Cryptophycin failed to advance through clinical trials due to lack of activity and severe toxicities.[2,14]

2.2.2. *Colchicine site-binding agents*

Colchicine was first extracted from *Colchicum autumnale* Linn (Liliaceae). It binds to tubulin at the interface of unpolymerized α/β-tubulin heterodimers,[14] thereby inhibiting tubulin assembly and blocking the formation of the mitotic spindle.[6] Colchicine has limited medicinal application because of its toxicity profile. Nevertheless, it can still be used as a lead compound for the generation of other potent anti-cancer agents.[6,23]

Combrestatins have been isolated from the bark of the African willow tree Combretum caffrum in the late 1980s.[24] They were shown to induce vascular disruption and inhibition of proliferation due to tubulin polymerization.[2] One of these agents is combrestatin A-4-phosphate (CA4P), which shows little effect on normal vasculature and is being evaluated for use in combination therapy with other chemotherapeutics, e.g. paclitaxel, against solid tumors.[14] Combretastatin A-4 phosphate is a prodrug. *In vivo*, it is dephosphorylated to its active metabolite, combretastatin A-4 (**8**). Compound **8** is a combretastatin, a type of stilbenoid (Fig. 2).

Many natural products bind to the colchicine site, notably podophyllotoxin (**9**) and flavonoids.[6] Other colchicine site-binding agents currently under investigation are ABT-751, which has shown activity against solid tumors in xenograft studies; NPI-2358 and SSR97225, which break down tumor vasculature; and a potent inhibitor of angiogenesis-2-Methoxyestradiol.[14] However, promising results are slightly distant as no objective responses have been observed so far and it is still unclear whether these 'new colchicines' demonstrate any advantage over approved drugs.[2]

3. TOPOISOMERASE INHIBITORS

Topoisomerases are important intranuclear enzymes responsible for the steric configuration of DNA molecules.[25] They have crucial functions in many aspects of DNA metabolism including DNA replication, RNA transcription, chromosome condensation and segregation at mitosis.[26–28] For these reasons, many tumor cells over-express topoisomerases to enhance cellular proliferation.[25,28]

Two classes can be distinguished: Type I topoisomerases (TOP1) induce transient single strand breaks to relax DNA sections that become supercoiled during transcription and replication.[28] Supercoiling describes a form of DNA in which the double helix is further twisted about itself, forming a tightly coiled structure.[29] Type II topoisomerases (TOP2) prevent the daughter molecule from being entangled as they cut both strands of the double helix to let another unbroken DNA strand pass through this gap, thus enabling chromatin rearrangements. Contrary to TOP1, the TOP2 reaction is ATP consuming and cell cycle dependent.[27–29] Inactivation of gene function is due to alterations at the chromatin after topoisomerase inhibition.[26] Any events that influence the topoisomerase-DNA complex will disturb cellular proliferation as it is the central intermediate in the catalytic cycle.[6] Topoisomerases are therefore among the most vital and pre-eminent anticancer agents.[6]

3.1. Topoisomerase I Inhibitors

During the catalytic relaxation of the supercoiled DNA mediated by topoisomerase I, a TOP1-DNA binary complex is formed. Inhibitors of DNA TOP1 target this central intermediate and induce reversible single strand breaks. When the trapped "cleavable complex" collides with the replication fork or DNA polymerase, double strand breaks are generated.[6,26,30,31]

Camptothecin (**10**)(CPT) is a quinoline alkaloid that was first isolated from the Chinese tree *Camptotheca acuminata* in 1958.[32] The mechanism of action involves TOP1, which is the only known target of camptothecin.[31] CPT targets TOP1 by trapping the catalytic intermediate of the topoisomerase I-DNA reaction, the cleavage complex,[33] by inhibiting the religation step.[27] Thereby, it generates single strand breaks, which are transformed into double strand breaks when a DNA polymerase collides with the trapped TOP1-DNA complex.[30] CPT exerts its cytotoxicity exclusively in the S-phase of the cell cycle.[28,34] There is an equilibrium between the active lactone form of CPTs and the inactive open ring carboxylate form.[35] Camptothecin shows remarkable antitumor activity against leukemia cells, liver carcinoma, head and neck tumors,[6] lung, ovary, breast, pancreas, and stomach cancers.[36] However, CPTs have several limitations as these drugs cause serious side effects, e.g. leucopenia.[31] Additionally, they are poorly water-soluble. Many derivatives have been synthesized to overcome these problems. Some of them are in various stages of preclinical and clinical development. Two representatives, topotecan (Hycamtin®) and irinotecan (Camptosar®), have successfully entered the market and are used for the treatment of ovarian and colorectal cancer[37] as well as brain tumors.[38] Irinotecan has predominantly GI and haematological toxicities, topotecan is principally haematologically toxic with little extramedullary toxicity.[39] Other camptothecin derivatives include lurtotecan, 9-aminocamptothecin (9-AC) and 9-nitrocamptothecin (9-NC), which are currently in different stages of clinical development.[37] In addition to camptothecins, a growing list of non-CPT TOP1 poisoning compounds has been identified,[27] notably indolocarbazoles, phenanthroline derivatives and indenoisoquinoline.[31]

3.2. Topoisomerase II Inhibitors

TOP2 has crucial functions, including DNA replication, transcription and chromosome segregation.[30] DNA TOP2 inhibitors trap an enzyme intermediate, the covalent complex, and induce DNA double strand breaks that lead to chromosomal aberrations.[26,30] This reaction is cell cycle independent.[26] Hence, TOP2 inhibitors can be highly cytotoxic even in the absence of active replication.[31]

Two broad classes of these drugs can be distinguished: TOP2 poisons and TOP2 catalytic inhibitors. TOP2 poisons stabilize the covalent DNA topoisomerase II complex ("cleavable complex").[40] They generate DNA strand breaks, protein-DNA-compounds, and thus block transcription and replication subsequently followed by apoptotic cell death. This class can be further subdivided into intercalating and non-intercalating poisons.[28] Anthracyclines like doxorubicin as well as other compounds like mitoxantrone, mAMSA and ellipticine belong to the chemically diverse group of intercalators. The ability of these drugs to trap TOP2 is only explainable by their ability to intercalate in DNA. They have many effects that are independent of their action against TOP2. Therefore it is doubtful that TOP2 is the most important target of the intercalators.[30] Non-intercalating TOP2 poisons including the epipodophyllotoxins etoposide and teniposide, and flouroquinolones do not act strongly with the DNA, but induce protein-drug interactions.[30]

Agents acting on any of the other steps in the catalytic cycle are called **catalytic inhibitors.**[40] These drugs kill cells through the elimination of the essential enzymatic activity of TOP2.[30] They target TOP2 within the cell and inhibit various genetic processes.[41] This heterogeneous group of compounds can interfere with the binding between DNA and TOP2 (aclarubicin, suramin), stabilize noncovalent DNA-TOP2 complexes (merbarone, ICRF-187, and structurally related bisdioxopiperazine derivatives), or inhibit ATP binding (novobiocin).[40] The novel drug novobiocin has several targets besides TOP2 and is therefore described elsewhere.

In the clinic it is common to use TOP2 poisons solely for their antitumor activities, whereas catalytic inhibitors are utilized for a variety of reasons, including their activity as antineoplastic agents (aclarubicin), cardioprotectors (ICRF-187), or modulators (novobiocin).[40] The two distinct classes of TOP2 inhibitors seem to act by separable mechanisms and it was found that catalytic inhibitors antagonize the toxicity of TOP2 poisons.[30,41]

Combination of TOP2 targeting drugs with other agents is a safe and effective applied strategy today.

An important problem in the clinic is that treatment with drugs that target TOP2 can lead to the formation of secondary malignancies. Also resistance was observed due to reduced expression of TOP2[30] and it is suggested that inhibition of one type of topoisomerase results in an increase of other topoisomerase activity to compensate the inhibitory effect.[28]

3.2.1. *Anthracyclines*

Anthracyclines are important fungal antibiotics[42] that were discovered in screening programs that tested antibiotics for their cytotoxic activity.[43] Cyclophosphamide, doxorubicin (**11**) (DOX), daunorubicin (**12**), epirubicin, idarubicin, mitoxantrone, and valrubicin are classified under this therapeutic category.[44–46] First generation anthracyclines, compounds **11** and **12**, were isolated from *Streptomyces peucetius* in the 1960s.[44] Their mechanism of action bases on stabilization of the reaction intermediate (cleavage complex) through inhibition of TOP2,[45] while their planar ring system is important for intercalation into dsDNA. The external (non-intercalating) moieties of the molecule seem to play an important role in the formation and stabilization of the ternary complex.[44] Taken together, these mechanisms cause structural changes,[45] which interfere with crucial cellular processes. Anthracyclines have a range of effects on cells that are independent of their action against TOP2 as **11** and other anthracyclines produce free radicals, cause membrane damage and induce protein-DNA crosslinks.[30] It is also proposed that p53 could play an important role as it activates p53-DNA binding resulting in apoptosis. Inhibition of TOP1 may play an ancillary role in anthracycline activity too.[44]

Anthracylines are active against a wide range of hematopoietic malignancies and different solid tumors,[42,47] notably nonlymphocytic leukemia, soft tissue sarcomas, breast and ovarian carcinoma, non-Hodgkin's lymphoma and small cell lung cancer.

Compounds **11** and **12** are the most commonly applied anthracyclines. The main use for doxorubicin is the treatment of solid tumors, especially breast cancer and aggressive lymphoma. Daunorubicin is utilized for acute lymphoblastic or myeloblastic leukemias.[44,47]

To improve activity and cardiac tolerability a number of analogs have been synthesized. Amongst them, epirubicin (EPI) and idarubicin (IDA) enjoy popularity as useful alternatives to **11** or **12**, respectively. Epirubicin is an **11** analog with equal activity but reduced cardiotoxicity.[44] Idarubicin, an **12** analog, is highly lipid soluble and therefore able to cross the blood-brain barrier.[45] The increased lipophilicity results in enhanced cellular uptake and improved stabilization of the drug-TOP2-DNA complex.[44] It can be administered orally and finds application in the treatment of acute leukemias and solid tumors.[48]

Menogaril and pirarubicin are other anthracycline derivates with less relevance at present.[49]

The use of anthracycline antibiotics is generally limited by cardiotoxicity,[30] congestive heart failure (CHF),[50] and development of resistance.

Pixantrone (BBR2778) is another novel anthracycline derivative for the treatment of non-Hodgkin lymphomas. It exhibits lower cardiac toxicity and better activity in comparison with alternative anthracyclines.[51] This is also true for Mitoxantrone.[45]

Combination therapy, e.g. taxanes with anthracyclines, is another strategy that is widely used in the clinic to reduce cardiotoxic effects,[44] but remains challenging due to an overlap of other toxicities.

3.2.2. *Bisdioxopiperazine*

TOP2 is the molecular target of the catalytic inhibitor bis-(2,6-dioxopiperazine).[41] ICRF-187 (dexrazoxane) and structurally related dioxopiperazine derivatives stabilize noncovalent DNA TOP2 complexes[40] and thus inhibit various genetic processes.[53] An effect on TOP1 could not be observed.[41] It was found that dioxopiperazine compounds, notably ICRF-159, ICRF-187 and MST-16, affect only cells in transition from G2 to M phase where they lock the enzyme in a closed clamp form, thus inhibiting ATPase activity[41] and blocking enzyme turnover.[30]

Dioxopiperazines have modest anti-tumour activity,[30] but they have shown potential to overcome cardiotoxicity of doxorubicin and other antitumor anthracycline antibiotics.[41,54] Thus, the main area of application for dioxopiperazines is cardioprotection. ICRF-187 is an (+)-enantiomer of racemic ICRF-159 and is already used to reduce cardiotoxicity, especially in Europe.[40] Otherwise, dioxopiperazines are supposed to influence the induction of secondary leukemia.[30]

3.2.3. *Podophyllotoxines*

Podophyllotoxin (**9**) was first extracted in the 1880s from the roots of plants of Podophyllum species like *Podophyllum peltatum, P. emodi, P. hexandrum* or others.[55] Compound **9** exerts interesting cytostatic properties against several types of neoplasms.[56] The substance belongs to the group of lignans, which are diphenolic and widely distributed compounds involved in plant defense mechanisms.[57] Due to problems concerning podophyllotoxin supply,[58] drug resistance, and water-solubility, diverse analogues were developed.[55] The three most important semi-synthetic derivatives of podophyllotoxin are etoposide, teniposide and etoposide phosphate (etopophose).[57] The primary molecular target of these anticancer drugs is TOP2.[59] The high level of DNA damage results from stabilization of the cleavage complex,[60] the ternary complex between the drug, protein and DNA.[30] Inhibition of microtubule assembly is a second mechanism of action of Podophyllotoxines, although etoposide and teniposide do not exert this one, but affect cell division in the late S or G2 phase through TOP2 inhibition.[6] In general, the podophyllotoxin analogues show efficiency in germ cell tumors, lung cancers, Hodgkin's disease, non-Hodgkin's lymphoma, and others.[55] Etoposide was introduced in 1971 and is currently used for the treatment of small cell lung cancer, testicular cancer, lymphomas and leukemia.[61] Teniposide is effective in neuroblastoma, all non-Hodgkin's lymphoma and brain tumors of children.[6] Drug resistance and several side effects such as myleosupression, neutropenia, and nausea limit the clinical use of these drugs.[55] There is also evidence that etoposide and other agents can lead to specific types of leukemia.[59]

3.2.4. *Ellipticine*

The natural plant alkaloid ellipticine (**13**) (Fig. 3) was first isolated in 1959 from *Ochrosia elliptica labil*, an Australian evergreen tree of the Apocynaceae family.[62] Besides ellipticine and Olivacine, 9-methoxyellipticine is the most important naturally occurring derivate as it possesses a relatively broad spectrum of activity.[6,62] Different ellipticine analogs have been synthesized to reduce toxicities and improve anticancer activities,

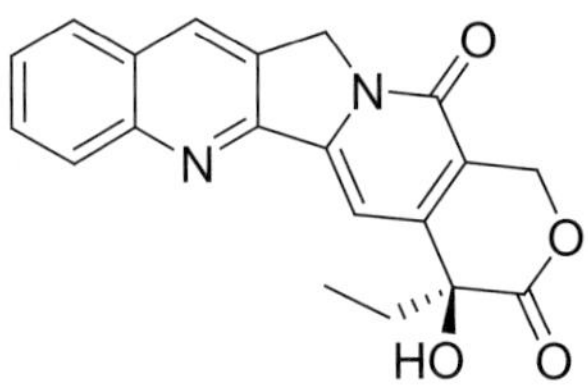

10

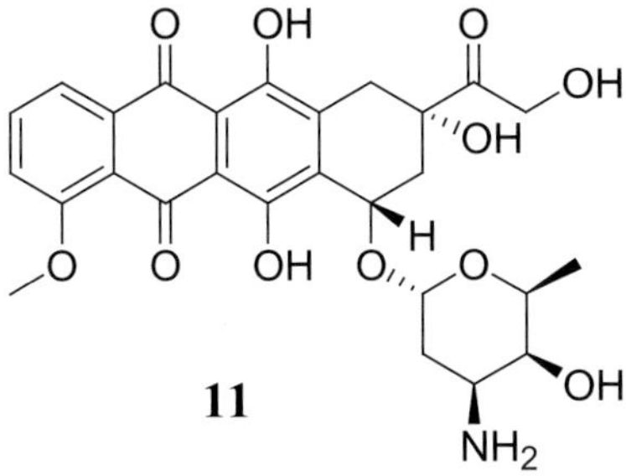

11

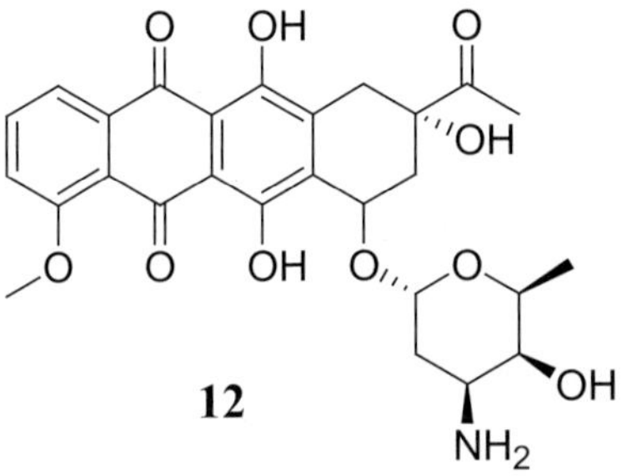

12

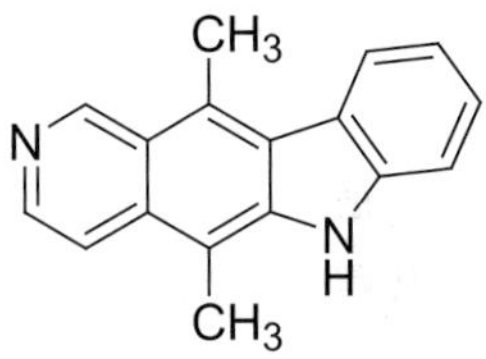

13

Figure 3 Chemical structures of four naturally occurring topoisomerase inhibitors.

administration, and biodistribution. 9-Hydroxyellipticine was less successful due to solubility problems. N-2-methyl-9-hydroxyellipticinium (NMHE) entered clinical trials as Celiptium in 1977. It reached phase I and II trials, but in spite of encouraging responses, there were several important toxicities that were displayed. There are also a number of hybrid molecules, like azatoxin, that were designed to achieve more effective targeting.[62]

DNA is thought to be a major cellular target of ellipticines. The drugs interact with DNA through intercalation and exhibit extremely tight binding. The structure is suitable for intercalation as it possesses a planar pyridocarbazole ring system with protonatable ring nitrogens.[6,62] The mode of action is also based on inhibition of TOP2 activity by preventing the enzyme from resealing transient DNA breaks.[62] Further mechanisms of action include interactions with membrane barriers and cytochrome P450, and the generation of free radicals by oxidative bioactivation.[62–64] These alterations lead to chromosomal abnormalities, inhibition of cell cycle at the G2 phase and subsequent cell death.[62]

Ellipticine improves the conditions of patients with advanced breast cancer refractory to all other treatment.[65] Activity against anaplastic thyroid carcinoma and ovarian carcinoma,[65] leukemias, melanoma and colon tumors was also reported.[62]

A major problem of ellipticine and its analogs is that they exhibit significant effects on the cardiovascular system.[62] Other observed side effects are kidney failure, nausea and vomiting, hypertension, muscular cramp, fatigue, mouth dryness, and mycosis of the tongue and esophagus.[65]

4. HEAT SHOCK PROTEIN 90 INHIBITORS

Heat shock proteins (Hsps) are molecular chaperones required for the stability and function of numerous proteins. They are integrally involved in cell signaling, proliferation, and survival and are therefore ubiquitously expressed.[66,67] Stress to the cell, including elevated temperature, non-physiolgical pH, nutrient deprivation and malignancy results in a significant overexpression of heat shock proteins to minimize the number of denatured proteins.[66] Also, many oncogenic proteins are dependent on this protein folding machinery as it is responsible for the conformational maturation of nascent polypeptides and refolding of denatured proteins into biologically active structures.[66,68] Especially tumor cells are able to successfully adapt to cellular stress, e.g. in case of drug application by increasing synthesis of several chaperones.[66,69]

Overexpression or accumulation of misfolded proteins is responsible for diseases such as cancer, Alzheimer's, Parkinson's, motor impairments, and multiple sclerosis.[68]

Heat shock proteins are classified by their molecular weights. There are several major families of chaperones with critical cellular roles including Hsp90, Hsp70, Hsp60 and the small Hsps.

One chaperone in particular has emerged as being of prime importance to the survival of cancer cells. This is the 90 kDa heat shock protein (Hsp90) which is expressed at 2- to 10-fold higher levels in tumor cells.[69] Hsp90 is a chaperone with over 100 identified client proteins including telomerase, mutated p53, Bcr-Abl, Raf-1, Akt, HER2/Neu (ErbB2), mutated B-Raf, mutated EGF receptor, and HIF-1alpha.[70] Therefore, Hsp90 has emerged as a promising target (not only) for cancer therapy as destabilization and eventual degradation of multiple Hsp90 client proteins[67] leads to tumor growth arrest. Hsp90 exists as a homodimer, separated in three highly evolutionary conserved domains. Hsp90 function depends upon the ability of the 25 kDa N-terminal domain to bind and hydrolyze ATP. This site also binds the natural products geldanamycin and radicicol. The 35 kDa middle domain is highly charged and has affinity for client proteins and co-chaperones. A second ATP-binding site is suggested in the C-terminal nucleotide binding pocket, a 12 kDa dimerization domain. Novobiocin, cisplatin, EGCG and taxol also bind to this site. The novobiocin-binding pocket seems to be adjacent to or overlapping with the radicicol-binding site, even though they are not identical.[71]

In the human proteome, four isomers of Hsp90 have been identified. Two isoforms, that are predominantly located in the cytosol, include the major, inducible Hsp90α and the minor, constitutive Hsp90β. Heat shock can induce alterations of the Hsp90 α/β ratio. Other isoforms include the 94 kDa glucose-regulated protein (GRP94), which is expressed in the endoplasmic reticulum, and the Hsp75/tumor necrosis factor receptor associated protein 1 (TRAP-1) in the mitochondrial matrix.[66,68]

The mechanisms of action of Hsp90 under normal conditions can be described as follows: ATP binds to the open chaperone conformation and Hsp90 clamps around the client protein substrate forming a closed molecular clamp.[66] Crystal structures clearly outline interactions between all three domains during this clamped conformation.[68] At this stage of the folding process, an inhibitor, instead of ATP, can bind competitively to the multiprotein complex, thus blocking the ability to fold or stabilize client proteins. As a consequence, unfolded proteins become part of an unproductive heteroprotein complex and subsequently degradation by the ubiquitin-proteasome pathway follows.[68] Inhibitors of Hsp90 are very effective as they simultaneously inhibit multiple signaling pathways by targeting only one specific molecule.[70] Additionally, Hsp90 inhibitors show high selectivity for malignant cells. Hsp90 is frequently overexpressed and predominantly activated in cancer cells, which leads to increased accumulation of inhibitors in these cells.[72–75] Taken together, this makes inhibitors of Hsp90 an interesting molecular target for anti-cancer drug development.[70]

4.1. Inhibitors of the N-Terminal ATP Binding Site

4.1.1. *Ansamycins*

Geldanamycin (**14**) (Fig. 4) was first isolated in 1970 from *Streptomyces hygroscopicus.*[68,76,77] The structure of this natural product includes a benzoquinone ring fused to a macrocyclic ansa-ring.[75] In 1997, the co-crystal structure of Hsp90 bound to **14** was reported[78,79] and it revealed that **14** binds specifically and with high affinity[80] to the N-terminal ATP binding site and thus reduces ATPase activity.[68] Compound **14** inhibits formation of an active src–Hsp90 heteroprotein complex.[68] Thereby, the chaperone's association with its client proteins is disrupted and subsequently leads to their destabilization and proteolysis.[71,73] Despite compound **14**'s promising anti-cancer activity, there are several problems in clinical application. It was especially poor solubility, stability, and significant hepatotoxicity[73,81,82] that

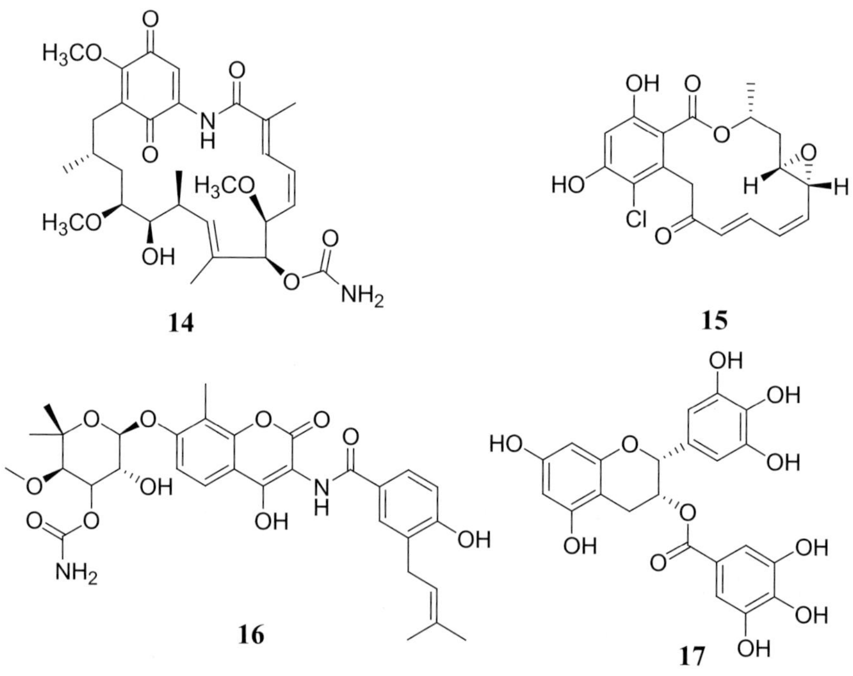

Figure 4 Chemical structures of geldanamycin (**14**), radicicol (**15**), novobiocin (**16**) and epigallocatechin-3-gallate (**17**).

led to the development of improved compound **14** derivatives. It was found that the C-17 methoxy group is responsible for these drawbacks. Replacement of the methoxy group with other substituents produced compounds with better tolerance and higher selectivity as well as improved *in vitro* activity over **14**.[68,73,80]

Allylamino-17-demethoxygeldanamycin (17-AAG, KOS-953)[85] is a more soluble analogue with similar anti-cancer effects as GA whereas it is less toxic.[80,82] 17-AAG was the first **14** derivate that entered phase II clinical trials.[67] It is useful against a variety of solid tumors and leukemia,[80] and several preclinical studies have shown that it may enhance the efficacy of conventional chemotherapy[82] and also radiation therapy.[84] Therefore, interest in combining 17-AAG with other anti-cancer therapies is increasing.

The 17-AAG prodrug, 17-allylamino-17-demethoxygeldanamycin hydroquinone hydrochloride (IPI-504) is a stable hydrochloride salt with reduced propensity to auto-oxidation. It exerts favorable pharmacological properties, especially very good water solubility and high affinity. For this reason, it was immediately taken for phase I/II clinical trials.[77,85,86]

Another **14** analogue that is currently in phase I clinical trials[80] is 17-(dimethylamino-ethylamino)-17-demethoxygeldanamycin (17-DMAG, KOS-1022).[87] It has been shown to be less toxic, more water soluble, and orally bioavailable than 17-AAG.[68]

4.1.2. *Macrolide: Radicicol and its analogues*

The natural product radicicol (**15**) is the most potent Hsp90 inhibitor[68,80] and was first isolated in 1953 from the fungus *Monosporium bonorden*.[80,88] It is a 14-member macrocyclic lactone antibiotic[73,89] that inhibits Hsp90 by tightly binding to the N-terminal ATP binding pocket.[68,71,73,78] Compound **15** mimics the ATP shape as also **14** does, but these two compounds are not similar in structure and they interact with Hsp90 differently.[77,80] However, the cellular effects of **15** are similar to those of ansamycins; **15** exerts its anticancer activity through inhibition of the Hsp90 protein folding machinery.[89] Compound **15** inhibits signal transduction pathways via depletion of Hsp90 client signaling molecules and induces ubiquitination and proteasomal degradation of oncogene products, such as K-ras and v-Src.[90] The molecular structure contains a resorcinol ring that provides key binding interactions with the protein.[89] Moreover, it was found that the stereochemistry of the oxime moiety is important for biological activity. The oxime derivative, KF58333, which contains the **15** oxime *E* isomer is more potent than its *Z* stereoisomer both *in vitro* and *in vivo* against breast cancer cells.[91]

Although **15** is less hepatotoxic[90] and more affine than **14** for the N-terminal ATP binding site, it lacks activity *in vivo* due to its rapid conversion into inactive metabolites.[68,80] It shows poor chemical and metabolic stability and a consequent decrease of anti-tumor potency *in vivo*.[77,92,93] Hence, considerable efforts have been undertaken to modify this agent. Important derivatives of **15**[90,94] include an oxime-radicicol series[92,93,95] and cycloproparadicicol.[96] The latter was synthesized by modifications to the **15** epoxide ring.[96] Also, the oxime-derivatives, e.g. KF25706, exert potent anti-tumor activity while showing no liver or renal toxicity in xenograft models.[92] Based on the observation that the L-shape conformation of **15** is bioactive, a number of analogues were analyzed. It was found that pochonin family members, e.g. Pochonin D[94] and A,[97] are nearly as potent as **15**[80] and bind Hsp90 with high affinity. A novel class of Hsp90 inhibitors is represented by chimeric molecules of GA and **15**. These are composed of **15** resorcinol ring and GA quinone connected by amide or ester linkages.[98] Radamide was the first chimeric compound synthesized.[89] Other chimeras with increased activity[89,99] include radamide[100] and radester.[98]

4.2. Inhibitors of the C-Terminal ATP Binding Pocket

4.2.1. *Novobiocin*

The coumarin antibiotics novobiocin (**16**), chlorobiocin and coumermycin were first isolated from several streptomyces strains and show activity against Gram-positive bacteria. They bind TOP2, including DNA gyrase, and thus inhibit enzyme-catalyzed hydrolysis of ATP.[66] As potent gyrase inhibitors[68] these agents were initially attractive for the treatment of infections with multi-resistant gram-positive bacteria.[66,101] Novobiocin as a clinically utilized antibiotic with tolerable toxicity[71] was the first and remains the most studied C-terminal Hsp90 inhibitor. It interacts with a previously unrecognized carboxy-terminal portion of Hsp90.[68,71,101] and thereby competes with ATP (and also other coumarines) for binding the Hsp90 nucleotide binding pocket.[71,101] Novobiocin inhibits Hsp90 and induces degradation of the Hsp90 client proteins via the ubiquitin-proteasome pathway.[66,101] This interference with the chaperone function finally leads to depletion of malignant cells.[71] Interestingly, binding of compound **16** to the C-terminus displaces inhibitors bound to the N-terminus.[71,102,103] It was proposed that this kind of substrate release could be induced by conformational changes.[104] Unfortunately, **16** binds only weakly to the Hsp90 C-terminal nucleotide binding site and has therefore weak clinical potential. For this reason, different structural modifications were synthesized that led to analogues with greater efficacy against various cancer cell lines.[66]

Although there is no reported co-crystal structure of Hsp90 bound to any inhibitor yet,[66] previous structure-activity relationship studies identified key moieties that appear important for Hsp90 inhibitory activity.[103,105] These studies highlighted a crucial role of the C-4 and/or C-7 positions of the coumarin ring for biological activity.[71,101,106] Interesting compounds for the development of more potent novobiocin analogues are 4TCNA and 7TCNA, which lack the noviose moiety and represent a new class of highly potent Hsp90 inhibitors.[101] Compound **16** represents an alternative to other Hsp90-targeting drugs that are not as well tolerated. The unique interaction with Hsp90 led to the identification of a previously uncharacterized C-terminal ATP binding site in the chaperone thus revealing an additional target on this protein for pharmacologic interference.[70,71] Therefore, **16** and the other coumarin antibiotics are going to be further investigated for their antitumor activity and potential in combination therapy.

4.2.2. *Epigallocatechin-3-gallate*

Epigallocatechin-3-gallate (**17**) (EGCG) is the most active and major compound of green tea.[68,107] This catechin is supposed to bind at or near the Hsp90 C-terminus[108] similar like the coumarin antibiotic novobiocin. It has proteasome-inhibitory effects and modifies the association of Hsp90 with several co-chaperones.[66,108,109] EGCG also inhibits tumour formation in oral-digestive tract carcinogenesis, prostate and breast cancers,[110] and all stages of lung cancer[111] as it suppresses oncogenic signaling pathways and induces cell cycle arrest or apoptosis.[112] It exerts anticancer effects by acting as a Hsp90 inhibitor,[108] and mitigates cellular damage arising from oxidative stress.[110] The antioxidant activity of **17** is due to the polyphenolic structure that interacts with reactive oxygen species and quenches free radicals.[111]

4.3. Histone Deacetylase Inhibitors (HDACs)

Epigenetic modulations of gene expression are important regulatory processes besides genetic alterations. The acetylation status of histones regulates access of transcription factors to DNA and thereby strongly influences regulatory processes within the cell. Histone acetylation is determined by the dynamic interaction of histone acetylases and histone deacetylases (HDACs).[113–116] These enzymes are involved in cell-cycle progression, differentiation, and contribute a major influence on the level of cellular protein modification.[123,126] In normal cells there is a balance between histone acetylation and deacetylation. An imbalance in acetylation levels of histones is caused by dysfunction and deregulation of HDAC enzymes and is associated with a variety of human diseases including cancer.[114,117–120] In tumors, a high expression level of HDAC isoenzymes and corresponding hypoacetylation of histones is common. This causes compaction of the DNA/histone complex. In turn, this blocks gene transcription, inhibits differentiation,[115,121] and thus alters pathways that promote proliferation, angiogenesis, and survival in support of cancer cells.[115]

HDACs catalyze the hydrolysis of *N*-acetyl lysine residues of the histone protein in chromatin[122] and also from acetylated sites in many other nuclear and cytoplasmic proteins,[117,118] including cytoskeletal proteins like α-tubulin, transcription factors like p53,[123] E2F,[124] c-Myc,[125] nuclear factor κB,[126] hypoxia-inducible factor 1α,[127] hormone receptors like estrogen receptor α26,[128] and chaperone proteins like the important heat shock protein 90 (Hsp90).[115,129] Hsp90 normally protects cellular and cancer-related proteins from degradation.[115] The activity of this critical protein chaperone is regulated by posttranscriptional modulation and Hsp90 inhibition can thereby be indirectly achieved through increased acetylation caused by HDAC inhibitors (HDACs).[80,130] This hyperacetylation of Hsp90 results in loss of the chaperone's function.[115]

As HDACs can effect a wide array of cellular proteins, this group of naturally occurring substances has stimulated interest in clinical trials and development of HDACs is at the center of attention. HDACs are a relatively new group of epigenetic anticancer agents that alter the acetylation status of chromatin and also non-histone proteins by shifting the balance between acetylation and deacetylation, resulting in hyperacetylation of core histones.[115,118,119] Structure-activity relation studies revealed that HDACs act by binding a critical Zn^{2+}-ion required for catalytic function of the HDAC enzyme.[114] HDAC inhibition induces cell death, apoptosis, differentiation, and cell cycle arrest of many tumor types *in vitro* and *in vivo*. However, the effects on normal tissue are minimal even though the increased susceptibility of transformed cells to HDACs is still poorly understood.[115,117,118] Presumably for this reason HDACs are generally well tolerated.

Currently, numerous new HDACs are being investigated in phase I, II and III clinical trials with variable efficacy.[118,131] Due to their different mechanism of action compared to conventional anti-neoplastic agents, HDACs are a treatment option for refractory cancers.[117] Several HDACs have entered the clinic against hematologic malignancies and others are effective against solid tumors as single agents.[115,118]

The first HDAC was hexamethylene bisacetamide.[115] A related natural product, trichostatin A, was isolated from a *Streptomyces hygroscopicus* strain.[132] The small molecule suberoylanilide hydroxamic acid (SAHA, vorinostat) is the only FDA-approved HDAC yet. It has activity against hematologic and solid tumors and is approved for the treatment of relapsed or refractory cutaneous T-cell lymphoma (CTCL).[133,155] The cyclic tetrapeptide depsipeptide and the benzamide MGCD0103 exert promising antitumour activities against lymphoma, leukemia and solid tumors.[115,134] Depsipeptide has shown modest clinical efficacy and the drug was relatively well tolerated.[115] MGCD0103 is an orally bioavailable HDAC with activity in hematologic malignancies.[115,135,136] It appears tolerable and exhibits favorable pharmacokinetic and pharmacodynamic profiles.[134] The hydroxamate panobinostat (LBH589) and the benzamide entinostat (SNDX-275) have both attractive preclinical and phase I safety and efficacy profiles. In ongoing trials these drugs show activity against transformed cells in culture.[115] Other HDACs include valproic acid, romidepsin (FK-228), and

LBH589. In general, HDACs show a favorable safety profile and have only minimal adverse effects, although cardiac toxicity needs to be further evaluated.[115] Some HDACs also have potential in the treatment of neurodegenerative disorders.[113]

Drug resistance to HDACs can be mediated by clusterin (CLU), a key contributor to chemoresistance to anticancer agents.[137] In late stage tumors, increased expression of the cytoplasmic/secretory clusterin form (sCLU) was found. Also, the acquisition of multidrug resistance[138] is a critical issue and a lot of effort is invested in overcoming these difficulties. A promising strategy to avoid resistance to HDACs is the application of combination therapies; e.g. simultaneously targeting HDACs and DNA methylation, or HDACs and Hsp90, demonstrates favorable clinical outcome.[124] Interestingly, HDACs also enhance tumor cell radiosensitivity.[116] Different studies and clinical data support the use of HDACs in combination with other anticancer agents including epigenetic or chemotherapeutic agents.[118,130] However, this requires a better understanding of the mode of action of each administered agent.[118]

Clinical trials have demonstrated activity of HDACs but further investigation is necessary to improve efficiency, reduce toxic side effects, and overcome drug resistance.

5. NATURAL PRODUCTS AFFECTING CELL CYCLE AND PROTEIN SYNTHESIS

5.1. Homoharringtonine

Homoharringtonine (4 methy-2-hydroxy-4 methylpentyl butanedioate) is a cephalotaxus alkaloid that can be obtained by alcoholic extraction from various *Cephalotaxus* species, such as *C. harrintonia, C. hainanensis*, and *C. qinensis.*[139,140] These evergreen trees are coniferous shrubs with yew-like leaves[139] native to southern China and Japan.[141] Cephalotaxine itself is inactive,[139,140] but there are four active cephalotaxine esters: harringtonine, isoharringtonine, homoharringtonine (**18**) (Fig. 5), and doxyharringtonine.[142] Compound **18** and its analogs are primarily inhibitors of protein synthesis in a dose- and time-dependent

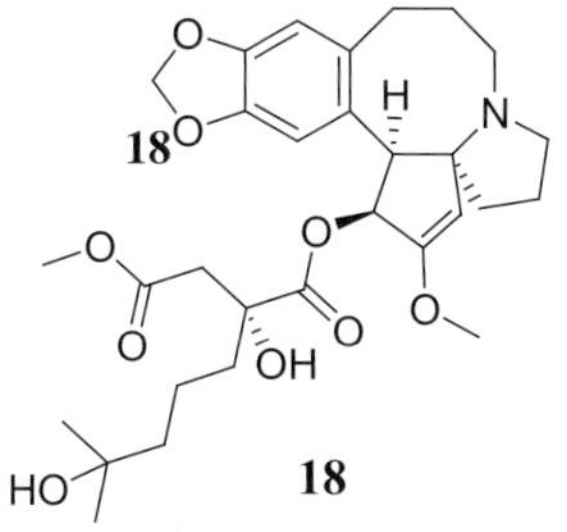

18

Figure 5 Chemical structures of homoharringtonine (**18**).

manner.[139–141] Additionally, they seem to have effects on DNA.[140] It is proposed that **18** inhibits the elongation cycle of translation by suppressing the substrate binding to the receptor site on the 60S ribosome subunit,[140,143,144] thus inducing differentiation and promoting apoptosis.[139,140] Cytotoxicity of this natural product is cell-cycle specific and preferentially affects cells in G1 and G2 phase.[139,141,145]

Controlled clinical studies have shown that **18** exerts significant antileukemic activity.[1,140] Initial trials were conducted in the 1970s in China and later in the US with encouraging results.[140,146,147] Compound **18** shows high response rates in acute myeloid leukemia (AML), acute promyelocytic leukemia (APL), chronic myeloid leukemia (CML) and myelodysplastic syndrome (MDS).[140,141,148,149] In acute lymphoblastic leukemia (ALL) and solid tumors it has very little effect.[140,141,150] Rapidly growing tumors tend to be more sensitive.[151] Moreover, the potential use of **18** for the treatment of non-malignant diseases, such as malaria has been discussed.[140,152]

As HHT exerts promising activity in hematologic malignancies, it drew the attention of scientists in the 1990s.[139] In the following years the development of **18** decelerated[139] due to the success of tyrosine kinase inhibitors (TKI), notably imatinib mesylate,[153] in the treatment of imatinib-resistant CML.[139,154–156] However, in recent years interest in **18**-based therapy has reinvigorated because of positive results in patients who failed second-generation TKIs and interferon-alpha (IFN-α) therapy.[139,157]

Further results from Phase II studies suggest that omacetaxine mepesuccinate, a semisynthetic form of **18**, is also active in patients with imatinib-resistant CML including in patients who harbor highly resistant BCR-ABL1 kinase domain mutations. Omacetaxine mepesuccinate is subcutaneously bioavailable and seems to be a valuable option in the management of leukemia. Since 2006, it is approved for the treatment of CML in all phases after failure of imatinib therapy.[139,158] Unfortunately, **18** belongs to the category of multidrug resistance (MDR)-related drugs. Cells resistant to **18** are often times cross-resistant to anthracyclines, vinca alkaloids, epipodophyllotoxin derivatives and taxol.[140,141,159,160] Due to increased expression of the MDR P170-glycoprotein,[159,161] the effect of **18** is poor especially in refractory and relapsed acute leukemia and in the blastic phase of CML.[159] The tumor suppressor p53 is another important factor of chemoresistance. It has been shown that cephalotoxin esters are more active in cell lines without p53 mutation.[162]

Like most neoplastic agents, **18** has several dose-related adverse effects. Primarily, it is toxic to hematopoietic, cardiac and gastrointestinal systems,[140,146] and induces hypotension as well as myelosuppression doses at high.[150] It has been found that intravenous infusion often provokes cardiovascular disturbances,[171] whereas subcutaneous **18** is well tolerated.[140,157] Future efforts aim at reducing these side effects and elucidate broader or different antitumor profiles. This may be done by further manipulations in the side chain of **18** and production of more second-generation analogues like cytarabine and gemcitabine.[140]

5.2. Mistletoe

Mistletoe (*Viscum album* L.) is a parasitic plant that attaches itself to and grows within the branches of different trees and shrubs. Extracts and preparations of mistletoe are widely used in cancer treatment as monotherapy or concomitantly with chemo- or radiotherapy.[163–165] Different biologically active substances of the plant were identified including the lectins I, II, and III (ML-I to -III); viscotoxins; polysaccharides; alkaloids and Vester proteins.[163] Especially the lectins viscum album agglutinin (VAA)-I and -II are expected to play an important role in the efficacy of mistletoe therapy as they belong to the group of ribosome-inactivating proteins.[175] Another constituent, the galactoside-specific lectin, enhances the secretion of proinflammatory cytokines such as IL-6.[167] Several preclinical and *in vitro* studies have shown a cytotoxic, proapoptotic, and immunostimulating effect of mistletoe extracts. However, translation of these findings into the clinic remains problematic.[163,166,168]

So far, results from clinical trials were mixed and no direct anticancer action could be clearly demonstrated.[169] Nevertheless, there is some evidence that mistletoe application may improve the quality of life and can reduce side effects of chemotherapy and radiation,[165,168] whereas increased survival is still a subject under discussion.[164]

To further investigate clinical efficacy and also toxicity and pharmacokinetic profiles of mistletoe extracts, well-planned clinical trials should be encouraged.[120]

6. OTHER TARGETS

6.1. Inhibitors for the p53-MDM2 Interaction

Protein-protein interactions (PPIs) have essential functions in many biological processes such as cell growth and differentiation, intracellular signaling, and apoptosis.[170] The interfaces have become a novel target for medical intervention, also in the treatment of cancer. PPI targets are more complex than enzyme targets. However, many small molecule disruptors have been identified. These so-called "hot spots" are small regions at the protein-protein interface with over 80 percent responsible binding energy.[170] In half of all human cancer types the genes of the DNA-binding domain of p53 are mutated or deleted.[171] The transcription factor p53 ("guardian of the genome") normally plays an important role in cell cycle regulation, DNA repair, and apoptosis. As a result, the development of cancer is accelerated in the case of a mutation. In other human cancer types the tumor suppressor function of the intact p53 is blocked by the MDM2 (murine double minute 2) oncoprotein. The corresponding human isoform of this oncogen HDM2 (human double minute 2) is over-expressed in a few human cancer types.[170] Due to the fact that the interaction between these proteins is understood, the development of inhibitors of this PPI is the approach that is followed for the treatment of cancer. The two proteins mutually maintain their low cellular concentrations through an auto-regulatory feedback loop.[170] When p53 is activated and stabilized the transcription of the *mdm2* gene is triggered. Increased MDM2 inhibits p53 protein activity in three ways: (1) by physically binding to the N-terminal domain of p53 and blocking its transcriptional activity, (2) by inducing nuclear export, and (3) by provoking the degradation of the E3 ubiquitin ligase/proteasome pathway. The residues 19-102 of the N-terminus of MDM2 were determined as the p53 binding domain. The crystal structure of the PPI indicates that a deep, hydrophobic cavity on the MDM2 surface is responsible for the binding. The 15-amino acid peptide sequence at the N-terminal transactivation domain of p53 was identified as the MDM2 binding domain. More precisely, only three of these amino residues (Phe, Trp and Leu) link to the hydrophobic cavity of MDM2.[170]

The validation of MDM2 as a drug target occurred with the help of antisense oligonucleotides and monoclonal antibodies. It showed that disturbance of MDM2 and following severance of the PPI results in activation of p53.[171] Many small molecules have been determined and two natural products have also been identified to disrupt the p53-MDM2 PPI. The first one is the secondary fungal metabolite *Chlorofusin*, isolated from the *Microdochium caespitosum* species, which has a modified cyclic peptide structure and an IC_{50} of 4.6 μM. The second one, *(−)-Hexylitaconic acid*, is a non-peptide natural product inhibitor of MDM2. It was isolated from *Arthrinium* sp., also a fungus, but derived from a marine sponge collection in Toyama Bay in the Japan Sea. It has a modest activity with an IC_{50} of 233 μM, but shows dose-dependent inhibition of PPI.[173]

6.2. Inhibitors of Hypoxia-Inducible Factor-1 (HIF-1)

Hypoxia (insufficient oxygen supply) appears in consequence of interruption of blood flow, reduction in oxygen tension, decreased oxygen carrying capacity, or failure to transport oxygen from the microvasculature to cells where oxygen is needed by mitochondria.[174] Cells of solid tumors are often hypoxic in contrast to healthy cells, because of rapid tumor growth.[174] Even though hypoxia triggers tumor angiogenesis, the newly shaped blood vessels often do not come to maturity. With the result that tumor regions often stay under hypoxic stress, the extent correlates positively with advanced stages and poor prognosis. Hypoxia brings along higher resistance to radiation treatment and chemotherapy. It is also an important contributor to disease relapse. To take advantage of hypoxia, hypoxia-selective therapies have been developed. Equally, sensitizers to increase the sensitivity of hypoxic cells to radiation have been developed.[174,175]

But hypoxia also has indirect effects on tumor cells by inducing the expression of genes that promote hypoxic adaptation and survival. The transcription factor hypoxia-inducible factor-1 (HIF-1) regulates more than 70 of these genes, and the list is growing rapidly. HIF-1 is activated in many biological processes (including tumorigenesis, vascular remodeling, inflammation, and hypoxia/ischemia-related tissue damage) and also by numerous factors such as cytokines, growth factors, hormones, activated oncogenes, or inactivated tumor suppressors.[174]

HIF-1 is a heterodimer of the basic helix-loop-helix (bHLH) PER-ARNT-SIM (PAS) proteins HIF-1α and HIF-1β/aryl hydrocarbon receptor nuclear translocator (ARNT). The HIF-1α subunit is stabilized under hypoxic conditions, and degraded rapidly under normoxic conditions, whereas the HIF-1β subunit is permanently existent. The activity of HIF-1α is the decisive factor for the bioactivity of HIF-1.[174,175] Over-expression of HIF-1α is crucial for advanced disease stages and poor prognosis among cancer patients. Preclinical studies in several animal models has shown that HIF-1 inhibition suppresses angiogenesis, retards tumor growth, and enhances treatment outcomes when used in combination with radiation or chemotherapy. Studies have proved that the outcome of HIF-1 inhibition is connected to the tumor microenvironment, so even HIF-1α deficiency could enhance tumor growth if located in high vascularized brain parenchyma.[174,175] In contrast, HIF-1α deficiency in low vascularized subcutaneous regions retards tumor growth. Numerous HIF-1 inhibitors have been identified from synthetic and from natural origin. The natural product-derived compounds are described subsequently according to their mode of action.[174]

6.2.1. *Inhibitors of HIF-1α protein synthesis*

The HIF-1α inhibitors may function through inhibition of transcription, degradation of HIF-1α mRNA, and decrease of translation.[175] Picroliv, a purified iridoid glycoside fraction from the roots of *Picrorhiza kurrooa* from the Indian traditional medicine showed a reduction of HIF-1α and vascular endothelial growth factor (VEGF) mRNA levels *in vitro*.[176] Leading compounds are the iridoid glycosides *picroside-I* and *kutkoside*.[177] The precise substances for inhibition and also the exact mode of action have not been determined yet.[175] Some soy-derived sphingolipids, most notably of the 4,8-sphingadiene glycosylceramide-type, led to reduced HIF-1α mRNA levels in the intestinal mucosa cells by more than 50 percent.[178] *Cycloheximide*, isolated from *Strepromyces griseus*, blocked hypoxia-induced HIF-1α protein accumulation.[175] *Cycloheximide* as well the *Streptomyces parvullus*-derived compound *actinomycin D* do not selectively inhibit gene expression. Consequently, the treatment-associated toxicity of these substances cast a shadow on their potency to inhibit cancer growth. Microtubule disrupting agents (MDA) can also have an effect on HIF-1α expression, either inhibiting or promoting effects. The natural estradiol metabolite *2-methoxyestradiol* suppresses HIF-1α synthesis in human prostate PC-3 and breast MDA-MB-231 carcinoma cells.[175] *Taxol (Taxus brevifolia)* and *vincristine (Catharanthus roseus)* also inhibit hypoxia-induced HIF-1α protein accumulation, whereas *vinblastine* and *colchicine* activate HIF-1 in several cell lines.[175] The PI3K inhibitor *wortmannin* (isolated from *Penicillium fumiculosum*) inhibits both hypoxia and non-hypoxic HIF-1 activation by inactivating targets that regulate the translation of HIF-1α mRNA.[175]

6.2.2. *Natural products that promote HIF-1α protein degradation*

Post-translational modifications of HIF-1α under normoxic conditions lead to degradation.[179] Some HIF-1 inhibitors use stimuli to contribute to degradation. As studies demonstrate, the heat shock protein 90 (hsp90) binds to the HIF-1α PAS domain at the same binding site as the HIF-1β/ARNT subunit and stabilizes the HIF-1α protein.[175] So, the use of pharmacological agents that inhibit hsp90 is effective in provoking HIF-1α degradation.

The hsp90 inhibitor geldanamycin (**14**) promotes pVHL- independent (Hippel-Lindau protein-independent) proteasomal degradation of HIF-1α protein under normoxic and hypoxic conditions.[175] Due to its hepatotoxicity, a derivative *17-N-allylamino-17-demethoxygeldanamycin* (*17-AAG*) is in clinical phase I trial for the treatment of cancer. In preclinical models

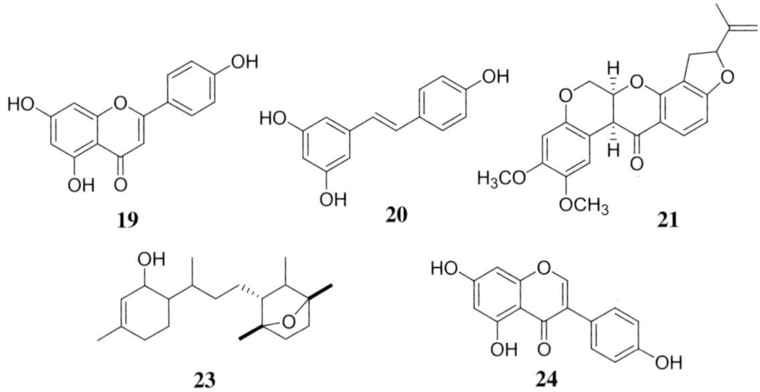

Figure 6 Chemical structures of some natural inhibitors of hypoxia-inducible factor-1.

it had delayed tumor growth and intensified the impact of radiation treatment on cancer cells.[175]

Flavone apigenin or 4′,5,7-trihydroxyflavone (**19**) (Fig. 6) can be found in many fruits and vegetables and shows anti-tumor activity in several *in vitro* and *in vivo* models. It is also expected to have chemopreventive properties.[175] The HIF-1 inhibitory effect is achieved by various different pathways including the PI3K/AKT/p70S6K1, hsp90, HDM2/p53 pathways, as well as the cellular targets nuclear factor-κ B (NFκB) and human epidermal growth factor receptor 2 (HER2/neu), like numerous studies have demonstrated.[175]

Resveratrol or trans-3,4,5′-trihydroxystilben (**20**) is a phytoalexin produced in consequence of environmental stress like fungal infection and injury.[175] It is found in many types of plants including grapes and peanuts. *Resveratrol* inhibits numerous cellular targets that are linked with tumor initiation, promotion, and progression. *Resveratrol* suppresses HIF-1α protein synthesis and promotes proteasome-mediated HIF-1α protein degradation. Mostly tested in concentrations from 10 to 100 μM, this is far more than consumable through food.[175]

The natural products rotenone (**21**) isolated from *Lonchocarpus spp.* and myxothiazol (**22**) derived from the myxobacteria *Myxococcus fulvus/Stigmatella aurantiaca* showed inhibition of hypoxia-induced HIF-1α protein accumulation, HIF-1 activity, and expression of HIF-1 target genes.[175] Laurenditerpenol (**23**), a secondary metabolite of the red alga *Laurencia intricata*, restricts oxygen consumption of mitochondria at the concentrations that inhibit hypoxia-induced HIF-1 activation.[175] Compound **23** is therefore the first member of a new class of marine natural product-based mitochondrial inhibitors.[175]

6.2.3. *Compounds that prevent HIF-1α protein accumulation*

The fact that redox status regulates HIF-1 activity led to the screening of redox factors. It was found that redox proteins such as redox factor 1 (REF-1) and thioredoxin enhance the activation of HIF-1 by exalting the interaction between HIF-1α and transcriptional coactivators CBP/p300 and SRC-1.[174] In MCF-7 human breast tumor cells the over-expression of thioredoxin-1 promoted the accumulation of HIF-1α protein, HIF-1 activation, and expression of HIF-1 target genes without changing the level of HIF-1α mRNA.[174] The natural compound *pleurotin*, an inhibitor of thioredoxin-1 reductase, derived from the mushroom *Pleurotus griseus*, is able to inhibit HIF-1α protein accumulation under normoxic and hypoxic conditions. The effect was certified with several other cancer cells; nevertheless the mode of action remains unclear.

The broad-spectrum tyrosine kinase inhibitor genistein (**24**), a plant isoflavone, blocks the induction of HIF-1α protein.[175] Other plant derived flavonoids and homoisoflavonoids like isorhamnetin, leuteolin, isoquercetin, quercetin and methylophiopogonanone B are also weak inhibitors of hypoxia-induced HIF-1 activation, which was shown in CHO (A4-4 clone) cells.

The dineolignans manassantin B and 4-*O*-demethylmanassantin B extracted from the aquatic plant *Saururus cernuus* L. *(Saururaceae)* are the most potent small molecule HIF-1 inhibitors discovered.[175] Another study affirmed that manassantin B blocks selectively the induction of HIF-1α protein. Furthermore, both compounds are involved in blocking the induction of VEGF.[175]

6.2.4. *Inhibitors of the interaction between HIF-1α and coactivators*

These inhibitors impede HIF-1 activation without lowering the HIF-1α protein concentration. The fungal natural product chetomin isolated from

Chaetomium cochliodes, links to the CH1 domain of coactivator p300 and breaks the tertiary structure.[175] Consequently, the interaction between HIF-1α CTAD domain and p300 is no longer possible. Both, hypoxia- and iron chelator-induced HIF-1 activation was repressed *in vitro* as well as *in vivo*.[175] Only the high toxicity, which was shown in animal models could reduce its therapeutic potential.

6.3. Kinases

Kinases are members belonging to the family of transferase-enzymes. They transfer phosphate groups from a donor molecule, such as ATP, to a specific substrate. In this way, other enzymes can become activated or inactivated. There are some kinases, which are overexpressed in tumor cells. Newer strategies for cancer treatment include development of small-molecules inhibitors of specific kinases.[180] The most common approach for the discovery and development of candidates for active agents against protein kinases has been to target the ATP-binding sites.[180] In this case, the problem is the close resemblance between the ATP-binding pockets[181] with the consequence of difficulties in the development of selective inhibitors. Most of the drug candidates are binding on multiple targets. Approach to a solution could be the differences in flexibility and volume of similar ATP binding sites.

Polo-like kinases (PLKs) play an important role in mitotic entry and regulation of mitosis.[180] They are members of the serine/threonine kinase family. Four subtypes of PLKs are established: Human PLK1, PLK2 (aka SNK), PLK3 (also known as PRK and FNK) and PKL4 (aka SAK). All of them are structurally homologous.

One can find PLK1, PLK2 and PLK3 in all tissues. They have many different tasks, whereas PKL4 controls unique physiological roles and just appears in certain tissues, e.g. testes and thymus.

Scytonemin (**25**) (Fig. 7) is a biological pigment synthesized by many strains of cyanobacteria, including *Calothrix* sp., *Lyngbya aestuarii*, and others.[182,183] The compound acts as PLK inhibitor, but does not contain a high selectivity.[180] Besides Polo-like kinases, other Serin/Threonin and Thyrosin/Threonin cell-cycle kinases are inhibited as well. For a long time, **25** was the only published small-molecule PKL inhibitor. Other kinase inhibitors of natural source include plant flavonoids morin (**26**)

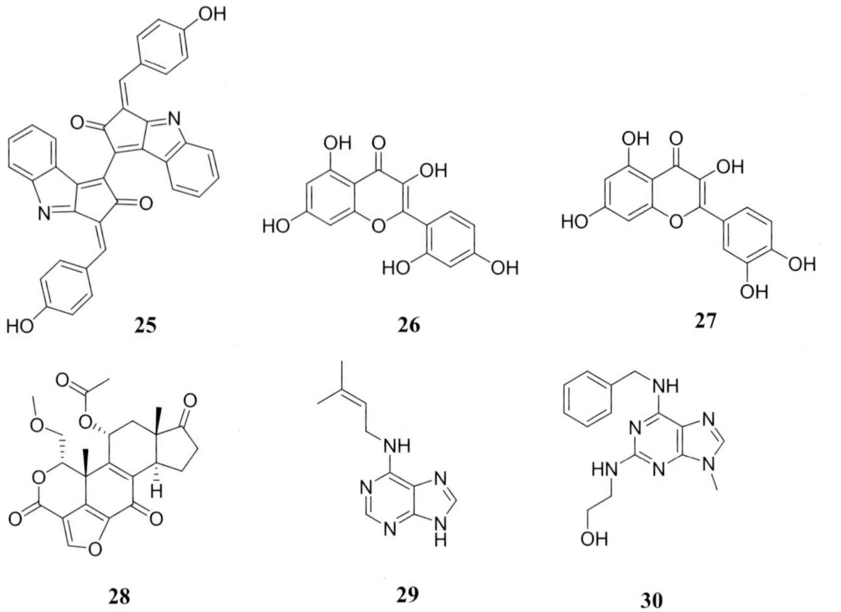

Figure 7 Chemical structures of some natural kinase inhibitors.

and quercetin (**27**) which inhibit PKL1, as well as the fungi (*Penicillium funiculosum*) metabolite, wortmannin (**28**).[184]

Mitotic histone Hi kinase, also known as cyclin-dependent kinase 1/cyclin B, plays a very important role in the control of the cell division cycle. Cyline B takes control of the G2/M-checkpoint, together with Cyclin A/Cdc2-complex. This checkpoint is important to repair DNA damages. It was noted that in many cancer types the cycline-dependent kinases are controlled incorrectly. Cyclin B is overexpressed in the early-stage of non-small cell lung cancer, leukemia, breast, and colon cancer.[185] This kinase can be inhibited by isopentenyl adenine (**29**), a constituent of *Castanea* sp. (Chestnut Tree).[185]

The secondary metabolite isolated from the radish *Raphanus sativus* L., olomucine (**30**), also inhibits the Mitotic histone Hi kinase specifically.[185] Further investigations of the structure of compound **30** came up with roscovitine known as CYC202 and seliciclib, which is currently in phase II clinical trials.[185]

REFERENCES

1. Efferth T, Li PC, Konkimalla VS, Kaina B. (2007) From traditional Chinese medicine to rational cancer therapy. *Trends Mol Med* **13**: 353–361.
2. Carlson RO. (2008) New tubulin targeting agents currently in clinical development. *Expert Opin Investig Drugs* **17**: 707–722.
3. Pellegrini F, Budman DR. (2005) Review: tubulin function, action of antitubulin drugs, and new drug development. *Cancer Invest* **23**: 264–273.
4. Edelman MJ. (2009) Novel taxane formulations and microtubule-binding agents in non-small-cell lung cancer. *Clin Lung Cancer* **10** Suppl 1: S30–S34.
5. Jordan MA, Horwitz SB, Lobert S, Correia JJ. (2008) Exploring the mechanisms of action of the novel microtubule inhibitor vinflunine. *Semin Oncol* **35**: S6–S12.
6. Mukherjee AK, Basu S, Sarkar N, Ghosh AC. (2001) Advances in cancer therapy with plant based natural products. *Curr Med Chem* **8**: 1467–1486.
7. Abal M, Andreu JM, Barasoain I. (2003) Taxanes: microtubule and centrosome targets, and cell cycle dependent mechanisms of action. *Curr Cancer Drug Targets* **3**: 193–203.
8. Morris PG, Fornier MN. (2008) Microtubule active agents: beyond the taxane frontier. *Clin Cancer Res* **14**: 7167–7172.
9. Barton-Burke M, Berg D. (1996) Cancer chemotherapy: a nursing process approach.
10. He L, Jagtap PG, Kingston DG, *et al.* (2000) A common pharmacophore for Taxol and the epothilones based on the biological activity of a taxane molecule lacking a C-13 side chain. *Biochemistry* **39**: 3972–3978.
11. Miller MC, III, Johnson KR, Willingham MC, Fan W. (1999) Apoptotic cell death induced by baccatin III, a precursor of paclitaxel, may occur without G(2)/M arrest. *Cancer Chemother Pharmacol* **44**: 444–452.
12. Saloustros E, Mavroudis D, Georgoulias V. (2008) Paclitaxel and docetaxel in the treatment of breast cancer. *Expert Opin Pharmacother* **9**: 2603–2616.
13. Dumontet C, Jordan MA, Lee FF. (2009) Ixabepilone: targeting beta III-tubulin expression in taxane-resistant malignancies. *Mol Cancer Ther* **8**: 17–25.
14. Risinger AL, Giles FJ, Mooberry SL. (2009) Microtubule dynamics as a target in oncology. *Cancer Treat Rev* **35**: 255–261.
15. Miele E, Spinelli GP, Miele E, *et al.* (2009) Albumin-bound formulation of paclitaxel (Abraxane ABI-007) in the treatment of breast cancer. *Int J Nanomedicine* **4**: 99–105.

16. Regina A, Demeule M, Che C, *et al.* (2008) Antitumour activity of ANG1005, a conjugate between paclitaxel and the new brain delivery vector Angiopep-2. *Br J Pharmacol* **155**: 185–197.
17. Trivedi M, Budihardjo I, Loureiro K, *et al.* (2008) Epothilones: a novel class of microtubule-stabilizing drugs for the treatment of cancer. *Future Oncol* **4**: 483–500.
18. Kuppens IE. (2006) Current state of the art of new tubulin inhibitors in the clinic. *Curr Clin Pharmacol* **1**: 57–70.
19. Pazdur R, Keegan P. (2009) FDA Approval for Ixabepilone.
20. Vahdat L. (2008) Ixabepilone: a novel antineoplastic agent with low susceptibility to multiple tumor resistance mechanisms. *Oncologist* **13**: 214–221.
21. Burris HA, III. (2008) Preclinical investigations with epothilones in breast cancer models. *Semin Oncol* **35**: S15–S21.
22. Kruczynski A, Hill BT. (2001) Vinflunine, the latest Vinca alkaloid in clinical development. A review of its preclinical anticancer properties. *Crit Rev Oncol Hematol* **40**: 159–173.
23. Bhattacharyya B, Panda D, Gupta S, Banerjee M. (2008) Anti-mitotic activity of colchicine and the structural basis for its interaction with tubulin. *Med Res Rev* **28**: 155–183.
24. Pettit GR, Singh SB, Hamel E, *et al.* (1989) Isolation and structure of the strong cell growth and tubulin inhibitor combretastatin A-4. *Experientia* **45**: 209–211.
25. Denny WA, Baguley BC. (2003) Dual topoisomerase I/II inhibitors in cancer therapy. *Curr Top Med Chem* **3**: 339–353.
26. Degrassi F, Fiore M, Palitti F. (2004) Chromosomal aberrations and genomic instability induced by topoisomerase-targeted antitumour drugs. *Curr Med Chem Anticancer Agents* **4**: 317–325.
27. Adams VR, Burke TG, editors. (2005) *Camptothecins in Cancer Therapy.* Totowa: Humana Press Inc., pp. 229–262.
28. Malonne H, Atassi G. (1997) DNA topoisomerase targeting drugs: mechanisms of action and perspectives. *Anticancer Drugs* **8**: 811–822.
29. A Dictionary of Biology. (2009) Supercoiling.
30. Nitiss JL. (2009) Targeting DNA topoisomerase II in cancer chemotherapy. *Nat Rev Cancer* **9**: 338–350.
31. Pommier Y. (2006) Topoisomerase I inhibitors: camptothecins and beyond. *Nat Rev Cancer* **6**: 789–802.
32. Basili S, Moro S. (2009) Novel camptothecin derivatives as topoisomerase I inhibitors. *Expert Opin Ther Pat* **19**: 555–574.

33. Hartmann JT, Lipp HP. (2006) Camptothecin and podophyllotoxin derivatives: inhibitors of topoisomerase I and II-mechanisms of action, pharmacokinetics and toxicity profile. *Drug Saf* **29**: 209–230.
34. D'Arpa P, Liu LF. (1989) Topoisomerase-targeting antitumor drugs. *Biochim Biophys Acta* **989**: 163–177.
35. Parchment RE, Pessina A. (1998) Topoisomerase I inhibitors and drug resistance. *Cytotechnology* **27**: 149–164.
36. Sriram D, Yogeeswari P, Thirumurugan R, Bal TR. (2005) Camptothecin and its analogues: a review on their chemotherapeutic potential. *Nat Prod Res* **19**: 393–412.
37. Herben VM, Ten Bokkel Huinink WW, Schellens JH, Beijnen JH. (1998) Clinical pharmacokinetics of camptothecin topoisomerase I inhibitors. *Pharm World Sci* **20**: 161–172.
38. Feun L, Savaraj N. (2008) Topoisomerase I inhibitors for the treatment of brain tumors. *Expert Rev Anticancer Ther* **8**: 707–716.
39. Seiter K. (2005) Toxicity of the topoisomerase I inhibitors. *Expert Opin Drug Saf* **4**: 45–53.
40. Larsen AK, Escargueil AE, Skladanowski A. (2003) Catalytic topoisomerase II inhibitors in cancer therapy. *Pharmacol Ther* **99**: 167–181.
41. Andoh T, Ishida R. (1998) Catalytic inhibitors of DNA topoisomerase II. *Biochim Biophys Acta* **1400**: 155–171.
42. Gray JW, Darzynkiewicz Z. (1987) Techniques in Cell Cycle Analysis. Clifton, NJ: Humana Press.
43. Pratt WB, Ruddon RW, Ensminger WD, Maybaum J. (1994) *The Anticancer Drugs*, 2nd edn. New York, USA: Oxford University Press.
44. Minotti G, Menna P, Salvatorelli E, *et al.* (2004) Anthracyclines: molecular advances and pharmacologic developments in antitumor activity and cardiotoxicity. *Pharmacol Rev* **56**: 185–229.
45. DeAngelis LM, Posner JB. (2009) Side effects of chemotherapy. In: *Neurologic Complications of Cancer*, 2nd edn. New York: Oxford University Press, p. 447.
46. Klein TE, Chang JT, Cho MK, *et al.* (2009) Integrating genotype and phenotype information: an overview of the PharmGKB project. *The Pharmacogenomics Journal* **2001**: 167–170.
47. Chabner BA, Longo DL. (2005) *Cancer Chemotherapy and Biotherapy: Principles and Practice.* Lippincott Williams & Wilkins.
48. Borchmann P, Hubel K, Schnell R, Engert A. (1997) Idarubicin: a brief overview on pharmacology and clinical use. *Int J Clin Pharmacol Ther* **35**: 80–83.

49. Strohl WR. (1997) Industrial antibiotics: today and the future. In: Strohl WR (ed.), *Biotechnology of Antibiotics*, 2nd edn. New York: Marcel Dekker, pp. 1–47.
50. Cortes-Funes H, Coronado C. (2007) Role of anthracyclines in the era of targeted therapy. *Cardiovasc Toxicol* 7: 56–60.
51. El-Helw LM, Hancock BW. (2009) Pixantrone: a promising drug in the treatment of non-Hodgkin lymphomas. *Future Oncol* 5: 445–453.
52. von Minckwitz G. (2007) Docetaxel/anthracycline combinations for breast cancer treatment. *Expert Opin Pharmacother* **8**: 485–495.
53. Andoh T. (1998) Bis(2,6-dioxopiperazines), catalytic inhibitors of DNA topoisomerase II, as molecular probes, cardioprotectors and antitumor drugs. *Biochimie* **80**: 235–246.
54. Hasinoff BB, Hellmann K, Herman EH, Ferrans VJ. (1998) Chemical, biological and clinical aspects of dexrazoxane and other bisdioxopiperazines. *Curr Med Chem* **5**: 1–28.
55. You Y. (2005) Podophyllotoxin derivatives: current synthetic approaches for new anticancer agents. *Curr Pharm Des* **11**: 1695–1717.
56. Xu H, Lv M, Tian X. (2009) A review on hemisynthesis, biosynthesis, biological activities, mode of action, and structure-activity relationship of podophyllotoxins: 2003–2007. *Curr Med Chem* **16**: 327–349.
57. Lamblin F, Hano C, Fliniaux O, *et al.* (2008) Interest of lignans in prevention and treatment of cancers. *Med Sci (Paris)* **24**: 511–519.
58. Farkya S, Bisaria VS, Srivastava AK. (2004) Biotechnological aspects of the production of the anticancer drug podophyllotoxin. *Appl Microbiol Biotechnol* **65**: 504–519.
59. Baldwin EL, Osheroff N. (2005) Etoposide, topoisomerase II and cancer. *Curr Med Chem Anticancer Agents* **5**: 363–372.
60. Montecucco A, Biamonti G. (2007) Cellular response to etoposide treatment. *Cancer Lett* **252**: 9–18.
61. Meresse P, Dechaux E, Monneret C, Bertounesque E. (2004) Etoposide: discovery and medicinal chemistry. *Curr Med Chem* **11**: 2443–2466.
62. Garbett NC, Graves DE. (2004) Extending nature's leads: the anticancer agent ellipticine. *Curr Med Chem Anticancer Agents* **4**: 149–172.
63. Stiborova M, Rupertova M, Schmeiser HH, Frei E. (2006) Molecular mechanisms of antineoplastic action of an anticancer drug ellipticine. *Biomed Pap Med Fac Univ Palacky Olomouc Czech Repub* **150**: 13–23.
64. Stiborova M, Sejbal J, Borek-Dohalska L, *et al.* (2004) The anticancer drug ellipticine forms covalent DNA adducts, mediated by human cytochromes P450, through metabolism to 13-hydroxyellipticine and ellipticine N2-oxide. *Cancer Res* **64**: 8374–8380.

65. Paoletti C, Le Pecq JB, Dat-Xuong N, *et al.* (1980) Antitumor activity, pharmacology, and toxicity of ellipticines, ellipticinium, and 9-hydroxy derivatives: preliminary clinical trials of 2-methyl-9-hydroxy ellipticinium (NSC 264-137). *Recent Results Cancer Res* **74**: 107–123.
66. Donnelly A, Blagg BS. (2008) Novobiocin and additional inhibitors of the Hsp90 C-terminal nucleotide-binding pocket. *Curr Med Chem* **15**: 2702–2717.
67. Koga F, Kihara K, Neckers L. (2009) Inhibition of cancer invasion and metastasis by targeting the molecular chaperone heat-shock protein 90. *Anticancer Res* **29**: 797–807.
68. Chaudhury S, Welch TR, Blagg BS. (2006) Hsp90 as a target for drug development. *Chem Med Chem* **1**: 1331–1340.
69. Isaacs JS, Xu W, Neckers L. (2003) Heat shock protein 90 as a molecular target for cancer therapeutics. *Cancer Cell* **3**: 213–217.
70. Neckers L. (2006) Using natural product inhibitors to validate Hsp90 as a molecular target in cancer. *Curr Top Med Chem* **6**: 1163–1171.
71. Marcu MG, Schulte TW, Neckers L. (2000) Novobiocin and related coumarins and depletion of heat shock protein 90-dependent signaling proteins. *J Natl Cancer Inst* **92**: 242–248.
72. Whitesell L, Shifrin SD, Schwab G, Neckers LM. (1992) Benzoquinonoid ansamycins possess selective tumoricidal activity unrelated to src kinase inhibition. *Cancer Res* **52**: 1721–1728.
73. Taldone T, Sun W, Chiosis G. (2009) Discovery and development of heat shock protein 90 inhibitors. *Bioorg Med Chem* **17**: 2225–2235.
74. Chiosis G, Neckers L. (2006) Tumor selectivity of Hsp90 inhibitors: the explanation remains elusive. *ACS Chem Biol* **1**: 279–284.
75. Kamal A, Thao L, Sensintaffar J, *et al.* (2003) A high-affinity conformation of Hsp90 confers tumour selectivity on Hsp90 inhibitors. *Nature* **425**: 407–410.
76. DeBoer C, Meulman PA, Wnuk RJ, Peterson DH. (1970) Geldanamycin, a new antibiotic. *J Antibiot* (Tokyo) **23**: 442–447.
77. Sgobba M, Rastelli G. (2009) Structure-based and in silico design of Hsp90 inhibitors. *Chem Med Chem* **4**: 1399–1409.
78. Prodromou C, Roe SM, O'Brien R, *et al.* (1997) Identification and structural characterization of the ATP/ADP-binding site in the Hsp90 molecular chaperone. *Cell* **90**: 65–75.
79. Stebbins CE, Russo AA, Schneider C, *et al.* (1997) Crystal structure of an Hsp90-geldanamycin complex: targeting of a protein chaperone by an antitumor agent. *Cell* **89**: 239–250.

80. Xiao L, Lu X, Ruden DM. (2006) Effectiveness of hsp90 inhibitors as anticancer drugs. *Mini Rev Med Chem* **6**: 1137–1143.
81. Supko JG, Hickman RL, Grever MR, Malspeis L. (1995) Preclinical pharmacologic evaluation of geldanamycin as an antitumor agent. *Cancer Chemother Pharmacol* **36**: 305–315.
82. Sharp S, Workman P. (2006) Inhibitors of the HSP90 molecular chaperone: current status. *Adv Cancer Res* **95**: 323–348.
83. Schulte TW, Neckers LM. (1998) The benzoquinone ansamycin 17-allylamino-17-demethoxygeldanamycin binds to HSP90 and shares important biologic activities with geldanamycin. *Cancer Chemother Pharmacol* **42**: 273–279.
84. Bisht KS, Bradbury CM, Mattson D, *et al.* (2003) Geldanamycin and 17-allylamino-17-demethoxygeldanamycin potentiate the in vitro and in vivo radiation response of cervical tumor cells via the heat shock protein 90-mediated intracellular signaling and cytotoxicity. *Cancer Res* **63**: 8984–8995.
85. Ge J, Normant E, Porter JR, *et al.* (2006) Design, synthesis, and biological evaluation of hydroquinone derivatives of 17-amino-17-demethoxygeldanamycin as potent, water-soluble inhibitors of Hsp90. *J Med Chem* **49**: 4606–4615.
86. Sydor JR, Normant E, Pien CS, *et al.* (2006) Development of 17-allylamino-17-demethoxygeldanamycin hydroquinone hydrochloride (IPI-504), an anticancer agent directed against Hsp90. *Proc Natl Acad Sci USA* **103**: 17408–17413.
87. Jez JM, Chen JC, Rastelli G, *et al.* (2003) Crystal structure and molecular modeling of 17-DMAG in complex with human Hsp90. *Chem Biol* **10**: 361–368.
88. Delmotte P, Motte-Plaque J. (1953) A new antifungal substance of fungal origin. *Nature* **171**: 344.
89. Hadden MK, Lubbers DJ, Blagg BS. (2006) Geldanamycin, radicicol, and chimeric inhibitors of the Hsp90 N-terminal ATP binding site. *Curr Top Med Chem* **6**: 1173–1182.
90. Soga S, Shiotsu Y, Akinaga S, Sharma SV. (2003) Development of radicicol analogues. *Curr Cancer Drug Targets* **3**: 359–369.
91. Soga S, Sharma SV, Shiotsu Y, *et al.* (2001) Stereospecific antitumor activity of radicicol oxime derivatives. *Cancer Chemother Pharmacol* **48**: 435–445.
92. Agatsuma T, Ogawa H, Akasaka K, *et al.* (2002) Halohydrin and oxime derivatives of radicicol: synthesis and antitumor activities. *Bioorg Med Chem* **10**: 3445–3454.

93. Ikuina Y, Amishiro N, Miyata M, *et al.* (2003) Synthesis and antitumor activity of novel O-carbamoylmethyloxime derivatives of radicicol. *J Med Chem* **46**: 2534–2541.
94. Moulin E, Zoete V, Barluenga S, *et al.* (2005) Design, synthesis, and biological evaluation of HSP90 inhibitors based on conformational analysis of radicicol and its analogues. *J Am Chem Soc* **127**: 6999–7004.
95. Soga S, Neckers LM, Schulte TW, *et al.* (1999) KF25706, a novel oxime derivative of radicicol, exhibits in vivo antitumor activity via selective depletion of Hsp90 binding signaling molecules. *Cancer Res* **59**: 2931–2938.
96. Yamamoto K, Garbaccio RM, Stachel SJ, *et al.* (2003) Total synthesis as a resource in the discovery of potentially valuable antitumor agents: cycloproparadicicol. *Angew Chem Int Ed Engl* **42**: 1280–1284.
97. Moulin E, Barluenga S, Winssinger N. (2005) Concise synthesis of pochonin A, an HSP90 inhibitor. *Org Lett* **7**: 5637–5639.
98. Shen G, Blagg BS. (2005) Radester, a novel inhibitor of the Hsp90 protein folding machinery. *Org Lett* **7**: 2157–2160.
99. Jadhav VD, Duerfeldt AS, Blagg BS. (2009) Design, synthesis, and biological activity of bicyclic radester analogues as Hsp90 inhibitors. *Bioorg Med Chem Lett* **19**: 6845–6850.
100. Clevenger RC, Blagg BS. (2004) Design, synthesis, and evaluation of a radicicol and geldanamycin chimera, radamide. *Org Lett* **6**: 4459–4462.
101. Messaoudi S, Peyrat JF, Brion JD, Alami M. (2008) Recent advances in Hsp90 inhibitors as antitumor agents. *Anticancer Agents Med Chem* **8**: 761–782.
102. Sullivan WP, Owen BA, Toft DO. (2002) The influence of ATP and p23 on the conformation of hsp90. *J Biol Chem* **277**: 45942–45948.
103. Marcu MG, Chadli A, Bouhouche I, *et al.* (2000) The heat shock protein 90 antagonist novobiocin interacts with a previously unrecognized ATP-binding domain in the carboxyl terminus of the chaperone. *J Biol Chem* **275**: 37181–37186.
104. Allan RK, Mok D, Ward BK, Ratajczak T. (2006) Modulation of chaperone function and cochaperone interaction by novobiocin in the C-terminal domain of Hsp90: evidence that coumarin antibiotics disrupt Hsp90 dimerization. *J Biol Chem* **281**: 7161–7171.
105. Burlison JA, Neckers L, Smith AB, *et al.* (2006) Novobiocin: redesigning a DNA gyrase inhibitor for selective inhibition of hsp90. *J Am Chem Soc* **128**: 15529–15536.
106. Radanyi C, Le BG, Marsaud V, *et al.* (2009) Antiproliferative and apoptotic activities of tosylcyclonovobiocic acids as potent heat shock protein 90 inhibitors in human cancer cells. *Cancer Lett* **274**: 88–94.

107. Tachibana H. (2009) Molecular basis for cancer chemoprevention by green tea polyphenol EGCG. *Forum Nutr* **61**: 156–169.
108. Yin Z, Henry EC, Gasiewicz TA. (2009) (−)-Epigallocatechin-3-gallate is a novel Hsp90 inhibitor. *Biochemistry* **48**: 336–345.
109. Yang H, Zonder JA, Dou QP. (2009) Clinical development of novel proteasome inhibitors for cancer treatment. *Expert Opin Investig Drugs* **18**: 957–971.
110. Butt MS, Sultan MT. (2009) Green tea: nature's defense against malignancies. *Crit Rev Food Sci Nutr* **49**: 463–473.
111. Yang CS, Wang X, Lu G, Picinich SC. (2009) Cancer prevention by tea: animal studies, molecular mechanisms and human relevance. *Nat Rev Cancer* **9**: 429–439.
112. Yamauchi R, Sasaki K, Yoshida K. (2009) Identification of epigallocatechin-3-gallate in green tea polyphenols as a potent inducer of p53-dependent apoptosis in the human lung cancer cell line A549. *Toxicol In Vitro* **23**: 834–839.
113. Chuang DM, Leng Y, Marinova Z, *et al.* (2009) Multiple roles of HDAC inhibition in neurodegenerative conditions. *Trends Neurosci* **32**: 591–601.
114. Finnin MS, Donigian JR, Cohen A, *et al.* (1999) Structures of a histone deacetylase homologue bound to the TSA and SAHA inhibitors. *Nature* **401**: 188–193.
115. Lane AA, Chabner BA. (2009) Histone deacetylase inhibitors in cancer therapy. *J Clin Oncol* **27**: 5459–5468.
116. Camphausen K, Tofilon PJ. (2007) Inhibition of histone deacetylation: a strategy for tumor radiosensitization. *J Clin Oncol* **25**: 4051–4056.
117. Marson CM. (2009) Histone deacetylase inhibitors: design, structure-activity relationships and therapeutic implications for cancer. *Anticancer Agents Med Chem* **9**: 661–692.
118. Ma X, Ezzeldin HH, Diasio RB. (2009) Histone deacetylase inhibitors: current status and overview of recent clinical trials. *Drugs* **69**: 1911–1934.
119. Marchion D, Munster P. (2007) Development of histone deacetylase inhibitors for cancer treatment. *Expert Rev Anticancer Ther* **7**: 583–598.
120. Jones PA, Baylin SB. (2007) The epigenomics of cancer. *Cell* **128**: 683–692.
121. Gray SG, Ekstrom TJ. (2001) The human histone deacetylase family. *Exp Cell Res* **262**: 75–83.
122. Newkirk TL, Bowers AA, Williams RM. (2009) Discovery, biological activity, synthesis and potential therapeutic utility of naturally occurring histone deacetylase inhibitors. *Nat Prod Rep* **26**: 1293–1320.
123. Gu W, Roeder RG. (1997) Activation of p53 sequence-specific DNA binding by acetylation of the p53 C-terminal domain. *Cell* **90**: 595–606.

124. Martinez-Balbas MA, Bauer UM, Nielsen SJ, *et al.* (2000) Regulation of E2F1 activity by acetylation. *EMBO J* **19**: 662–671.
125. Patel JH, Du Y, Ard PG, *et al.* (2004) The c-MYC oncoprotein is a substrate of the acetyltransferases hGCN5/PCAF and TIP60. *Mol Cell Biol* **24**: 10826–10834.
126. Chen L, Fischle W, Verdin E, Greene WC. (2001) Duration of nuclear NF-κB action regulated by reversible acetylation. *Science* **293**: 1653–1657.
127. Jeong JW, Bae MK, Ahn MY, *et al.* (2002) Regulation and destabilization of HIF-1alpha by ARD1-mediated acetylation. *Cell* **111**: 709–720.
128. Wang C, Fu M, Angeletti RH, *et al.* (2001) Direct acetylation of the estrogen receptor alpha hinge region by p300 regulates transactivation and hormone sensitivity. *J Biol Chem* **276**: 18375–18383.
129. Kovacs JJ, Murphy PJ, Gaillard S, *et al.* (2005) HDAC6 regulates Hsp90 acetylation and chaperone-dependent activation of glucocorticoid receptor. *Mol Cell* **18**: 601–607.
130. Reikvam H, Ersvaer E, Bruserud O. (2009) Heat shock protein 90 — a potential target in the treatment of human acute myelogenous leukemia. *Curr Cancer Drug Targets* **9**: 761–776.
131. Piekarz RL, Sackett DL, Bates SE. (2007) Histone deacetylase inhibitors and demethylating agents: clinical development of histone deacetylase inhibitors for cancer therapy. *Cancer J* **13**: 30–39.
132. Miller TA, Witter DJ, Belvedere S. (2003) Histone deacetylase inhibitors. *J Med Chem* **46**: 5097–5116.
133. Mann BS, Johnson JR, Cohen MH, *et al.* (2007) FDA approval summary: vorinostat for treatment of advanced primary cutaneous T-cell lymphoma. *Oncologist* **12**: 1247–1252.
134. Le TC, Siu LL. (2008) Promising antitumor activity with MGCD0103, a novel isotype-selective histone deacetylase inhibitor. *Expert Opin Investig Drugs* **17**: 1247–1254.
135. Garcia-Manero G, Assouline S, Cortes J, *et al.* (2008) Phase 1 study of the oral isotype specific histone deacetylase inhibitor MGCD0103 in leukemia. *Blood* **112**: 981–989.
136. Fournel M, Bonfils C, Hou Y, *et al.* (2008) MGCD0103, a novel isotype-selective histone deacetylase inhibitor, has broad spectrum antitumor activity in vitro and in vivo. *Mol Cancer Ther* **7**: 759–768.
137. Djeu JY, Wei S. (2009) Clusterin and chemoresistance. *Adv Cancer Res* **105**: 77–92.
138. Dedes KJ, Dedes I, Imesch P, *et al.* (2009) Acquired vorinostat resistance shows partial cross-resistance to 'second-generation' HDAC inhibitors and

correlates with loss of histone acetylation and apoptosis but not with altered HDAC and HAT activities. *Anticancer Drugs* **20**: 321–333.

139. Quintas-Cardama A, Kantarjian H, Cortes J. (2009) Homoharringtonine, omacetaxine mepesuccinate, and chronic myeloid leukemia circa 2009. *Cancer* **115**: 5382–5393.
140. Kantarjian HM, Talpaz M, Santini V, *et al.* (2001) Homoharringtonine: history, current research, and future direction. *Cancer* **92**: 1591–1605.
141. Zhou DC, Zittoun R, Marie JP. (1995) Homoharringtonine: an effective new natural product in cancer chemotherapy. *Bull Cancer* **82**: 987–995.
142. Smith CR, Jr., Powell RG, Mikolajczak KL. (1976) The genus Cephalotaxus: source of homoharringtonine and related anticancer alkaloids. *Cancer Treat Rep* **60**: 1157–1170.
143. Tujebajeva RM, Graifer DM, Karpova GG, Ajtkhozhina NA. (1989) Alkaloid homoharringtonine inhibits polypeptide chain elongation on human ribosomes on the step of peptide bond formation. *FEBS Lett* **257**: 254–256.
144. Fresno M, Jimenez A, Vazquez D. (1977) Inhibition of translation in eukaryotic systems by harringtonine. *Eur J Biochem* **72**: 323–330.
145. Baaske DM, Heinstein P. (1977) Cytotoxicity and cell cycle specificity of homoharringtonine. *Antimicrob Agents Chemother* **12**: 298–300.
146. Grem JL, Cheson BD, King SA, *et al.* (1988) Cephalotaxine esters: antileukemic advance or therapeutic failure? *J Natl Cancer Inst* **80**: 1095–1103.
147. Luo CY, Tang JY, Wang YP. (2004) Homoharringtonine: a new treatment option for myeloid leukemia. *Hematology* **9**: 259–270.
148. (1977) Harringtonine in acute leukemias. Clinical analysis of 31 cases. *Chin Med J (Engl)* **3**: 319–324.
149. Anonymous (1976) Cephalotaxine esters in the treatment of acute leukemia. A preliminary clinical assessment. *Chin Med J (Engl)* **2**: 263–272.
150. O'Dwyer PJ, King SA, Hoth DF, *et al.* (1986) Homoharringtonine–perspectives on an active new natural product. *J Clin Oncol* **4**: 1563–1568.
151. Takemura Y, Ohnuma T, Chou TC, *et al.* (1985) Biologic and pharmacologic effects of harringtonine on human leukemia-lymphoma cells. *Cancer Chemother Pharmacol* **14**: 206–210.
152. Whaun JM, Brown ND. (1990) Treatment of chloroquine-resistant malaria with esters of cephalotaxine: homoharringtonine. *Ann Trop Med Parasitol* **84**: 229–237.
153. Druker BJ, Tamura S, Buchdunger E, *et al.* (1996) Effects of a selective inhibitor of the Abl tyrosine kinase on the growth of Bcr-Abl positive cells. *Nat Med* **2**: 561–566.

154. Hughes TP, Branford S. (2009) Monitoring disease response to tyrosine kinase inhibitor therapy in CML. *Hematology Am Soc Hematol Educ Program* 2009: 477–487.
155. Goldman JM. (2009) Initial treatment for patients with CML. *Hematology Am Soc Hematol Educ Program* 2009: 453–460.
156. Pinilla-Ibarz J, Quintas-Cardama A. (2009) New agents in the treatment of chronic myelogenous leukemia. *J Natl Compr Canc Netw* 7: 1028–1037.
157. Quintas-Cardama A, Kantarjian H, Garcia-Manero G, *et al.* (2007) Phase I/II study of subcutaneous homoharringtonine in patients with chronic myeloid leukemia who have failed prior therapy. *Cancer* **109**: 248–255.
158. Quintas-Cardama A, Cortes J. (2008) Homoharringtonine for the treatment of chronic myelogenous leukemia. *Expert Opin Pharmacother* **9**: 1029–1037.
159. Russo D, Michelutti A, Melli C, *et al.* (1995) MDR-related P170-glycoprotein modulates cytotoxic activity of homoharringtonine. *Leukemia* **9**: 513–516.
160. Tebbi CK, Chervinsky D, Baker RM. (1991) Modulation of drug resistance in homoharringtonine-resistant C-1300 neuroblastoma cells with cyclosporine A and dipyridamole. *J Cell Physiol* **148**: 464–471.
161. Zhou DC, Ramond S, Viguie F, *et al.* (1996) Sequential emergence of MRP- and MDR1-gene over-expression as well as MDR1-gene translocation in homoharringtonine-selected K562 human leukemia cell lines. *Int J Cancer* **65**: 365–371.
162. Efferth T, Sauerbrey A, Halatsch ME, *et al.* (2003) Molecular modes of action of cephalotaxine and homoharringtonine from the coniferous tree Cephalotaxus hainanensis in human tumor cell lines. *Naunyn Schmiedebergs Arch Pharmacol* **367**: 56–67.
163. Mansky PJ. (2002) Mistletoe and cancer: controversies and perspectives. *Semin Oncol* **29**: 589–594.
164. Horneber MA, Bueschel G, Huber R, *et al.* (2008) Mistletoe therapy in oncology. *Cochrane Database Syst Rev* CD003297.
165. Kienle GS, Kiene H. (2007) Complementary cancer therapy: a systematic review of prospective clinical trials on anthroposophic mistletoe extracts. *Eur J Med Res* **12**: 103–119.
166. Hajto T, Hostanska K, Saller R. (1999) Mistletoe therapy from the pharmacologic perspective. *Forsch Komplementarmed* **6**: 186–194.
167. Gabius S, Gabius HJ. (1999) Immunomodulating mistletoe therapy by lectin standardization: a double-edged sword? *Versicherungsmedizin* **51**: 128–136.
168. Stauder H, Kreuser ED. (2002) Mistletoe extracts standardised in terms of mistletoe lectins (ML I) in oncology: current state of clinical research. *Onkologie* **25**: 374–380.

169. Hamre HJ, Kiene H, Kienle GS. (2009) Clinical research in anthroposophic medicine. *Altern Ther Health Med* **15**: 52–55.
170. Dudkina AS, Lindsley CW. (2007) Small molecule protein-protein inhibitors for the p53-MDM2 interaction. *Curr Top Med Chem* **7**: 952–960.
171. Bai L, Zhu W-G. (2006) p53: Structure, function and therapeutic applications. *J Cancer Molecules* **2**: 141–153.
172. Duncan SJ, Grueschow S, Williams DH, *et al.* (2001) Isolation and structure elucidation of Chlorofusin, a novel p53-MDM2 antagonist from a Fusarium sp. *J Am Chem Soc* **123**: 554–560.
173. Tsukamoto S, Yoshida T, Hosono H, *et al.* (2006) Hexylitaconic acid: a new inhibitor of p53-HDM2 interaction isolated from a marine-derived fungus, Arthrinium sp. *Bioorg Med Chem Lett* **16**: 69–71.
174. Martínez-Sánchez G, Giuliani A. (2007) Cellular redox status regulates hypoxia inducible factor-1 activity. Role in Tumor development. *J Exp Clin Cancer Res* **26**: 39–50.
175. Nagle DG, Zhou YD. (2006) Natural product-based inhibitors of hypoxia-inducible factor-1 (HIF-1). *Curr Drug Targets* **7**: 355–369.
176. Gaddipati JP, Madhavan S, Sidhu GS, *et al.* (1999) Picroliv, a natural product protects cells and regulates the gene expression during hypoxia/reoxygenation. *Mol Cell Biochem* **194**: 271–281.
177. Chander R, Kapoor NK, Dhawan BN. (1992) Picroliv, picroside-I and kutkoside from Picrorhiza kurrooa are scavengers of superoxide anions. *Biochem Pharmacol* **44**: 180–183.
178. Symolon H, Schmelz EM, Dillehay DL, *et al.* (2004) Dietary soy sphingolipids suppress tumorigenesis and gene expression in 1,2-dimethylhydrazine-treated CF1 mice and ApcMin/+ mice. *J Nutr* **134**: 1157–1161.
179. Park JH, Kim TY, Jong HS, *et al.* (2003) Gastric epithelial reactive oxygen species prevent normoxic degradation of hypoxia-inducible factor-1alpha in gastric cancer cells. *Clin Cancer Res* **9**: 433–440.
180. Zhang J, Yang PL, Gray NS. (2009) Targeting cancer with small molecule kinase inhibitors. *Nat Rev Cancer* **9**: 28–39.
181. Ochi T, Fujiwara H, Yasukawa M. (2009) Aurora-A kinase: a novel target both for cellular immunotherapy and molecular target therapy against human leukemia. *Expert Opin Ther Targets* **13**: 1399–1410.
182. Dillon JG, Castenholz RW. (2003) The synthesis of the UV-screening pigment, scytonemin, and photosynthetic performance in isolates from closely related natural populations of cyanobacteria *(Calothrix sp.). Environ Microbiol* **5**: 484–491.

183. Balskus EP, Case RJ, Walsh CT. (2011) The biosynthesis of cyanobacterial sunscreen scytonemin in intertidal microbial mat communities. *FEMS Microbiol Ecol* 77: 322–332.
184. McInnes C, Mezna M, Fischer PM. (2005) Progress in the discovery of polo-like kinase inhibitors. *Curr Top Med Chem* 5: 181–197.
185. Newman DJ. (2008) Natural products as leads to potential drugs: an old process or the new hope for drug discovery? *J Med Chem* 51: 2589–2599.

Index